GOD AND DESIGN

Is there reason to think a supernatural designer made our world?

Recent discoveries in physics, cosmology, and biochemistry have captured the public imagination and made the design argument – the theory that God created the world according to a specific plan – the object of renewed scientific and philosophical interest. Terms such as "cosmic fine-tuning," the "anthropic principle," and "irreducible complexity" have seeped into public consciousness, increasingly appearing within discussion about the existence and nature of God. This accessible and serious introduction to the design problem brings together both sympathetic and critical new perspectives from prominent scientists and philosophers including Paul Davies, Richard Swinburne, Sir Martin Rees, Michael Behe, Elliott Sober, and Peter van Inwagen.

Questions raised include:

- What is the logical structure of the design argument?
- How can intelligent design be detected in the Universe?
- What evidence is there for the claim that the Universe is divinely fine-tuned for life?
- Does the possible existence of other universes refute the design argument?
- Is evolutionary theory compatible with the belief that God designed the world?

God and Design probes the relationship between modern science and religious belief, considering their points of conflict and their many points of similarity. Is God the "master clockmaker" who sets the world's mechanism on a perfectly enduring course, or a miraculous presence continually intervening in and altering the world we know? Are science and faith, or evolution and creation, really in conflict at all? Expanding the parameters of a lively and urgent contemporary debate, *God and Design* considers the ways in which perennial questions of origin continue to fascinate and disturb us.

Neil A. Manson is Visiting Assistant Professor of Philosophy at Virginia Commonwealth University in Richmond, and a former Gifford Research Fellow in Natural Theology at the University of Aberdeen. He has a long standing interest in the science and religion debate.

GOD AND DESIGN

The teleological argument and modern science

Neil A. Manson

LONDON AND NEW YORK

First published 2003
by Routledge
11 New Fetter Lane, London EC4P 4EE

Simultaneously published in the USA and Canada
by Routledge
29 West 35th Street, New York, NY 10001

Routledge is an imprint of the Taylor & Francis Group

© 2003 Neil A. Manson for selection and editorial material; individual contributors for their contributions

Typeset in Times by Taylor & Francis Books Ltd
Printed and bound in Great Britain by MPG Books Ltd, Bodmin

All rights reserved. No part of this book may be reprinted or reproduced or utilized in any form or by any electronic, mechanical, or other means, now known or hereafter invented, including photocopying and recording, or in any information storage or retrieval system, without permission in writing from the publishers.

British Library Cataloguing in Publication Data
A catalogue record for this book is available from the British Library

Library of Congress Cataloging in Publication Data
A catalog record for this book has been requested

ISBN 0–415–26343–3 (hbk)
ISBN 0–415–26344–1 (pbk)

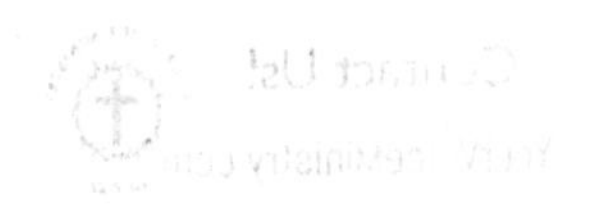

CONTENTS

ILLUSTRATIONS

Tables

Figures

CONTRIBUTORS

Michael Behe is Professor of Biological Sciences at Lehigh University in Bethlehem, Pennsylvania, and the author of *Darwin's Black Box: The Biochemical Challenge to Evolution.*

Robin Collins is Associate Professor of Philosophy at Messiah College in Grantham, Pennsylvania. He has written widely on the argument from fine-tuning and is working on a book tentatively entitled *The Well-tempered Universe: God, Fine-tuning, and the Laws of Nature.*

Simon Conway Morris is Professor of Evolutionary Paleobiology at the University of Cambridge and a Fellow of the Royal Society. He is author of *The Crucible of Creation* and of *Life's Solution.* His search for fossils has taken him to many parts of the world, including China, Mongolia, Australia, and Greenland.

William Lane Craig is Research Professor of Philosophy at the Talbot School of Theology in La Mirada, California. He is author of many journal articles and several books, including *Reasonable Faith* and *God, Time, and Eternity.*

Paul Davies is Honorary Professor at the Australian Centre for Astrobiology at Macquarie University of Sydney. A winner of the Templeton Prize, he has published numerous research papers in specialist journals, in the fields of cosmology, gravitation, and quantum field theory, with particular emphasis on black holes and the origin of the Universe. His many popular books include *The Mind of God* and *The Fifth Miracle.*

William Dembski, a mathematician and philosopher, is Associate Research Professor in the Conceptual Foundations of Science at Baylor University in Waco, Texas, and Senior Fellow with the Center for the Renewal of Science and Culture of the Discovery Institute in Seattle. He is author of *The Design Inference* and *No Free Lunch: Why Specified Complexity Cannot be Purchased without Intelligence.*

John Leslie is University Professor Emeritus at the University of Guelph, Ontario. A philosopher who works in metaphysics, philosophy of religion, and philosophy of science, his publications include *Universes* and *Infinite Minds*.

Timothy McGrew is Associate Professor of Philosophy at Western Michigan University in Kalamazoo. He is author of *Foundations of Knowledge* and "Direct inference in the problem of induction," in *The Monist* 84. His wife **Lydia McGrew** has authored several articles, including "Agency and the metalottery fallacy," in *Australasian Journal of Philosophy* 80.

Neil A. Manson is Visiting Assistant Professor of Philosophy at Virginia Commonwealth University in Richmond. He held the positions of Gifford Research Fellow in Natural Theology at the University of Aberdeen, Scotland, and Postdoctoral Research Associate at the Center for Philosophy of Religion at the University of Notre Dame in Indiana. His interests include metaphysics, philosophy of religion, and philosophy of science.

D.H. Mellor is Emeritus Professor of Philosophy at the University of Cambridge and a Fellow of both Darwin College and of the British Academy. His books include *Matters of Metaphysics*, *The Facts of Causation*, and *Real Time II*.

Kenneth R. Miller is a cell biologist and Professor of Biology at Brown University in Providence, Rhode Island. In addition to publishing articles that have appeared in scientific journals such as *Nature*, *Scientific American*, and *Cell*, he is author of *Finding Darwin's God: A Scientist's Search for Common Ground between God and Evolution*.

Jan Narveson is Professor of Philosophy at the University of Waterloo, Ontario. A member of the Royal Society of Canada, he is author of *The Libertarian Idea* and *Moral Matters*.

Robert O'Connor is Associate Professor of Philosophy at Wheaton College in Illinois. His current research interests are in the area of theological realism, and he has authored a number of papers on the connections between science and religion.

Del Ratzsch is Professor of Philosophy at Calvin College in Grand Rapids, Michigan. He is author of *The Philosophy of Science* and *Nature, Design, and Science*.

Martin Rees Great Britain's Astronomer Royal, is Royal Society Professor at Cambridge University and a Fellow of King's College. His research interests are in cosmology and astrophysics. Apart from his extensive research publications, he is the author of many general articles and of several books, including *Before the Beginning*, *Just Six Numbers*, and *Our Cosmic Habitat*.

Michael Ruse is Professor of Philosophy at Florida State University in Tallahassee. He is author of numerous books, including *The Darwinian Revolution*, *Taking Darwin Seriously*, and *Can a Darwinian be a Christian?*

Elliott Sober is Hans Reichenbach Professor and Henry Vilas Research Professor of Philosophy at the University of Wisconsin, Madison. His research is mainly in the philosophy of science, particularly in the philosophy of evolutionary biology. He has published many papers and several books, including *Philosophy of Biology* and *From a Biological Point of View*.

Richard Swinburne is Nolloth Professor of the Philosophy of the Christian Religion at the University of Oxford and a Fellow of the British Academy. His major writings include a trilogy on the philosophy of theism (including *The Coherence of Theism* and *The Existence of God*), a tetralogy on the philosophy of Christianity, *The Evolution of the Soul*, and – most recently – *Epistemic Justification*.

Peter van Inwagen is the John Cardinal O'Hara Professor of Philosophy at the University of Notre Dame in Indiana. He is author of numerous books, including *An Essay on Free Will*, *Material Beings*, *Metaphysics*, and *God, Knowledge, and Mystery*.

Eric Vestrup is Assistant Professor in the Department of the Mathematical Sciences at DePaul University in Chicago. His interests include measure theory, probability, and foundational questions in mathematics and statistics.

Roger White is Assistant Professor of Philosophy at New York University. His research focuses on epistemological issues in the philosophy of science, particularly those having to do with probability and explanation.

PREFACE

As the contents of this book indicate, there has been a tremendous resurgence of interest in the design argument in recent years. Unfortunately, discussions of the design argument (particularly in North America) have tended to generate as much heat as light. This is largely due to the association of the biological version of the design argument with the controversial matter of the content of public school science curricula. Those who make a design argument from biological evidence are likely to be accused of religious fundamentalism and of belief in creationism. Cosmic design arguments, on the other hand, are far more respected, at least in so far as those who make them are not so apt to be subject to *ad hominem* attack. As a philosopher, I fail to see why cosmic design arguments should be treated so differently from biological ones. Certainly cosmic design arguments are no less astounding for locating supernatural activity at the very beginning of the Universe rather than in more recent history. That is why I have included in this volume papers on both sorts of design argument. I also made a point of including papers on what is emerging as the primary naturalistic alternative to the design hypothesis: the "multiverse" hypothesis that there are many universes in addition to our own. My hope is that, by bringing together up-to-date papers on these diverse strands, I have, with respect to the design argument, given philosophers, theologians, scientists, and interested laypeople a sense of where the action is.

The publishers and I wish to thank the following for permission to reprint copyright material in this book: Blackwell Publishers, for Roger White's "Fine-tuning and multiple universes," in *Nous* 34 (2000), pp. 260–76; Blackwell Publishers again, for Elliott Sober's "The design argument," in William Mann (ed.) *The Blackwell Guide to Philosophy of Religion* (forthcoming); the New York Academy of Sciences and the American Association for the Advancement of Science, for John Leslie's "The meaning of design," in Jim Miller (ed.) *Cosmic Questions,* NYAS Annals 950 (2001) pp. 128–38; Oxford University Press, for Timothy McGrew, Lydia McGrew, and Eric Vestrup's "Probabilities and the fine-tuning argument: A skeptical view," in

Mind 110 (no. 440), pp. 1,027–38; and the Evangelical Philosophical Society, for Michael Behe's "The modern intelligent design hypothesis: Breaking rules," in *Philosophia Christi*, (2001) vol. 3, no. 1 (2001), pp. 165–79.

In addition to thanking all of the contributors to this volume – they are the ones who will make this book worth reading – I would like to thank various individuals and institutions for their help in my getting exposed to and coming to understand the design argument. First and foremost I thank Peter van Inwagen for directing my research; he is the person who got me interested in the design argument in the first place. I am also grateful to the Department of Philosophy at the University of Aberdeen for funding my research fellowship there, and also for supporting the Gifford Bequest International Conference in Aberdeen in May 2000. Many of the contributors to this book participated in that conference. I profited greatly from my time spent in Aberdeen and I thank all the members of the department there, particularly Gordon Graham. Through its Seminars in Christian Scholarship, Calvin College enabled me to participate in the summer of 2000 in a six-week seminar on the design argument: "Design, Self-organization, and the Integrity of Creation." My thanks go to Calvin, to its administrative staff, and to all the seminar participants. The Center for Philosophy of Religion at the University of Notre Dame also supported my work through a postdoctoral fellowship; it was there that I completed this book. The members of the Center engaged me in numerous productive conversations and reviewed my introduction in the Center's weekly discussion group. I thank all of the participants in that group, particularly Tom Flint, Marcin Iwanicki, Ernan McMullin, Christian Miller, John Mullen, and Alvin Plantinga. Adolf Grünbaum also provided extensive comments on an earlier draft of my introduction; he has my gratitude. My thoughts on the design argument over the last several years have been greatly clarified through conversations and correspondence with a number of other people, including but not limited to the following: Jose Benardete, Nick Bostrom, Robin Collins, Andrew Cortens, William Lane Craig, William Dembski, Timothy Kenyon, John Leslie, Lydia McGrew, Timothy McGrew, Brent Mundy, Graham Oppy, Del Ratzsch, Alasdair Richmond, Jack Smart, Quentin Smith, Richard Swinburne, Michael Thrush, Roger White, and David Woodruff. Sherry Levy-Reiner long ago took me on as an editorial intern at the Association of American Colleges and Universities; without the training I got from her the process of editing this book would have been vastly more difficult. I thank the staff at Routledge, particularly Clare Johnson, Vanessa Winch, Celia Tedd and Tony Nixon. Lastly, I would like to thank my family – particularly my parents, Bill and Shirley Manson – for giving me unqualified love and support.

INTRODUCTION

Neil A. Manson

This introduction has two functions. First, it apprises readers of some of the basic data, terminology, and formalisms used in contemporary discussions of the design argument while also giving a sense of the argument's history. Other pieces in this anthology – particularly those of Elliott Sober, John Leslie, Paul Davies, and Michael Ruse – cover some of the same ground. Second, it gives readers some idea of what the various contributors will say and why their contributions are important for understanding the design argument.[1] Though I will raise my own concerns at various points, I will (so far as I can) leave the philosophical and scientific heavy lifting to the distinguished contributors.

Classifying the design argument

Design arguments involve reasoning from seemingly purposeful features of the observable world to the existence of at least one supernatural designer. Because of this appeal to purpose, design arguments are teleological (from the Greek word "*telos*", meaning "goal" or "end"). Though design arguments almost always are mounted for the ultimate purpose of proving the existence of God (as opposed to some other being), in most versions of the argument the inference is not directly to God, but rather just to the existence of some supernatural designer(s) or other; further arguments are needed to identify the supernatural designer(s) with God.[2] Since the design argument relies on a premise that can be known only through observation of the empirical world, it counts as an *a posteriori* argument. In this way it contrasts with *a priori* arguments for the existence of God – arguments all of the premises of which can be known to be true independent of sense experience. Ontological arguments are the most notable example of *a priori* arguments for the existence of God.

Cosmological arguments for the existence of God are also *a posteriori*. The causal version of the cosmological argument moves from the existence of causal sequences in the observable world to the existence of a first cause. The contingency version of the cosmological argument moves from the existence of things that might not have existed to the existence of a necessary being. Cosmological arguments differ from the design argument, however, in that their *a posteriori* premises are highly general and apparently incorrigible. The passage of time and the development of scientific knowledge will presumably provide neither more nor less reason to believe that there are sequences of cause-and-effect relationships or that there are things that might not have existed.

The eutaxiological argument (from the Greek word "*eutaxia*", meaning "good order") moves from the lawful regularity and comprehensibility of the world to the existence of an ordering being.[3] In addition to the wealth of historical and scientific facts it displays, the argument by Paul Davies is interesting for its strong eutaxiological flavor – its emphasis on the fact that "the physical world is both ordered and intelligible." Unlike the cosmological argument, the eutaxiological argument's *a posteriori* premise can get support from science. As science progresses, the world does seem to become more orderly and comprehensible, at least in so far as phenomena that were previously thought to be unrelated (e.g. electricity and magnetism) come to be seen as related. Yet the sort of empirical evidence in favor of the eutaxiological argument is not nearly so detailed as the sort of evidence offered in favor of the design argument. Orderliness, lawfulness, and comprehensibility are quite general features of the world. The features of the world to which proponents of the design argument point are much more specific and the allure of the argument depends very much on the state of scientific knowledge during a particular slice of history.[4]

The resurgence of the design argument in the late twentieth century

As of half a century ago, that allure was minimal. Darwin's theory of evolution by natural selection, articulated in *On the Origin of Species* (1859), was thought to have robbed proponents of what Elliott Sober calls the "organismic" design argument of the move from apparent design to a designer. The detailed biological observations in William Paley's *Natural Theology* (1802) and in the Bridgewater Treatises of the early nineteenth century were widely thought to be explicable in terms of evolution through natural selection, with no need for a designer. The origin of life, meanwhile, was not even seen as an important scientific problem; this was largely due to the underlying assumption (discussed below) that the Universe is temporally and spatially infinite. The story of the decline of the organismic design argument subsequent to Darwin is a fascinating one, but since Sober, Ruse, and (more briefly) Michael Behe all recount that story in their chapters, I will not repeat it here.

The prospects for a "cosmic" design argument, meanwhile, seemed nonexistent so long as prevailing attitudes towards the cosmos held sway. The discoveries of modern physical cosmology have permeated contemporary intellectual sensibilities so thoroughly that some of us have a hard time remembering the wariness with which cosmology and its object were viewed even as recently as the 1960s. The tradition of suspicion dates at least to Kant, who claimed in his First Antinomy in the *Critique of Pure Reason* that any talk of the Universe as a unified object (comparable to, say, Jupiter) led to contradiction.[5] Moving ahead to the early twentieth century, logical positivists advanced the verificationist theory of meaning, according to which statements the truth of which cannot be verified are meaningless. Verificationism prompted doubts about whether the Universe was a legitimate object of scientific inquiry (or even an object at all). Is it a unique whole? If it is unique, does it make any sense to talk of it following laws? Can we ever observe it (as opposed to observing just part of it)? Can we even have a meaningful concept of the Universe – a concept that succeeds, or could fail to succeed, in picking out something? Like a nominee at a Senate confirmation hearing, the Universe had its ontological candidacy rejected on the grounds that too many questions could be raised.

Because of this aura of disrepute popular among both philosophers and scientists, cosmology was largely disregarded. It was simply assumed that the Universe is eternal and infinite, and that otherwise there is nothing for scientists (or philosophers) to say about it. These assumptions (and the atheism with which they are consonant) were deeply entrenched, which explains the tremendous surprise and hostility with which the Big Bang model was greeted. To get a sense of the reaction, consider the following account C.F. von Weizsäcker gave of a conversation he had in 1938 with Nobel Prize-winning physical chemist Walther Nernst. Nernst had reacted to von Weizsäcker's presentation of some calculations he had made regarding the age of the Universe:

> He said, the view that there might be an age of the universe was not science. At first I did not understand him. He explained that the infinite duration of time was a basic element of all scientific thought, and to deny this would mean to betray the very foundations of science. I was quite surprised by this idea and I ventured the objection that it was scientific to form hypotheses according to the hints given by experience, and that the idea of an age of the universe was such a hypothesis. He retorted that we could not form a scientific hypothesis which contradicted the very foundations of science. He was just angry, and thus the discussion, which was continued in his private library, could not lead to any result.
>
> (von Weizsäcker 1964: 151)

Nowadays one can hardly surf the Internet or peruse the science section of a chain bookstore without stumbling across a website, article, or book about the design argument. (You chose wisely!) What accounts for this change in fortunes? The answer lies in the spectacular growth in the middle of the last century of (1) physical cosmology and (2) the closely related fields of molecular biology, cell biology, and biochemistry. A series of breakthroughs in physics and observational astronomy led to the development of the Big Bang model and the discovery that the Universe is highly structured, with precisely defined parameters such as age, mass, entropy (degree of disorder), curvature, temperature, density, and rate of expansion. Using clever experimentation and astounding instrumentation, physical cosmologists were able to determine the values of these parameters to remarkably precise degrees. The specificity of the Universe prompted theoretical exploration of how the Universe would have been if the values of its parameters had been different. This led to the discovery of numerous "anthropic coincidences" and supported the claim that the Universe is fine-tuned for life – that is, that the values of its parameters are such that, if they differed even slightly, life of any sort could not possibly have arisen in the Universe. Furthermore, the temporal and spatial finitude of the Universe meant that there were not unlimited opportunities for life to originate by chance. So the discovery of the Big Bang did not just resurrect the possibility of mounting a cosmic design argument. It also created an opening for biological design arguments as well.

That opening was widened as the inner workings of the cell were made accessible due to the introduction (beginning in the post-Second World War period) of powerful new tools and experiments. Prior to that time not much was known about the cell. Though it was acknowledged to be the most basic form of life and to contain within it the key to reproduction, it was generally regarded as quite simple. Cells were viewed as hunks of protoplasm – things that could have arisen easily enough from an inorganic, prebiotic soup of the sort that presumably covered the earth billions of years ago. As Michael Behe notes (1996: 6–13), the development of electron microscopy, X-ray crystallography, and nuclear magnetic resonance (NMR) imaging in the latter half of the twentieth century caused this conception of the cell to topple. The cell itself and the mechanisms for replication contained within it were now seen to be powered by molecular structures of tremendous complexity – a complexity that Behe argues is (in some cases) "irreducible."

Clearly, science has not run its course; just as the fortunes of the design argument rose, they may fall again. Indeed, some will claim that they are *now* falling. As we will see shortly when the topic turns to the multiverse hypothesis, some scientists think the theory of a unique Big Bang and the temporal singularity it implies can and should be discarded. But until these scientific revolutions occur, we cannot fault design proponents for drawing from currently accepted scientific facts and theories. The data in support of the claims of fine-tuning and irreducible complexity – and criticisms of these interpretations of the data – are presented with admirable clarity by several of the contributors to this volume. Robin Collins and William Lane

Craig both document extensively the array of force strengths, mass ratios, and other fundamental constants that seem to be fine-tuned, with Collins going into considerable technical detail to provide six solid cases of fine-tuning. Michael Behe, Kenneth Miller, and Michael Ruse, meanwhile, all apply their biological expertise to supporting or debunking claims that particular cellular mechanisms could not have arisen via natural selection. Since the scientific details of the design argument are so ably explained by the aforementioned authors, I will devote the remainder of this introduction to highlighting the philosophical issues the design argument raises.

The logic of the design argument

Design arguments nowadays typically employ a probabilistic logical apparatus. This distinguishes contemporary design arguments from earlier analogical versions of the sort that Hume criticized in *Dialogues Concerning Natural Religion* (1779). To say that modern design arguments employ a probabilistic logical apparatus, however, still leaves much room for disagreement as to their precise logical structure. For example, are design arguments Bayesian? Bayes's theorem is a formula in the probability calculus. This theorem provides us with Bayes's rule – a rule that shows how one might revise, in the light of new evidence, the probabilities one initially assigned to competing hypotheses. Many contemporary philosophers think one should always evaluate the impact of new evidence in conformity with Bayes's rule. These "Bayesians" think Bayes's rule is a crucial constraint on scientific reasoning.[6]

To get a sense of Bayesian reasoning, suppose you are tracking the leader board of a men's professional golf tournament in which there are 100 contestants, one of whom you know is Tiger Woods. Unlike the typical golf tournament, however, the players are not identified by name on the leader board, but rather by number. Now consider the following hypothesis (T), the following datum (L), and the following statement of background knowledge (B):

T = Contestant 93 is Tiger Woods.
L = Contestant 93 is leading by eight strokes.
B = There are 100 numbered contestants, all of them are golf professionals, and one of them is Tiger Woods.

What is the relationship between the hypothesis, the datum, and the background knowledge? To answer this, it will be helpful to use the following notation: $P(x|y)$ stands for the probability of x given that ("conditional on") y. Now, knowing nothing else about contestant 93 except that he is one of 100 professional golfers, you think he has one chance in a hundred of being Tiger. So prior to getting any information about how contestant 93 is doing, you think

$$P(T|B) = 0.01.$$

But now you see on the leader board that contestant 93 is leading by eight strokes. This is an extremely large lead for a professional golf tournament; a golfer who could build such a lead would have to be much, much better even than the typical professional golfer. So while you think the probability of an eight-stroke lead being built by contestant 93 conditional on his being a contestant other than Tiger is extremely low (say, one in ten thousand), you think the probability of an eight-stroke lead's being built by contestant 93 conditional on his being Tiger is fairly high (say, one in a hundred). (Of course, these epistemic probability assignments are artificially precise, but for the purposes of this example we will ignore this problem.) So

$P(L|{\sim}T \,\&\, B) = 0.0001$ (where "~" stands for "it is not the case that")

and

$$P(L|T \,\&\, B) = 0.01.$$

In light of the evidence L, you realize you should assign a much higher probability to T than you did prior to looking at the leader board. Bayes's rule, many philosophers would think, tells you exactly how much higher. Bayes's rule says the probability that golfer 93 is Tiger given that golfer 93 leads by eight strokes is the *particular probability* – the probability that golfer 93 is Tiger and leads by eight strokes – divided by the *total probability* – the probability that any one of the 100 golfers leads by eight strokes. So

$$P(T|L \,\&\, B) = \{P(L|T \,\&\, B) \bullet P(T|B)\}/\{P(L|{\sim}T \,\&\, B) \bullet P({\sim}T|B) + P(L|T \,\&\, B) \bullet P(T|B)\}$$
$$= \{0.01 \bullet 0.01\}/\{0.0001 \bullet 0.99 + 0.01 \bullet 0.01\}$$
$$= 0.0001/0.000199$$
$$= 0.5025.$$

That is, evaluating the evidence in light of Bayes's rule, there is about a one in two chance that golfer 93 is Tiger Woods.

Given our best scientific knowledge, say many contemporary proponents of the design argument, we see that certain special features of the Universe are extremely unlikely if the Universe is not the product of design but are quite likely if it is. In order to get what they say to fit the Bayesian format, they (or we, on their behalf) must articulate three specific propositions: a proposition concerning the relevant scientific background data (K); a proposition about the Universe's having a certain special feature (E); and a proposition identifying a particular design hypothesis (D). For example, a Bayesian design argument might involve the following propositions:

K_1 = Many of the initial conditions and free parameters of a universe need to be finely tuned in order for the development of life in that universe to be possible.
E_1 = The Universe is such that the development of life in it is possible.
D_1 = There is at least one supernatural designer.

The proponent of this sample Bayesian design argument would then make the following claims:

(1) $P(E_1|K_1 \&{\sim}D_1)$ is extremely low.
(2) $P(E_1|K_1 \& D_1)$ is quite high.
(3) $P(D_1|K_1)$ is considerably greater than $P(E_1|K_1 \&{\sim}D_1)$.

Claims (1)–(3) provide all of the necessary ingredients for a Bayesian inference. Using Bayes's rule, proponents of this sample design argument reach a profound conclusion:

(4) $P(D_1|E_1 \& K_1)$ is quite high.

That is, the existence of at least one supernatural designer is quite high given that life is possible in the Universe and given what we know about how the Universe must be if life is to be possible in it. Notice that, for this argument to work, something must be said about $P(D_1|K_1)$ relative to $P(E_1|K_1 \&{\sim}D_1)$. That is, something must be said about how the probability of the design hypothesis compares to the probability that life is possible in the Universe given the denial of the design hypothesis. The proponent of a Bayesian design argument cannot remain silent on the issue of the prior probability of the design hypothesis.

Richard Swinburne's version of the design argument is robustly Bayesian. For Swinburne, the relevant background data K is provided by contemporary physical cosmology and life science. The proposition E for Swinburne is that the Universe permits the existence of embodied agents that are sentient, intelligent, and free. The design hypothesis D just is that God exists. Swinburne argues that the prior probability that God exists is quite high – something near 0.5 – because God is the metaphysically simplest being we can conceive. In light of this high prior probability and in light of the restrictions the possibility of embodied agents puts on a universe, Swinburne argues, the posterior probability of theism is very high indeed.

As Elliott Sober presents it, however, the modern design argument is not Bayesian, but is rather an argument from likelihoods. The design arguments Sober considers are silent on the question of the prior probability of the design hypothesis, and so they are incapable of producing the conclusion that the posterior probability of the design hypothesis is high. They are only meant to show that the probability of a designer – whatever that probability is – is raised by the evidence. William Lane Craig, meanwhile, employs the logical apparatus articulated in William Dembski's *The Design Inference* (1998). As can be seen from reading Craig's presentation of Dembski's Generic Chance Elimination Argument and Michael Ruse's presentation of Dembski's Explanatory Filter, Dembski is no Bayesian. Instead, his model of design inference is akin to Ronald Fisher's model of scientific inference. In developing his notion of "significance tests," Fisher (1959) explicitly rejects the Bayesian account of what is essential to scientific inference. Fisher says scientists routinely and rightly reject hypotheses for making the data too improbable, doing so without assigning prior probabilities to the hypotheses and without considering any alternative hypotheses.

As we can see, there is considerable disagreement regarding the best way to frame the design argument. Even so, Swinburne, Sober, Craig, and Dembski at least agree that the design argument is best presented as an inference that involves probabilities at some level. Del Ratzsch does not. His intriguing, and disruptive, suggestion is that design is perceived, not inferred. Drawing on some remarks by Scottish Enlightenment philosopher Thomas Reid, Ratzsch proposes that recognizing design is like seeing, smelling, or hearing. Ratzsch even sees signs of a Reidian view of design recognition in the work of Paley. As the range of positions indicates, what is the best framework for formulating the design argument is a matter of considerable philosophical interest.

Defining "fine-tuned" and "irreducibly complex"

Let us set aside Ratzsch's suggestion for now and consider versions of the design argument that do employ a probabilistic inferential apparatus. With such arguments the evidence of design – whether it be cosmic fine-tuning or biological complexity – must support claims of improbability, whether explicitly or implicitly. As D.H. Mellor argues persuasively, this evidence will have to be *physically* improbable, not just epistemically so. I have maintained elsewhere (Manson 2000b) that even if a cosmic parameter P is such that life could not have arisen had the numerical value of P been slightly different, that does not imply that it is physically improbable that P takes a value that permits life. Robin Collins, William Lane Craig, Richard Swinburne, and John Leslie all work with just such a "slight-difference" or "narrow-limits" definition of fine-tuning (though Collins also argues there are circumstances in which the actual value of a parameter could reasonably count as fine-tuned for life even if the life-permitting range of values for that parameter is not narrow). Without the introduction of further assumptions, however, statements about how things would be if other things were slightly different cannot be converted into statements about how physically probable it is that things are the way they are.

For example, a size 10 shoe would not fit its wearer if it were more than half a shoe size larger or smaller, but to move from this 10 percent window of (shoe) fitness to the conclusion that there is a 10 percent chance the shoe fits would be a bizarre *non sequitur*. To justify that conclusion, one would need to make very odd assumptions regarding the sizes the shoe could have had and, for each of those possible sizes, how likely it was that the shoe would be that size. Again, an approach shot by Tiger Woods would not land within twenty feet of the pin if any component of his swing were slightly different, but that does not make it improbable that an approach shot of his lands within twenty feet of the pin. Unlike the swing of the typical golfer, Tiger's actual swing is extremely unlikely to be more than the slightest bit different from his intended swing. This is true even though (due to his strength and flexibility) the range of possible swings for Tiger is considerably greater than the range of possible swings for the typical golfer.

As Timothy McGrew, Lydia McGrew, and Eric Vestrup argue, what proponents of arguments from fine-tuning need to provide is a normalizable measure of the space of values the cosmic parameters might take; that is, the regions of this space of possibilities must be capable of adding up to one. Only then can there be meaningful talk about the probability that the cosmic parameters lie within the life-permitting regions; by definition, probabilities lie in the interval [0,1]. To get a normalizable space, however, one must assume either that there are limits on the numerical values the cosmic parameters could have taken (rather like the limitations the length of his arms imposes on Tiger's possible swings) or that some possible values are more likely than others (in the way that Tiger's great skill makes good swings more likely for him than bad ones). As I contend in Manson (2000b), neither of these assumptions – that the possible values are bounded or that a density function should be imposed on the space of possibilities – is (or could be?) warranted by current physical theory.

It is precisely for this reason that the McGrews and Vestrup regard what Robert O'Connor calls "local design arguments" (and what they call "life support arguments") as more promising than cosmic (for O'Connor, "global") design arguments. By focusing on what is possible *within* the arena defined by the Universe as a whole, in principle there is the possibility of providing well-defined probabilities for the items to be used as evidence of design. Taking the Universe as a whole – including the fact that it is fine-tuned for life – as an unexplained given, proponents of local design arguments instead seek evidence of design in scientifically established contingencies. These include such facts as that life has arisen in the Universe (which may be very improbable even if the Universe is fine-tuned for life), that the earth is a climatologically appropriate distance from the Sun, that several gas giant planets serve to deflect most large asteroids from collision courses with the Earth, and so on.

However, whether local design arguments are, indeed, more promising than cosmic ones is not so clear. O'Connor casts doubt on the notion that such local design arguments are far less reliant on controversial philosophical premises than their global counterparts. On the contrary, he says, they presuppose the extra-scientific claims that any scientific explanations of scientifically established contingencies either are discoverable by us or would have been discovered already. O'Connor cautions that any design argument will presuppose disputed *a priori* philosophical and metaphysical principles. There is no such thing as a "strictly scientific" design argument, whether that design argument be cosmic or local.

One kind of local design argument is the sort advanced by Michael Behe and other advocates of "Intelligent Design Theory." They claim that certain biological structures could not have arisen within the Universe by Darwinian means and so must be explained supernaturally. (Note that, in saying this, they presume that there are no non-Darwinian natural means for the production of these biological structures; this may not be true.) Their arguments rest on the notion of irreducible complexity, but the definition of "irreducibly complex" is as much a matter of contention as that of "fine-tuned for life". Behe defines the phrase as follows:

> By irreducibly complex I mean a single system composed of several well-matched, interacting parts that contribute to the basic function, wherein the removal of any one of the parts causes the system to effectively cease functioning.
>
> (Behe 1996: 39)

What Behe is after is a definition such that, if a biological structure meets it, that biological structure could not have arisen by a Darwinian process. And it seems that if a biological structure is irreducibly complex in Behe's sense of the term, it indeed could not have been selected for by a Darwinian process. Evolution selects from functioning systems, yet any precursor to an irreducibly complex structure would, it seems, be non-functional. Given that such a biological system could not be explained in Darwinian terms, the next step in the inference to a designer would be to calculate the probability that the system arose by chance. And as may not be the case with fine-tuning, such calculations could make reference to scientifically established facts about the Universe regarding its age, the number of particles in it, the number of habitable planets in it, and so on in order to establish the number of opportunities that were available for the irreducibly complex structures to arise.

As Kenneth Miller notes, however, it is not enough for Behe's argument that he identify a biological structure the existence of which *currently lacks* a Darwinian explanation. According to Miller, Behe is trying to define the sort of biological structure the existence of which Darwin's theory *could not possibly* explain. Miller says Behe's definition fails to satisfy Behe's own criterion. Behe talks about *the* basic function of a biological system, when in order to tackle Darwinism on its own terms he should be talking about *some function or other* of a biological system. Biological functionality is defined only in the context of an environment, Miller insists. As an environment changes, the function of a system operating within it can change too. Miller argues that selectable functions do exist for the components of allegedly irreducibly complex systems (e.g. eubacterial flagella). Michael Ruse illustrates the same point when he discusses the energy-converting Krebs cycle. Ruse notes as well that Behe's definition does not take simplifying changes into consideration. A biological system could be irreducibly complex in Behe's sense yet be achievable via a Darwinian process if there existed a more complex precursor that was itself not irreducibly complex. Ruse uses the example of an arched stone bridge to illustrate this point. Once the keystone is placed, the bridge builders can remove the scaffolding. The stone bridge then becomes such that the removal of any one of its parts will cause the bridge to collapse, but that does not mean the stone bridge was not the product of a gradual process.

In considering these objections, however, we must not forget the key point of Behe's contribution to this volume. It should be possible to define a biological system such that, if it were to exist, its existence could not be explained in Darwinian fashion. If it is impossible to define such a biological system, then it will be impossible to formulate an empirical test that might disconfirm Darwin's theory. Darwinism's claim to be a genuine scientific theory would suffer a serious (if not mortal) blow. Darwin himself, Behe notes, recognized that he needed to provide a criterion for falsifying his theory. Yet Behe claims that in practice the defenders of Darwinism fail to admit the possibility of falsifying Darwinism. Meanwhile, they assert in the same breath both that Behe's "Intelligent Design" hypothesis is unfalsifiable and that there is evidence against it! (Miller does not make this mistake; his position is that Behe's hypothesis does make predictions and is falsifiable.) So even if the particular definition of 'irreducibly complex' Behe provides is inadequate (is such that the existence of a biological system which meets it would not necessarily disconfirm Darwinism), it might be in the interest of Darwinists to repair the definition to make it adequate.

Specifying for what the Universe is designed

The design argument involves the claim that the Universe, or some part of it, is designed for something. For example, design arguments from fine-tuning rest on the claim that the Universe is fine-tuned for something. But for what? A range of answers is given. William Lane Craig, Robin Collins, and John Leslie specify intelligent life as that for which the Universe is fine-tuned. Richard Swinburne's design argument from fine-tuning is framed in terms of the necessary conditions for the existence of embodied agents that are sentient, intelligent, and free. Timothy McGrew, Lydia McGrew, and Eric Vestrup speak of fine-tuning for carbon-based life, while D.H. Mellor and Martin Rees talk of fine-tuning for mere life.[7] Paul Davies, meanwhile, sees the Universe as set up for the production of complex, self-organizing systems, though he does also talk about consciousness being written into the laws of nature.

As I have noted elsewhere (Manson 2000a), the design argument is almost always characterized by its critics as involving anthropocentrism or (to use J.J.C. Smart's term) "psychocentrism" (Smart and Haldane 1996: 26–7). Hume, for example, claimed proponents of the design argument made the mistake of applying a particular mode of explanation – namely, explanation in terms of the possession of particular thoughts – to the Universe as a whole just because that mode of explanation often works with respect to humans. In doing this, he said, proponents of the design argument make humans "the model of the whole Universe."

> But allowing that we were to take the *operations* of one part of nature upon another for the foundation of our judgment concerning the *origin* of the whole (which never can be admitted), yet why select so minute, so weak, so bounded a principle as the reason and design of animals is found to be upon this planet? What peculiar privilege has this little agitation of the brain which we call *thought*, that we must thus make it the model of the whole universe? Our partiality in our own favor does indeed present it on all occasions, but sound philosophy ought carefully to guard against so natural an illusion.
>
> (Hume 1970 [1779]: 28)

Similarly, Bertrand Russell criticizes the design argument for resting on an inegalitarian ethical picture:

> Is there not something a trifle absurd in the spectacle of human beings holding a mirror before themselves, and thinking what they behold so excellent as to prove that a Cosmic Purpose must have been aiming at it all along? Why, in any case, this glorification of Man? How about lions and tigers? They destroy fewer animal or human lives than we do, and they are much more beautiful than we are.... Would not a world of nightingales and larks and deer be better than our human world of cruelty and injustice and war?
>
> (Russell 1961: 221)

As we can see, specifying for what the Universe is designed is not ethically unproblematic. For one thing, there is the risk of causing offense by leaving out of the specification important kinds of beings. For example, specifying the Universe as fine-tuned for intelligent life suggests that anything that is not both living and intelligent would not be a worthy end for a designer. Contemporary environmentalists and animal advocates would likely take exception to such a specification.

Proponents of the argument from fine-tuning could buy themselves some room for maneuver, however, if the probabilities on the chance of the Universe being suitable for life, intelligent life, carbon-based life, self-organizing complex systems, Gaia, nightingales, larks, deer, and so on were all effectively the same. This does seem to be the case. As indicated by the accounts of fine-tuning Collins and Craig provide, had any of the free cosmic parameters been the slightest bit different, the Universe would have been *radically* different. It would have lasted only a microsecond, or its matter would have been a billion times more diffuse, or its mean temperature would have been a million times greater. It appears the Universe would not have allowed for *any* of the beings specified if the values of its free parameters had been even slightly different.

In connection with this point, Simon Conway Morris argues that the chances of *human-like* life eventually arising in the Universe are effectively the same as the chances of life eventually arising in the Universe. He would agree with Paul Davies that "the emergence of life and consciousness somewhere and somewhen in the cosmos is...assured by the underlying laws of nature." The widespread phenomenon of convergent evolution suggests to Conway Morris that the eventual emergence in a biosphere of human-like biological properties is extremely likely given a reasonable amount of time. In taking this line he rejects the popular view that evolution is a "random walk" in which the evolution of humans is not to be expected. What is *really* not to be expected, says Conway Morris, is the existence of such a biosphere. Recent discoveries indicate such biospheres are (cosmically speaking) few and far between.

This suggests a picture of our place in the Universe that runs contrary to the "cosmic accident" view, according to which science – especially since Copernicus and Darwin – has shown our cosmic insignificance. Indeed, there are those for whom the teachings of Darwin and Copernicus are the organizing principles of an ethical cause on behalf of which they proselytize. Consider, again, Russell:

> Man, as a curious accident in a backwater, is intelligible: his mixture of virtues and vices is such as might be expected to result from a fortuitous origin. But only abysmal self-complacency can see in Man a reason which Omniscience could consider adequate as a motive for the Creator. The Copernican revolution will not have done its work until it has taught men more modesty than is to be found among those who think Man sufficient evidence of Cosmic Purpose.
>
> (Russell 1961: 222)

Russell is not the only one who sees science as putting humanity in its place. The idea that a heliocentric model of the solar system and an evolutionary account of humanity's existence have some sort of homiletic "work" to do has wide currency. Contrary to Russell's claim that we are "a curious accident in a backwater," however, one of the key discoveries of contemporary science is that our evolution depends crucially on the broad-scale features of the Universe and on specific phenomena such as star formation and star death. The chapters by Craig, Collins, and Conway Morris make this abundantly clear. So in so far as the "cosmic accident" view is mistaken – in so far as it is plausible to maintain the Universe is *for* something – the design argument has a chance of getting off the ground.

What should we expect from a supernatural designer?

Even if the existence of intelligent life now is radically contingent, does that necessarily disqualify the Universe and the intelligent life within it from counting as products of design? Kenneth Miller, Michael Ruse, John Leslie, and Peter van Inwagen say no. Van Inwagen searches for, but cannot find, a good reason for thinking God would not use the mechanism of natural selection to produce rational beings. He sees Darwin's account of evolution as wholly compatible with the claim that living beings (including rational beings such as ourselves) are the products of intelligent design, even though the evolution of intelligent life is not guaranteed in a universe with the laws and initial conditions of ours. Miller makes the same point in *Finding Darwin's God*, claiming that "the notion that we must find historical inevitability in a process in order to square it with the intent of a Creator makes absolutely no sense" (1999: 273):

> Can we really say that no Creator would have chosen an indeterminate, natural process as His workbench to fashion intelligent beings? Gould argues that if we were to go back to the Cambrian era and start over a second time, the emergence of intelligent life exactly 530 million years later would not be certain. I think he is right, but I also think this is less important than he believes. Is there some reason to expect that the God we know from Western theology had to preordain a timetable for our appearance? After 4.5 billion years, can we be sure He wouldn't have been happy to wait a few million longer? And, to ask the big question, do we have to assume that from the beginning He planned intelligence and consciousness to develop in a bunch of nearly hairless, bipedal, African primates? If another group of animals had evolved to self-awareness, if another creature had shown itself worthy of a soul, can we really say for certain that God would have been less than pleased with His new Eve and Adam? I don't think so.
>
> (Miller 1999: 274)

Leslie agrees with Miller, saying that the Universe is designed, not for our species in particular, but for intelligent life more generally.[8]

Likewise, Ruse claims not to see why God would not use evolution as His means for producing intelligent life. Like several of the contributors to this volume, Ruse finds *most* attractive a theological picture according to which God does not intervene in the Universe subsequent to bringing it into existence. Leslie opines that "any deity who supplemented laws of physics by life-forces and acts of interference would have produced a disappointingly untidy universe." Paul Davies agrees, saying he "would rather that nature can take care of itself." According to this line of thought (which is similar to deism, but is now often referred to as "theistic evolutionism"), it is to be expected that God would frontload into the Universe all that He wanted it eventually to produce. God, according to theistic evolutionists, would be expected to let evolution do the (dirty) work of bringing about the existence of intelligent life. Though theistic evolutionism is not the standard view regarding God's relationship to His creation, it is an increasingly popular one.

It is precisely with respect to the "Why?" and "How?" of creation, however, that skeptics such as Jan Narveson object to the design argument. The hypothesized designer of this universe will need a motive for having designed it, yet Narveson sees no reason for thinking there is such a motive. He takes the argued-for designer just to be God. Being absolutely perfect, however, speaks against God's having a motive. In saying this, Narveson echoes Spinoza, who thought that "if God acts with an end in view, he must necessarily be seeking something that he lacks" (Spinoza 1982 [1677]: 59) and hence must be incomplete and imperfect. Possessing omnipotence but lacking a motive, then, means God is no more likely to create one conceivable universe than any other. Yes, God *might* use evolution as a means to produce intelligent life (although skeptics will be quick to contend that evolution is an amazingly cruel and wasteful process – one that produces an amount of suffering no supremely good being would allow). God also might create the world in seven days, with humans being fashioned out of dust. He might create a universe hostile to life, then overcome that hostility and create beings like us. God might even create a universe that lasts a microsecond. Or God might simply not create anything at all. All of these are possible, but why think one is preferred? Proponents of the design argument are trying to argue from the way the world is to God, not just to reconcile the way the world is with God.[9] So unless proponents of the design argument can show why we should expect God to create our sort of universe, the hypothesis that God exists makes it no more likely that our universe exists. Because of this intractable problem, says Narveson, the design argument fails. Elliott Sober levels a similar criticism, saying "the assumption that God can do anything is part of the problem, not the solution. An engineer who is more limited would be more predictable."

As I note in Manson (2000a), one might think this objection can be avoided simply by refusing to identify straightaway the designer(s) with God. Most proponents of the design argument do just this, maintaining (as Michael Behe does explicitly at the beginning of his chapter) that the design hypothesis with which they operate is much weaker than theism:

> [W]hile I argue for design, the question of the identity of the designer is left open. Possible candidates for the role of designer include: the God of Christianity; an angel – fallen or not; Plato's demiurge; some mystical new age force; space aliens from Alpha Centauri; time travelers; or some utterly unknown intelligent being.

Behe thinks this more modest version of the design hypothesis keeps the design argument from falling afoul of issues such as the problem of evil and the paradox of omnipotence. But the increased plausibility of Behe's modest design hypothesis is purchased at the cost of explanatory power. At least the notion of moral perfection is included within the concept of God. To say, however, that a powerful supernatural being exists is to say nothing about that being's motivations, unless a set of preferences can somehow be teased out of the very concept of rationality. The prospects for doing so are dim if we accept the dominant view in contemporary philosophy of mind and action, according to which rationality just is effectiveness at using means to achieve desired ends. Being rational, according to this view, does not imply preferring the good. (Kantians will surely see this as a defect of purely means–ends accounts of rationality.) What all this shows is that proponents of design hypotheses weaker than theism will find themselves in deep philosophical waters when they try to explain why their hypotheses make the existence of a universe like ours more probable.

An analogy will be helpful in grasping this point. Compare the proponent of a non-theistic design hypothesis to a poker player who accuses the dealer of having fixed the deck on a particular hand. The allegation of cheating is credible when the dealer gets a valuable hand (e.g. a Royal Flush) but not when the dealer gets a worthless hand (e.g. the two of clubs, the five of diamonds, the seven of spades, the nine of hearts, and the queen of clubs). The ability to fix decks alone does not raise the probability that the dealer will get the worthless hand. The player could remedy this problem by attributing to the dealer a fetish for that particular worthless sequence of cards. But that move is no good either, for while it is highly probable that a dealer with such a fetish would deal herself just that sequence, it is highly improbable that any dealer has such a fetish. Likewise, the hypothesis that there exists a designer with the power to design a universe such as ours does not raise the probability of the existence of a universe such as ours unless the designer also has a motive for creating such a universe. Yet building enough of a motive into a non-theistic design hypothesis for it to make the existence of a universe like ours more probable risks driving down the prior probability of that hypothesis.

Unlike many proponents of the design argument, Richard Swinburne takes up the challenge of explaining in detail why God would be expected to create a universe like ours. In his chapter he argues that: (1) it follows from God's nature that He will try to bring about a great amount of the greatest sort of good; (2) bringing about a great amount of the greatest sort of good requires bringing about the existence of free beings; (3) free beings need an arena in which to develop morally and interact socially; and (4) this arena requires the creation of a fine-tuned and law-governed universe. Notice, however, that Swinburne begins by working with a theistic design hypothesis rather than a weaker design hypothesis of the sort Behe and others advocate. In giving reasons for expecting a designer to create a universe like ours, proponents of such design hypotheses cannot help themselves to the greater resources the theistic design hypothesis provides.

The much-maligned multiverse

Our discussion of the design argument would not be complete without mention of the multiverse hypothesis. "I really do believe that the case for design stands or falls upon whether we can find another explanation in terms of multiple universes," Paul Davies said in a recent interview (Davies 2002). According to the multiverse hypothesis, there are very many (if not infinitely many) things like the Universe. Though these huge physical systems share certain basic lawful structures (e.g. they all follow quantum-mechanical laws), the free cosmic parameters randomly take different values in the different universes. Given this multiverse, it is unsurprising that at least one universe in the vast ensemble is fit for the production of life. Furthermore, with respect to irreducible complexity and the origin of life, if vastly many universes in the ensemble are fit for life, then the "probabilistic resources" (to use William Dembski's term) for attributing the origin of life and the existence of irreducibly complex biological structures to chance might be inflated sufficiently to render appeal to the design hypothesis unnecessary. Thus the multiverse is (to use another of Dembski's terms) an "inflaton":

> [S]ome entity, process, or stuff outside the known universe that in addition to solving some problem also has associated with it numerous probabilistic resources as a byproduct. These resources in turn help to shore up chance when otherwise chance would seem unreasonable in explaining some event.

How, exactly, is the multiverse hypothesis supposed to explain fine-tuning? According to the weak version of what physicist Brandon Carter dubbed "the anthropic principle," observers should expect the Universe to meet whatever conditions are necessary for the existence of observers.[10] As Leslie notes, the anthropic principle calls to our attention an "observational selection effect" at work in cosmology – a feature of our methods of observation that systematically selects from only a subset of the set of observations we might have made.[11] To take an example from the social sciences, conducting a telephone poll introduces an observational selection effect. The method of telephone polling guarantees that one's survey will neglect certain segments of the population (e.g. those without telephones). So the multiverse hypothesis, when considered in light of the observational selection effect to which the anthropic principle calls our attention, is thought to provide a plausible naturalistic alternative to the claim that the apparent design in and of the Universe was produced by a supernatural designer. As Martin Rees suggests:

> [T]he cosmos maybe has something in common with an "off the shelf" clothes shop: if the shop has a large stock, we're not surprised to find one suit that fits. Likewise, if our universe is selected from a multiverse, its seemingly designed or fine-tuned features wouldn't be surprising.

As the chapters by D.H. Mellor, William Dembski, William Lane Craig, and Roger White indicate, there is considerable hostility towards the multiverse hypothesis. Perhaps the most common reactions to it are that it is *ad hoc* – "a sort of backhanded compliment to the design hypothesis," as Craig claims – and that it is metaphysically extravagant. The only motivation for believing it, goes the first complaint, is to avoid the obvious religious implications of the discovery of fine-tuning. The multiverse hypothesis is alleged to be the last resort for the desperate atheist. According to the second, the multiverse hypothesis violates Occam's razor, the philosophical injunction not to multiply entities beyond necessity when giving explanations. Assuming two hypotheses have the same explanatory power, Occam's razor dictates that we pick the simpler one. Swinburne and Craig claim the design hypothesis involves postulating a relatively simple entity. A multiverse, on the other hand, is (they claim) a vast, jumbled, arbitrary mess.

Regarding the first common objection, while it is certainly possible that what prompts some proponents of the multiverse hypothesis is a desire to avoid theism, it would be wrong to reject the multiverse hypothesis on that basis alone. The multiverse hypothesis may be false, but the fact (if it is a fact) that its originators developed it and its proponents defend it in order to avoid believing in God does not make it false. The key question is whether the multiverse hypothesis has independent support. Rees insists it could. The multiverse hypothesis, he says, is scientifically testable; those who deny this on the grounds that other universes are unobservable must explain why hypotheses about objects that lie beyond the detection of current telescopes or that cannot be detected during the current cosmic era are not likewise unscientific. And if there is independent scientific evidence for the multiverse hypothesis, says Rees, who could object to appealing to that hypothesis to explain the fine-tuning of the Universe for life?

Regarding the second common objection to the multiverse hypothesis, we should be wary of measuring simplicity too simplistically and of taking simplicity as the sole criterion of the merit of a hypothesis. The simplicity of a hypothesis is not merely a function of the raw number of entities it posits. The Standard Model in particle physics posits a small number of types of subatomic particle, but of course there are countless tokens (instances) of each of these types. Yet the Standard Model is rightly regarded as a good scientific explanation – one perfectly in accord with Occam's razor – because of its symmetry and because it invokes a small number of types. Depending on how it is fleshed out, the multiverse hypothesis, too, could exhibit simplicity in these regards. What multiverse critics need here is a comprehensive account of simplicity and clear, detailed statements of both the design and multiverse hypotheses before they deem the latter metaphysically extravagant.[12] With regard to this last point, skeptics like Narveson will retort that the design hypothesis is hardly simple if it just is the hypothesis that there exists an eternal, personal being of unlimited power, knowledge, and goodness. They find such a being incomprehensibly complex.

An increasingly popular objection to the multiverse hypothesis is that it fails to explain why *this* universe is fine-tuned. The "This Universe" objection is well expressed by Alan Olding:

> [T]he "world-ensemble" theory provides no explanatory comfort whatsoever. The situation is this. We have our own universe with planets occasionally, if not always, producing life; and, to escape explaining this fact, we surround it with a host of other universes, most limp and halting efforts and some, perhaps, bursting at the seam with creatures. But where is the comfort in such numbers? The logical situation is unchanged – *our* universe, the one that begat and nourished us, is put together with as unlikely a set of fine-tuned physical values whether it exists in isolation or lost in a dense scatter of worlds. So, then, by itself or surrounded by others, the existence of our universe still cries out for explanation.
>
> (Olding 1991: 123)

Craig, Dembski, Mellor, and Elliott Sober all raise the "This Universe" objection in their chapters in one form or another, but it is spelled out in a particularly detailed way by Roger White. He argues that the multiverse hypothesis ("M") merely "screens off" the probabilistic support that fine-tuning lends to the design hypothesis. That is, if there are many universes, then the probability that *this* one is life-permitting will be no greater on the supposition that there is a designer than on the supposition that there is not. To use the notation we introduced earlier, White says

$$P(E|D \,\&\, M \,\&\, K) = P(E|{\sim}D \,\&\, M \,\&\, K).$$

This is because there is no reason, White thinks, why a designer would single out *this* universe (as opposed to one of the others) to be the one that permits life. Despite this, the multiverse hypothesis fails to raise the probability that our universe is fine-tuned and so is not confirmed by the fact that our universe is fine-tuned. To appreciate this point, suppose for the sake of argument that 1 percent of all the universes that are possible within the multiverse scenario are such as to permit life, and that, according to the multiverse hypothesis, there are exactly 1,000 universes, all chosen at random from the set of possible universes. White would say that the probability that our universe permits life is still just 1 percent, because what goes on in the other 999 universes does not affect what goes on in ours. Of course, on this particular multiverse scenario, the probability that *some universe or other* permits life is much higher: 99.99 percent.[13] But that is not relevant, White would say.

Dembski gets at the same point when he asks us to consider the hypothesis that there are infinitely many Arthur Rubinstein lookalikes. If we postulate enough such impostors then we can be confident that somewhere in all of reality there is a Rubinstein impostor who by pure luck plunks down his fingers so that Liszt's "Hungarian Rhapsody" is played. How do we avoid the conclusion that the multiRubinsteins hypothesis explains our observing that a person who looks just like Arthur Rubinstein is performing Liszt's "Hungarian Rhapsody"? Dembski says we do this by demanding that our explanation make the performance likely on a local scale. That is, our explanation must make it likely that *this* person who looks like Rubinstein – the person in front of *us* – is performing Liszt's "Hungarian Rhapsody." The best explanation of that fact, Dembski urges, is that the performer *really is* Arthur Rubinstein. The multiRubinsteins hypothesis makes it likely that *some Rubinstein lookalike or another* is (by pure luck) giving a great performance, but makes it no more likely that *this* Rubinstein lookalike is doing so.

Michael Thrush and I (2003) see several problems with the "This Universe" objection. First, the sort of question to which its proponents demand an answer – "Why is *this* universe fit for life?" – is not asked with respect to comparable explanations in terms of great replicational resources. For example, when it comes to explaining the fitness of the Earth for life, accounts that appeal to the vast number of planets in our universe (and hence the vast number of chances for conditions to be just right) surely are not to be faulted for failing to explain why *this* planet is the fit one. One reason why is that, when we set aside all of the features of the Earth that are essential to its ability to produce living creatures (including relational properties such as distance from the right sort of star), there is otherwise nothing special about it. There *might* have been something special about the Earth. For example, it could have been that only from the vantage point of the Earth could an observer see that the constellations spell out "THIS UNIVERSE IS GOD'S HANDIWORK." But in the absence of such a special feature, there is no motivation for the demand to explain why this planet in particular is fit for life. So why think the "This Universe" objection is any more worrisome than the "This Planet" objection?[14]

Furthermore, Thrush and I argue, the "This Universe" objection helps itself to some non-obvious metaphysical assumptions, the most important of which is that the Universe could have taken different values for its free parameters. Yet whether the values of its free parameters are among the essential properties of a universe will depend, we think, on what a given multiverse theory says a universe is. In fairness to proponents of the "This Universe" objection, however, we acknowledge that multiverse proponents are generally silent on the identity conditions of the type of object they postulate. We conclude that much more scientific and philosophical groundwork must be laid before the multiverse hypothesis can rightly be regarded as explaining – or failing to explain – apparent design. Whether cosmic fine-tuning and biological complexity require any explanation at all is a question we leave for the reader.

Notes

1 Unless otherwise indicated, references to the works of the contributors are to their chapters in this volume. Thus, when you read that a contributor says so-and-so, take that to mean the contributor says so-and-so in his or her contribution to this book.

2 Richard Swinburne's argument is a notable exception to this rule.

3 See Barrow and Tipler (1986: 29) for a definition of "eutaxiological argument". Swinburne (1979: Ch. 8) provides another good example of the eutaxiological argument when he reasons to the existence of God from the "temporal order" of the world. G.K. Chesterton's story of Elfland (1936: Ch. 4) provides another good illustration of the argument.

4 To say that the design argument is distinct from the cosmological and eutaxiological arguments, however, is not to say there are no logical connections among those arguments. In "The poverty of theistic cosmology" (forthcoming in *Philo: The Journal of the Society of Humanist Philosophers*), Adolf Grünbaum claims that the theistic design and eutaxiological arguments take as their explanatory framework a theistic cosmological scenario of *creation ex nihilo*. According to the doctrine of *creation ex nihilo*, God is the creator of all logically contingent existing entities and of the laws that those entities follow; no concrete beings and no laws exist independently of Him. If theists are committed to the doctrine of *creation ex nihilo* and if the design and eutaxiological arguments are arguments for the existence of God, then the picture at work in those arguments cannot be one of God designing or imparting order to an independently existing world – to material for the existence of which He is not responsible. Grünbaum argues that the cosmological argument fails and that, because the theological explanation of

the laws is inseparable from the theological explanation of the contents of the world, proponents of the design argument are burdened with the need to support as well the *creation ex nihilo* framework on which their argument depends.

5 Throughout this book the term "the Universe" has been capitalized to indicate it is being used as a proper name for a unique astronomical object. The reason for doing so will be apparent later when we consider the multiverse hypothesis, according to which the Universe is just one instance or token of a particular natural kind. One exception to this rule is the chapter by D.H. Mellor; in it he uses "the Universe" to mean "everything that exists in some space–time or other."

6 For a comprehensive history, explication, and defense of Bayesianism, see Howson and Urbach (1993).

7 This is how Paley saw biological design. Throughout *Natural Theology* he appeals to the intricacy of life and the "works of nature," with nary a mention of intelligence or consciousness and with humanity mentioned only with respect to its anatomy. The evidence for Paley's argument would be just as strong were it drawn from cases more than several million years old – well before we humans came on the scene.

8 As Leslie points out, one sobering consequence of this view is that cosmic design is compatible with the eventual extinction of our species.

9 Hume's Philo makes precisely this point; see Hume (1970 [1779]: 94–5).

10 The term "anthropic" misleadingly suggests the principle refers to humans only, as opposed to observers more generally (e.g. Martians, Arcturans, or very smart dolphins); since the term "anthropic principle" is so entrenched, however, most people who write about cosmic fine-tuning continue to use it.

11 For a detailed and vigorous presentation and discussion of observational selection effects, see Bostrom (2002).

12 Swinburne tried to provide just such a criterion of simplicity in his 1997 Aquinas Lecture (Swinburne 1997).

13 The probability that at least one of the 1,000 universes permits life is equal to 1 minus $(0.99)^{1,000}$ – the probability that none of the 1,000 universes permits life.

14 On hearing news reports that a lone family in a remote Armenian village survived a devastating earthquake in December 1988 (nearly 50,000 Armenians were killed by that earthquake), a friend of mine said at the time "It's a miracle." When I noted that, given the size of the area, it wasn't unlikely that some family occupied a protected position in a fortified cellar at the time of the quake, she replied "Well, it's a miracle that *they* survived." When I retorted that this was (from her point of view) equivalent to saying "Well, it's a miracle that *the survivors* survived" and that there was nothing the least surprising about *that*, she made a few choice comments about how philosophers ruin everything.

References

Barrow, J.D. and Tipler, F.J. (1986) *The Anthropic Cosmological Principle*, Oxford: Oxford University Press.

Behe, M. (1996) *Darwin's Black Box: The Biochemical Challenge to Evolution*, New York: Free Press.

Bostrom, N. (2002) *Anthropic Bias: Observation Selection Effects in Science and Philosophy*, New York: Routledge.

Chesterton, G.K. (1936) *Orthodoxy*, New York: Doubleday/Image Books.

Darwin, C. (1859) *On the Origin of Species*, London: John Murray.

Davies, P. (2002) "God and time machines: A conversation with Templeton Prize-winning physicist Paul Davies," *Books & Culture* 8 (March/April): 29.
Dembski, W. (1998) *The Design Inference*, Cambridge: Cambridge University Press.
Fisher, R.A. (1959) *Statistical Methods and Scientific Inference*, London: Oliver & Boyd.
Howson, C. and Urbach, P. (1993) *Scientific Reasoning: The Bayesian Approach*, 2nd edn, Chicago: Open Court Press.
Hume, D. (1970 [1779]) *Dialogues Concerning Natural Religion*, ed. N. Pike, Indianapolis: Bobbs-Merrill Educational Publishing.
Manson, N.A. (2000a) "Anthropocentrism and the design argument," *Religious Studies* 36 (June): 163–76.
—— (2000b) "There is no adequate definition of 'fine-tuned for life'," *Inquiry* 43 (September): 341–51.
Manson, N.A. and Thrush, M. (2003) "Fine-tuning, multiple universes, and the 'this universe' objection," *Pacific Philosophical Quarterly* (forthcoming);84(1).
Miller, K. (1999) *Finding Darwin's God*, New York: Harper & Row.
Olding, A. (1991) *Modern Biology and Natural Theology*, New York: Routledge.
Paley, W. (1819 [1802]) *Natural Theology (Collected Works: IV)*, London: Rivington.
Russell, B. (1961) *Religion and Science*, Oxford: Oxford University Press.
Smart, J.J.C. and Haldane, J.J. (1996) *Atheism and Theism*, London: Blackwell.
Spinoza, B. (1982 [1677]) *The Ethics*, trans. S. Shirley and ed. S. Feldman, *The Ethics and Selected Letters*, Indianapolis: Hackett.
Swinburne, R. (1997) *Simplicity as Evidence of Truth*, Milwaukee: Marquette University Press.
—— (1979) *The Existence of God*, Oxford: Oxford University Press.
von Weizsäcker, C.F. (1964) *The Relevance of Science*, New York: Harper & Row.

Part I

GENERAL CONSIDERATIONS

1

THE DESIGN ARGUMENT

Elliott Sober

The design argument is one of three main arguments for the existence of God; the others are the ontological argument and the cosmological argument. Unlike the ontological argument, the design argument and the cosmological argument are *a posteriori*. And whereas the cosmological argument could focus on any present event to get the ball rolling (arguing that it must trace back to a first cause, namely God), design theorists are usually more selective.

Design arguments have typically been of two types – *organismic* and *cosmic*. Organismic design arguments start with the observation that organisms have features that adapt them to the environments in which they live and that exhibit a kind of *delicacy*. Consider, for example, the vertebrate eye. This organ helps organisms survive by permitting them to perceive objects in their environment. And were the parts of the eye even slightly different in their shape and assembly, the resulting organ would not allow us to see. Cosmic design arguments begin with an observation concerning features of the entire cosmos – the Universe obeys simple laws, it has a kind of stability, its physical features permit life and intelligent life to exist. However, not all design arguments fit into these two neat compartments. Kepler, for example, thought that the face we see when we look at the moon requires explanation in terms of intelligent design. Still, the common thread is that design theorists describe some empirical feature of the world and argue that this feature points towards an explanation in terms of God's intentional planning and away from an explanation in terms of mindless natural processes.

The design argument raises epistemological questions that go beyond its traditional theological context. As William Paley (1802) observed, when we find a watch while walking across a heath, we unhesitatingly infer that it was produced by an intelligent designer. No such inference forces itself upon us when we observe a stone. Why is explanation in terms of intelligent design so compelling in the one case, but not in the other? Similarly, when we observe the behavior of our fellow human beings, we find it irresistible to think that they have minds that are filled with beliefs and desires. And when we observe non-human organisms, the impulse to invoke mentalistic explanations is often very strong, especially when they look a lot like us. When does the behavior of an organism – human or not – warrant this mentalistic interpretation? The same question can be posed about machines. Few of us feel tempted to attribute beliefs and desires to hand calculators. We use calculators to help us add, but they don't literally figure out sums; in this respect, calculators are like pieces of paper on which we scribble our calculations. There is an important difference between a device that *we* use to help us think and a device that *itself* thinks. However, when a computer plays a decent game of chess, we may find it useful to explain and predict its behavior by thinking of it as having goals and deploying strategies (Dennett 1987b). Is this merely a useful fiction, or does the machine really have a mind? And if we think that present-day chess-playing computers are, strictly speaking, mindless, what would it take for a machine to pass the test? Surely, as Turing (1950) observed, it needn't look like us. In all these contexts, we face the problem of other minds (Sober 2000a). If we understood the ground rules in this general epistemological problem, that would help us think about the design argument for the existence of God. And, conversely, if we could get clear on the theological design argument, that might throw light on epistemological problems that are not theological in character.

What is the design argument?

The design argument, like the ontological argument, raises subtle questions about what the logical structure of the argument really is. My main concern here will not be to describe how various thinkers have presented the design argument, but to find the soundest formulation that the argument can be given.

The best version of the design argument, in my opinion, uses an inferential idea that probabilists call the *likelihood principle* (LP). This can be illustrated by way of Paley's (1802) example of the watch on the heath. Paley describes an observation that he claims discriminates between two hypotheses:

(W) O_1: the watch has features $G_1 \ldots G_n$.
W_1: the watch was created by an intelligent designer.
W_2: the watch was produced by a mindless chance process.

Paley's idea is that O_1 would be unsurprising if W_1 were true, but would be very surprising if W_2 were true. This is supposed to show that O_1 favors W_1 over W_2; O_1 supports W_1 more than it supports W_2. Surprise is a matter of degree; it can be captured by the concept of conditional probability. The probability of observation (O) given hypothesis (H) – $Pr(O|H)$ – represents how unsurprising O would be if H were true. LP says that comparing such conditional probabilities is the way to decide what the direction is in which the evidence points:

(LP) Observation O supports hypothesis H_1 more than it supports hypothesis H_2 if and only if $Pr(O|H_1) > Pr(O|H_2)$.

There is a lot to say on the question of why the likelihood principle should be accepted (Hacking 1965; Edwards 1972; Royall 1997; Forster and Sober 2003); for the purposes of this essay, I will take it as a given.

We now can describe the likelihood version of the design argument for the existence of God, again taking our lead from one of Paley's favorite examples of a delicate adaptation. The basic format is to compare two hypotheses as possible explanations of a single observation:

(E) O_2: the vertebrate eye has features $F_1 \ldots F_n$.
E_1: the vertebrate eye was created by an intelligent designer.
E_2: the vertebrate eye was produced by a mindless chance process.

We do not hesitate to conclude that the observations strongly favor Design over Chance in the case of argument (W); Paley claims that precisely the same conclusion should be drawn in the case of the propositions assembled in (E).[1]

Clarifications

Several points of clarification are needed here concerning likelihood in general and the likelihood version of the design argument in particular. First, I use the term "likelihood" in a technical sense. Likelihood is not the same as probability. To say that H has a high likelihood, given observation O, is to comment on the value of $Pr(O|H)$, not on the value of $Pr(H|O)$; the latter is H's *posterior probability*. It is perfectly possible for a hypothesis to have a high likelihood and a low posterior probability. When you hear noises in your attic, this confers a high likelihood on the hypothesis that there are gremlins up there bowling, but few of us would conclude that this hypothesis is probably true.

Although the likelihood of H (given O) and the probability of H (given O) are different quantities, they are related. The relationship is given by Bayes's theorem:

$$Pr(H|O) = Pr(O|H) \cdot Pr(H)/Pr(O).$$

Pr(H) is the hypothesis' *prior probability* – the probability that H has before we take the observation O into account. From Bayes's theorem we can deduce the following:

$$Pr(H_1|O) > Pr(H_2|O) \text{ if and only if } Pr(O|H_1) \cdot Pr(H_1) > Pr(O|H_2) \cdot Pr(H_2).$$

Which hypothesis has the higher posterior probability depends on how their likelihoods are related, but also on how their prior probabilities are related. This explains why the likelihood version of the design argument does not show that Design is more probable than Chance. To draw this further conclusion, we would have to say something about the prior probabilities of the two hypotheses. It is here that I wish to demur (and this is what separates me from card-carrying Bayesians). Each of us perhaps has some subjective degree of belief, before we consider the design argument, in each of the two hypotheses (E_1) and (E_2). However, I see no way to understand the idea that the two hypotheses have *objective* prior probabilities. Since I would like to restrict the design argument as much as possible to matters that are objective, I will not represent it as an argument concerning which hypothesis is more probable. However, those who have prior degrees of belief in (E_1) and (E_2) may use the likelihood argument to update their subjective probabilities. The likelihood version of the design argument says that the observation O_2 should lead you to increase your degree of belief in (E_1) and reduce your degree of belief in (E_2).

My restriction of the design argument to an assessment of likelihoods, not probabilities, reflects a more general point of view. Scientific theories often have implications about which observations are probable (and which are improbable), but it rarely makes sense to describe them as having objective probabilities. Newton's law of gravitation (along with suitable background assumptions) tells us that the return of Halley's comet was to be expected, but what is the probability that Newton's law is true? Hypotheses have objective probabilities when they describe possible outcomes of a chance process. But as far as anyone knows, the laws that govern our universe were not the result of a chance process. Bayesians think that *all* hypotheses have probabilities; the position I am advocating sees this as a special feature of *some* hypotheses.[2]

Not only do likelihood considerations leave open what probabilities one should assign to the competing hypotheses; they also don't tell you which hypothesis you should believe. I take it that belief is a dichotomous concept – you either believe a proposition or you do not. Consistent with this is the idea that there are three attitudes one might take to a statement – you can believe it true, believe it false, or withhold judgment. However, there is no simple connection of the matter-of-degree concept of probability to the dichotomous (or trichotomous) concept of belief. This is the lesson I extract from the lottery paradox (Kyburg 1961). Suppose 100,000 tickets are sold in a fair lottery; one ticket will win and each has the same chance of winning. It follows that each ticket has a very high probability of not winning. If you adopt the policy of believing a proposition when it has a high probability, you will believe of each ticket that it will not win. However, this conclusion contradicts the assumption that the lottery is fair. What this shows is that high probability does not suffice for belief (and low probability does not suffice for disbelief). It is for this reason that many Bayesians prefer to say that individuals have *degrees* of belief. The rules for the dichotomous concept are unclear; the matter-of-degree concept at least has the advantage of being anchored to the probability calculus.

In summary, likelihood arguments have rather modest pretensions. They don't tell you which hypotheses to believe; in fact, they don't even tell you which hypotheses are probably true. Rather, they evaluate how the observations at hand discriminate among the hypotheses under consideration.

I now turn to some details concerning the likelihood version of the design argument. The first concerns the meaning of the intelligent design hypothesis. This hypothesis occurs in (W_1) in connection with the watch and in (E_1) in connection with the vertebrate eye. In the case of the watch, Paley did not dream that he was offering an argument for the existence of *God*. However, in the case of the eye, Paley thought that the intelligent designer under discussion was God Himself. Why are these cases different? The bare bones of the likelihood arguments (W) and (E) do not say. What Paley had in mind is that building the vertebrate eye and the other adaptive features which organisms exhibit requires an intelligence far greater than anything that human beings could muster. This is a point that we will revisit at the end of this essay.

It is also important to understand the nature of the hypothesis with which the intelligent design hypothesis competes. I have used the term "chance" to express this alternative hypothesis. In large measure, this is because design theorists often think of chance as the alternative to design. Paley is again exemplary. *Natural Theology* is filled with examples like that of the vertebrate eye. Paley was not content to describe a few cases of delicate adaptations; he wanted to make sure that even if he got a few details wrong, the weight of evidence would still be overwhelming. For example, in Chapter 15 he considers the fact that our eyes point in the same direction as our feet; this has the convenient consequence that we can see where we are going. The obvious explanation, Paley (1802: 179) says, is intelligent design. This is because the alternative is that the direction of our eyes and the direction of our gait were determined by chance, which would mean that there was only a 1/4 probability that our eyes would be able to scan the quadrant into which we are about to step.

I construe the idea of chance in a particular way. To say that an outcome is the result of a *uniform chance process* means that it was one of a number of *equiprobable* outcomes. Examples in the real world that come close to being uniform chance processes may be found in gambling devices – spinning a roulette wheel, drawing from a deck of cards, tossing a coin. The term "random" becomes more and more appropriate as real-world systems approximate uniform chance processes. As R.A. Fisher once pointed out, it is not a "matter of chance" that casinos turn a profit each year, nor should this be regarded as a "random" event. The financial bottom line at a casino is the result of a large number of chance events, but the rules of the game make it enormously probable (though not certain) that casinos end each year in the black. All uniform chance processes are probabilistic, but not all probabilistic outcomes are "due to chance."

It follows that the two hypotheses considered in my likelihood rendition of the design argument are not exhaustive. Mindless uniform chance is one alternative to intelligent design, but it is not the only one. This point has an important bearing on the dramatic change in fortunes that the design argument experienced with the advent of Darwin's (1859) theory of evolution. The process of evolution by natural selection is not a uniform chance process. The process has two parts. Novel traits arise in individual organisms "by chance"; however, whether they then disappear from the population or increase in frequency and eventually reach 100 percent representation is anything but a "matter of chance." The central idea of natural selection is that traits which help organisms survive and reproduce have a better chance of becoming common than traits that hurt. The essence of natural selection is that evolutionary outcomes have *unequal* probabilities. Paley and other design theorists writing before Darwin did not and could not cover all possible mindless natural processes. Paley addressed the alternative of uniform chance, not the alternative of natural selection.[3]

Just to nail down this point, I want to describe a version of the design argument formulated by John Arbuthnot. Arbuthnot (1710) carefully tabulated birth records in London over eighty-two years and noticed that, in each year, slightly more sons than daughters were born. Realizing that boys die in greater numbers than girls, he saw that this slight bias in the sex ratio at birth gradually subsides until there are equal numbers of males and females at the age of marriage. Arbuthnot took this to be evidence of intelligent design; God, in his benevolence, wanted each man to have a wife and each woman to have a husband. To draw this conclusion, Arbuthnot considered what he took to be the relevant competing hypothesis – that the sex ratio at birth is determined by a uniform chance process. He was able to show that if the probability is 1/2 that a baby will be a boy and 1/2 that it will be a girl, then it is enormously improbable that the sex ratio should be skewed in favor of males in every one of the years he surveyed (Stigler 1986: 225–6).

Arbuthnot could not have known that R.A. Fisher (1930) would bring sex ratio within the purview of the theory of natural selection. Fisher's insight was to see that a mother's mix of sons and daughters affects the number of *grand*-offspring she will have. Fisher demonstrated that when there is random mating in a large population, the sex ratio strategy that evolves is one in which a mother invests equally in sons and daughters (Sober 1993: 17). A mother will put half her reproductive resources into producing sons and half into producing daughters. This equal division means that she should have more sons than daughters, if sons tend to die sooner. Fisher's model therefore predicts the slightly uneven sex ratio at birth that Arbuthnot observed.[4]

My point in describing Fisher's idea is not to fault Arbuthnot for living in the eighteenth century. Rather, the thing to notice is that what Arbuthnot meant by "chance" was very different from what Fisher was talking about when he described how a selection process might shape the sex ratio found in a population. Arbuthnot was right that the probability of there being more males than females at birth in each of eighty-two years is extremely low, if each birth has the same chance of producing a male as it does of producing a female. However, a male-biased sex ratio in the population is extremely probable, if Fisher's hypothesized process is doing the work. Showing that Design is more likely than Chance leaves it open that some third, mindless, process might still have a higher likelihood than Design. This is not a defect in the design argument, so long as the conclusion of that argument is not overstated. Here the modesty of the likelihood version of the design argument is a point in its favor. To draw a stronger conclusion – that the Design hypothesis is more likely than *any* hypothesis involving mindless natural processes – one would have to attend to more alternatives than just Design and (uniform) Chance.[5]

I now want to draw the reader's attention to some features of the likelihood version of the design argument (E) concerning how the observation and the competing hypotheses are formulated. First, notice that I have kept the observation (O_2) conceptually separate from the two hypotheses (E_1) and (E_2). If the observation were simply that "the vertebrate eye exists," then since (E_1) and (E_2) both entail this proposition, each would have a likelihood of unity. According to LP, this observation does not favor Design over Chance. Better to formulate the question in terms of explaining the properties of the vertebrate eye, not explaining why the eye exists. Notice also that I have not formulated the design hypothesis as the claim that God exists; this existence claim says nothing about the putative designer's involvement in the creation of the vertebrate eye. Finally, I should point out that it would do no harm to have the design hypothesis say that God created the vertebrate eye; this possible reformulation is something I'll return to later.

Other formulations of the design argument, and their defects

Given the various provisos that govern probability arguments, it would be nice if the design argument could be formulated deductively. For example, if the hypothesis of mindless chance processes entailed that it is *impossible* that organisms exhibit delicate adaptations, then a quick application of *modus tollens* would sweep that hypothesis from the field. However much design theorists might yearn for an argument of this kind, there apparently are none to be had. As the story about monkeys and typewriters illustrates, it is *not* impossible that mindless chance processes should produce delicate adaptations; it is merely very *improbable* that they should do so.

If *modus tollens* cannot be pressed into service, perhaps there is a probabilistic version of *modus tollens* that can achieve the same result. Is there a Law of Improbability that begins with the premise that Pr(O|H) is very low and concludes that H should be rejected? There is no such principle (Royall 1997: Ch. 3). The fact that you won the lottery does not, by itself, show that there is something wrong with the conjunctive hypothesis that the lottery was fair and a million tickets were sold and you bought just one ticket. And if we randomly drop a very sharp pin onto a line that is a thousand miles long, the probability of its landing where it does is negligible; however, that outcome does not falsify the hypothesis that the pin was dropped at random.[6]

The fact that there is no probabilistic *modus tollens* has great significance for understanding the design argument. The logic of this problem is essentially comparative. To evaluate the design hypothesis, we must know what it predicts and compare this with the predictions made by other hypotheses. The design hypothesis cannot win by default. The fact that an observation would be very improbable if it arose by chance is not enough to refute the chance hypothesis. One must show that the design hypothesis confers on the observation a higher probability, and even then the conclusion will merely be that the observation *favors* the design hypothesis, not that that hypothesis *must be true*.

In the continuing conflict (in the USA) between evolutionary biology and creationism, creationists attack evolutionary theory, but never take even the first step in developing a positive theory of their own. The three-word slogan "God did it" seems to satisfy whatever craving for explanation they may have. Is the sterility of this intellectual tradition a mere accident? Could intelligent design theory be turned into a scientific research program? I am doubtful, but the present point concerns the logic of the design argument, not its future prospects. Creationists sometimes assert that evolutionary theory "cannot explain" this or that finding (e.g. Behe 1996). What they mean is that certain outcomes are *very improbable* according to the evolutionary hypothesis. Even this more modest claim needs to be scrutinized. However, even if it were true, what would follow about the plausibility of creationism? In a word – *nothing*.

It isn't just defenders of the design hypothesis who have fallen into the trap of supposing that there is a probabilistic version of *modus tollens*. For example, the biologist Richard Dawkins (1986: 144–6) takes up the question of how one should evaluate hypotheses that attempt to explain the origin of life by appeal to strictly mindless natural processes. He says that an acceptable theory of this sort can say that the origin of life on Earth was somewhat improbable, but it cannot go too far. If there are N planets in the Universe that are "suitable" locales for life to originate, then an acceptable theory of the origin of life on Earth must say that that event had a probability of at least 1/N. Theories that say that terrestrial life was less probable than this should be rejected. This criterion may look plausible, but I think there is less to it than meets the eye. How does Dawkins obtain this lower bound? Why is the number of planets relevant? Perhaps he is thinking that if 1/N is the actual frequency of life-bearing planets among "suitable" planets (i.e. planets on which it is possible for life to evolve), then the true probability of life's evolving on Earth must also be 1/N. There is a mistake here, which we can uncover by examining how actual frequency and probability are related. With a small sample size, it is perfectly possible for these quantities to have different values (consider a fair coin that is tossed three times and then destroyed). However, Dawkins is obviously thinking that the sample size is very large, and here he is right that the actual frequency provides a good estimate of the true probability. It is interesting that Dawkins tells us to reject a theory if the probability it assigns is too low, but why doesn't he also say that it should be rejected if the probability it assigns is too high? The reason, presumably, is that we cannot rule out the possibility that Earth was not just suitable but *highly conducive* to the evolution of life. However, this point cuts both ways. Even if 1/N is the probability of a randomly selected suitable planet having life evolve on it, it still is possible that different suitable planets might have different probabilities – some may have values greater than 1/N while others may have values that are lower. Dawkins' lower bound assumes *a priori* that the Earth was above average; this is a mistake that might be termed the "Lake Wobegon Fallacy."

Some of Hume's (1779) criticisms of the design argument in his *Dialogues Concerning Natural Religion* depend on formulating the argument as something other than a likelihood inference. For example, Hume at one point has Philo say that the design argument is an argument from analogy, and that the conclusion of the argument is supported only very weakly by its premises. His point can be formulated by thinking of the design argument as follows:

Watches are produced by intelligent design.
Organisms are similar to watches to degree p.
p[══════════════════════════════
Organisms were produced by intelligent design.

Note that the letter "p" appears twice in this argument. It represents the degree of similarity of organisms and watches, and it represents the probability that the premises confer on the conclusion. Think of similarity as the proportion of shared characteristics. Things that are 0 percent similar have no traits in common; things that are 100 percent similar have all traits in common. The analogy argument says that the more similar watches and organisms are, the more probable it is that organisms were produced by intelligent design.

Let us grant the Humean point that watches and organisms have relatively few characteristics in common (it is doubtful that there is a well-defined totality consisting of all the traits of each, but let that pass). After all, watches are made of metal and glass, and go "tick tock"; organisms metabolize and reproduce and go "oink" and "bow wow." This is all true, but entirely irrelevant, if the design argument is a likelihood inference. It doesn't matter how similar watches and organisms are overall. With respect to argument (W), what matters is how one should explain the fact that watches are well adapted for the task of telling time; with respect to (E), what matters is how one should explain the fact that organisms are well adapted to their environments. Paley's analogy between watches and organisms is merely heuristic. The likelihood argument about organisms stands on its own (Sober 1993).

Hume also has Philo construe the design argument as an inductive argument, and then complain that the inductive evidence is weak. Philo suggests that for us to have good reason to think that our world was produced by an intelligent designer, we would have to visit other worlds and observe that all or most of them were produced by intelligent design. But how many other worlds have we visited? The answer is – not even one. Apparently, the design argument is an inductive argument that could not be weaker; its sample size is zero. This objection dissolves once we move from the model of inductive sampling to that of likelihood. You don't have to observe the processes of intelligent design and chance at work in different worlds to maintain that the two hypotheses confer different probabilities on the observations.

Three objections to the likelihood argument

There is another objection that Hume makes to the design argument, one that many philosophers apparently think is devastating. Hume points out that the design argument does not establish the attributes of the designer. The argument does not show that the designer who made the Universe, or who made organisms, is morally perfect, or all-knowing, or all-powerful, or that there is just one of him. Perhaps this undercuts some versions of the design argument, but it does not touch the likelihood argument we are considering. Paley, perhaps responding to this Humean point, makes it clear that his design argument aims to establish the *existence* of the designer, and that the question of the designer's *characteristics* must be addressed separately.[7] Does this limitation of the design argument make the argument trivial? Not at all – it is *not* trivial to claim that the adaptive contrivances of organisms are due to intelligent design. This supposed "triviality" would be *big* news to evolutionary biologists.

The likelihood version of the design argument consists of two premises – Pr(O|Chance) is very low and Pr(O|Design) is higher. Here O describes some observation of the features of organisms or some feature of the entire cosmos. The first of these claims is sometimes rejected by appeal to a theory that Hume describes under the heading of the Epicurean hypothesis. This is the monkeys-and-typewriters idea that if there are a finite number of particles that have a finite number of possible states, then, if they swarm about at random, they will eventually visit all possible configurations, including configurations of great order.[8] Thus, the order we see in our universe, and the delicate adaptations we observe in organisms, in fact had a high probability of eventually coming into being, according to the hypothesis of chance. Van Inwagen (1993: 144) gives voice to this objection and explains it by way of an analogy. Suppose you toss a coin twenty times and it lands heads every time. You should not be surprised at this outcome if you are one among millions of people who toss a fair coin twenty times. After all, with so many people tossing, it is all but inevitable that some people should get twenty heads. The outcome you obtained, therefore, was not improbable, according to the chance hypothesis.

There is a fallacy in this criticism of the design argument, which Hacking (1987) calls "the Inverse Gambler's Fallacy." He illustrates his idea by describing a gambler who walks into a casino and immediately observes two dice being rolled that land double-six. The gambler considers whether this result favors the hypothesis that the dice had been rolled many times before the roll he just observed or the hypothesis that this was the first roll of the evening. The gambler reasons that the outcome of double-six would be more probable under the first hypothesis:

$$\Pr(\text{double-six on this roll} \mid \text{there were many rolls}) > \Pr(\text{double-six on this roll} \mid \text{there was just one roll}).$$

In fact, the gambler's assessment of the likelihoods is erroneous. Rolls of dice have the *Markov property*: the probability of double-six on this roll is the same (1/36), regardless of what may have happened in the past. What is true is that the probability that a double-six will occur *at some time or other* increases as the number of trials is increased:

$$\Pr(\text{a double-six occurs sometime} \mid \text{there were many rolls}) > \Pr(\text{a double-six occurs sometime} \mid \text{there was just one roll}).$$

However, the *principle of total evidence* says that we should assess hypotheses by considering *all* the evidence we have. This means that the relevant observation is that *this* roll landed double-six; we should not focus on the logically weaker proposition that a double-six occurred *sometime*. Relative to the stronger description of the observations, the hypotheses have identical likelihoods.

If we apply this point to the criticism of the design argument that we are presently considering, we must conclude that the criticism is mistaken. There is a high probability (let us suppose) that a chance process will sooner or later produce order and adaptation. However, the relevant observation is not that these events occur at some time or other, but that they are true here and now

– *our* universe is orderly and the organisms here on Earth are well adapted. These events *do* have very low probability, according to the chance hypothesis, and the fact that a weaker description of the observations has high probability on the chance hypothesis is not relevant (see also White 2000).[9]

If the first premise in the likelihood formulation of the design argument – that Pr(O|Chance) is very low – is correct, then the only question that remains is whether Pr(O|Design) is higher. This, I believe, is the Achilles' heel of the design argument. The problem is to say how probable it is, for example, that the vertebrate eye would have features $F_1 \ldots F_n$ if the eye were produced by an intelligent designer. What is required is not the specification of a single probability value, or even a range of such. All that is needed is an argument that shows that this probability is indeed higher than the probability that Chance confers on the observation.

The problem is that the design hypothesis confers a probability on the observation only when it is supplemented with further assumptions about what the designer's goals and abilities would be if he existed. Perhaps the designer would never build the vertebrate eye with features $F_1 \ldots F_n$, either because he would lack the goals or because he would lack the ability. If so, the likelihood of the design hypothesis is zero. On the other hand, perhaps the designer would want to build the eye with features $F_1 \ldots F_n$ and would be entirely competent to bring this plan to fruition. If so, the likelihood of the design hypothesis is unity. There are as many likelihoods as there are suppositions concerning the goals and abilities of the putative designer. Which of these, or which class of these, should we take seriously?

It is no good answering this question by assuming that the eye was built by an intelligent designer and then inferring that he must have wanted to give the eye features $F_1 \ldots F_n$ and that he must have had the ability to do so since, after all, these are the features we observe. For one thing, this pattern of argument is question-begging. One needs *independent* evidence as to what the designer's plans and abilities would be if he existed; one can't obtain this evidence by *assuming* that the design hypothesis is true (Sober 1999). Furthermore, even if we assume that the eye was built by an intelligent designer, we can't tell from this what the probability is that the eye would have the features we observe. Designers sometimes bring about outcomes that are not very probable given the plans they have in mind.

This objection to the design argument is an old one; it was presented by Keynes (1921) and before him by Venn (1866). In fact, the basic idea was formulated by Hume. When we behold the watch on the heath, we know that the watch's features are not particularly improbable on the hypothesis that the watch was produced by a designer who has the sorts of *human* goals and abilities with which we are familiar. This is the deep and non-obvious disanalogy between the watchmaker and the putative maker of organisms and universes. We are invited, in the latter case, to imagine a designer who is radically different from the human craftsmen we know about. But if this designer is so different, why are we so sure that he would build the vertebrate eye in the form in which we find it?

This challenge is not turned back by pointing out that we often infer the existence of intelligent designers when we have no clue as to what they were trying to achieve. The biologist John Maynard Smith tells the story of a job he had during the Second World War inspecting a warehouse filled with German war *matériel*. He and his co-workers often came across machines whose functions were entirely opaque to them. Yet, they had no trouble seeing that these objects were built by intelligent designers. Similar stories can be told about archaeologists who work in museums; they often have objects in their collections that they know are artifacts, although they have no idea what the makers of these artifacts had in mind.

My claim is not that design theorists must have independent evidence that singles out a specification of the exact goals and abilities of the putative intelligent designer. They may be uncertain as to which of the goal-plus-abilities pairs $GA_1, GA_2, \ldots, GA_n$ is correct. However, since

$$\Pr(\text{the eye has } F_1 \ldots F_n | \text{Design}) = \Sigma_i \Pr(\text{the eye has } F_1 \ldots F_n | \text{Design \& } GA_i) \bullet \Pr(GA_i | \text{Design}),$$

they do have to show that

$$\Sigma_i [\Pr(\text{the eye has } F_1 \ldots F_n | \text{Design \& } GA_i) \bullet \Pr(GA_i | \text{Design})] > \Pr(\text{the eye has } F_1 \ldots F_n | \text{Chance}).$$

I think that Maynard Smith in his warehouse and archaeologists in their museums are able to do this. They aren't sure exactly what the intelligent designer was trying to achieve (e.g. they aren't certain that GA_1 is true and that all the other GA pairs are false), but they are able to see that it is not terribly improbable that the object should have the features one observes if it were made by a human intelligent designer. After all, the items in Maynard Smith's warehouse were symmetrical and smooth metal containers that had what appeared to be switches, dials, and gauges on them. And the "artifacts of unknown function" in anthropology museums likewise bear marks of human handiwork.

It is interesting in this connection to consider the epistemological problem of how one would go about detecting intelligent life elsewhere in the Universe (if it exists). The SETI (Search for Extraterrestrial Intelligence) project, funded until 1993 by the US National Aeronautics and Space Administration and now supported privately, dealt with this problem in two ways (Dick 1996). First, the scientists wanted to send a message into deep space that would allow any intelligent extraterrestrials who received it to figure out that it was produced by intelligent designers (namely, us). Second, they would scan the night sky hoping to detect signs of intelligent life elsewhere.

The message, transmitted in 1974 from the Arecibo Observatory, was a simple picture of our Solar System, a representation of oxygen and carbon, a picture of a double helix representing DNA, a stick figure of a human being, and a picture of the Arecibo telescope. How sure are we, if intelligent aliens find these clues, that they will realize that the clues were produced by intelligent designers? The hope is that this message will strike the aliens who receive it as evidence favoring the hypothesis of intelligent design over the hypothesis that some mindless physical process (not necessarily one involving uniform chance) was responsible. It is hard to see how the SETI engineers could have done any better, but still one cannot dismiss the possibility that they will fail. If extraterrestrial minds are very different from our own – either because they have different beliefs and desires or process information in different ways – it may turn out that their interpretation of the evidence differs profoundly from the interpretation that human beings would arrive at, were they on the receiving end. To say anything more precise about this, we would have to be able to provide specifics about the aliens' mental characteristics. If we are uncertain as to how the mind of an extraterrestrial will interpret this evidence, how can we be so sure that God, if he were to build the vertebrate eye, would endow it with the features we find it to have?

When SETI engineers search for signs of intelligent life elsewhere in the Universe, what are they looking for? The answer is surprisingly simple. They are looking for narrow-band radio emissions. This is because human beings build machines that produce these signals and, as far as we know, such emissions are not produced by mindless natural processes. The SETI engineers search for this signal, not because it is "complex" or fulfills some *a priori* criterion that would make it a "sign of intelligence," but simply because they think they know what sorts of mechanisms are needed to produce it.[10] This strategy may not work, but it is hard to see how the scientists could do any better. Our judgments about what counts as a sign of intelligent design must be based on empirical information about what designers often do and what they rarely do. As of now, these judgments are based on our knowledge of *human* intelligence. The more our hypotheses about intelligent designers depart from the human case, the more in the dark we are as to what the ground rules are for inferring intelligent design. It is imaginable that these limitations will subside as human beings learn more about the cosmos. But, for now, we are rather limited.

I have been emphasizing the fallibility of two assumptions – that we know what counts as a sign of extraterrestrial intelligence and that we know how extraterrestrials will interpret the signals we send. My point has been to shake a complacent assumption that figures in the design argument. However, I suspect that SETI engineers are on much firmer ground than theologians. If extraterrestrials evolved by the same type of evolutionary process that produced human intelligence, that may provide useful constraints on conjectures about the minds they have. No theologian, to my knowledge, thinks that God is the result of biological processes. Indeed God is usually thought of as a *super*natural being who is radically different from the things we observe *in* nature. The problem of extraterrestrial intelligence is therefore an intermediate case, lying somewhere between the watch found on the heath and the God who purportedly shaped the vertebrate eye (but much closer to the first). The upshot of this point for Paley's design argument is this: *Design arguments for the existence of human (and human-like) watchmakers are often unproblematic; it is design arguments for the existence of God that leave us at sea.*

I began by formulating the design hypothesis in argument (E) as the claim that an intelligent designer made the vertebrate eye. Yet, I have sometimes discussed the hypothesis as if it asserted that *God* is the designer in question. I don't think this difference makes a difference with respect to the objection I have described. To say that some designer or other made the eye is to state a disjunctive hypothesis. To figure out the likelihood of this disjunction, one needs to address the question of what each putative designer's goals and intentions would be.[11] The theological formulation shifts the problem from the evaluation of a disjunction to the evaluation of a disjunct, but the problem remains the same. Even supposing that God is omniscient, omnipotent, and perfectly benevolent, what is the probability that the eye would have features $F_1 \ldots F_n$ if God set his hand to making it? He *could* have produced those results if he had wanted. But why think that this is what he *would* have wanted to do? The assumption that God can do anything is part of the problem, not the solution. An engineer who is more limited would be more predictable.

There is another reply to my criticism of the design argument that should be considered. I have complained that we have no way to evaluate the likelihood of the design hypothesis, since we don't know which auxiliary assumptions about goal/ability pairs we should use. But why not change the subject? Instead of evaluating the likelihood of Design, why not evaluate the likelihood of various conjunctions – (Design & GA_1), (Design & GA_2), etc.? Some of these will have high likelihoods, others will have low, but it will no longer be a mystery what likelihoods these hypotheses possess. There are two problems with this tactic. First, it is a game that two can play. Consider the hypothesis that the vertebrate eye was created by the mindless process of electricity. If I simply get to invent auxiliary hypotheses without having to justify them independently, I can simply stipulate the following assumption – if electricity created the vertebrate eye, the eye must have features $F_1 \ldots F_n$. The electricity hypothesis is now a conjunct in a conjunction that has maximum likelihood, just like the design hypothesis. This is a dead end. My second objection is that it is an important part of scientific practice that conjunctions be broken apart (when possible) and their conjuncts scrutinized (Sober 1999, 2000b). If your doctor runs a test to see whether you have tuberculosis, you will not be satisfied if she reports that the conjunction "you have tuberculosis & auxiliary assumption 1" is very likely while the conjunction "you have tuberculosis & auxiliary assumption 2" is very unlikely. You want your doctor to address the first *conjunct*, not just various *conjunctions*. And you want her to do this by using a test procedure that is *independently* known to have small error probabilities. Demand no less of your theologian.

The relationship of the organismic design argument to Darwinism

Philosophers who criticize the organismic design argument often believe that the argument was dealt its death blow by Hume. True, Paley wrote after Hume, and the many Bridgewater Treatises elaborating the design argument appeared after Hume's *Dialogues* were published posthumously. Nonetheless, for these philosophers, the design argument after Hume was merely a corpse that could be propped up and paraded. Hume had taken the life out of it.

Biologists often take a different view. Dawkins (1986: 4) puts the point provocatively by saying that it was not until Darwin that it was possible to be an intellectually fulfilled atheist. The thought here is that Hume's skeptical attack was not the decisive moment; rather, it was Darwin's development and confirmation of a substantive scientific explanation of the adaptive features of organisms that really undermined the design argument (at least in its organismic formulation). Philosophers who believe that theories can't be rejected until a better theory is developed to take its place often sympathize with this point of view.

My own interpretation coincides with neither of these. As indicated above, I think that Hume's criticisms largely derive from an empiricist epistemology that is too narrow. However, seeing the design argument's fatal flaw does not depend on seeing the merits of Darwinian theory. True, LP says that theories must be evaluated comparatively, not on their own. But for this to be possible, each theory must make predictions. It is at this fundamental level that I think the design argument is defective.

Biologists often present two criticisms of creationism. First, they argue that the design hypothesis is untestable. Second, they contend that there is plenty of evidence that the hypothesis is false. Obviously, these two lines of argument are in conflict. I have already endorsed the first criticism, but I want to say a little about the second. A useful example is Stephen Jay Gould's (1980) widely read article about the panda's thumb. Pandas are vegetarian bears who have a spur of bone (a "thumb") protruding from their wrists. They use this device to strip bamboo, which is the main thing they eat. Gould says that the hypothesis of intelligent design predicts that pandas should not have this inefficient device. A benevolent, powerful, and intelligent engineer could and would have done a lot better. Evolutionary theory, on the other hand, says that the panda's thumb is what we should expect. The thumb is a modification of the wrist bones found in the common ancestor that pandas share with carnivorous bears. Evolution by natural selection is a tinkerer; it does not design adaptations from scratch, but modifies pre-existing features, with the result that adaptations are often imperfect.

Gould's argument, I hope it is clear, is a likelihood argument. I agree with what he says about evolutionary theory, but I think his discussion of the design hypothesis falls into the same trap that ensnared Paley. Gould thinks he knows what God would do if He built pandas, just as Paley thought he knew what God would do if He built the vertebrate eye. But neither of them knows this. Both help themselves to *assumptions* about God's goals and abilities. However, it is not enough to make assumptions about these matters; one needs independent evidence that these auxiliary assumptions are true. Paley's problem is also Gould's.

Anthropic reasoning and cosmic design arguments

Evolutionary theory seeks to explain the adaptive features of organisms; it has nothing to say about the origin of the Universe as a whole. For this reason, evolutionary theory conflicts with the organismic design hypothesis, but not with the cosmic design hypothesis. Still, the main criticism I presented of the first type of design argument also applies to the second. I now want to examine a further problem that cosmic design arguments sometimes encounter.[12]

Suppose I catch 50 fish from a lake, and you want to use my observations O to test two hypotheses:

O: All the fish I caught were more than ten inches long.
F_1: All the fish in the lake are more than ten inches long.
F_2: Only half the fish in the lake are more than ten inches long.

You might think that LP says that F_1 is better supported, since

(1) $\Pr(O|F_1) > \Pr(O|F_2)$.

However, you then discover how I caught my fish:

(A1) I caught the 50 fish by using a net that (because of the size of its holes) can't catch fish smaller than ten inches long.

This leads you to replace the analysis provided by (1) with the following:

(2) $\Pr(O|F_1 \,\&\, A_1) = \Pr(O|F_2 \,\&\, A_1) = 1$.

Furthermore, you now realize that your first assessment, (1), was based on the erroneous assumption that

(A_0) The fish I caught were a random sample from the fish in the lake.

Instead of (1), you should have written

$$Pr(O|F_1 \& A_0) > Pr(O|F_2 \& A_0).$$

This inequality is true; the problem, however, is that (A_0) is false.

This example, from Eddington (1939), illustrates the idea of an *observational selection effect* (an OSE). When a hypothesis is said to render a set of observations probable (or improbable), ask what assumptions allow the hypothesis to have this implication. The point illustrated here is that the procedure you use to obtain your observations can be relevant to assessing likelihoods.[13]

One version of the cosmic design argument begins with the observation that our universe is "fine-tuned." That is, the values of various physical constants are such as to permit life to exist, and, if they had been even slightly different, life would have been impossible. I'll abbreviate this fact by saying that "the constants are right." A design argument can now be constructed, one that claims that the constants being right should be explained by postulating the existence of an intelligent designer, one who wanted life to exist and who arranged the Universe so that this would occur (Swinburne 1990a). As with Paley's organismic design argument, we can represent the reasoning in this cosmic design argument as the assertion of a likelihood inequality:

(3) Pr(constants are right|Design) > Pr(constants are right|Chance).

However, there is a problem with (3) that resembles the problem with (1). Consider the fact that

(A_3) We exist, and if we exist the constants must be right.

We need to take (A_3) into account; instead of (3), we should have said:

(4) Pr(constants are right|Design & A_3) =
Pr(constants are right|Chance & A_3) = 1.0.

That is, given (A_3), the constants must be right, regardless of whether the Universe was produced by intelligent design or by chance.

Proposition (4) reflects the fact that our observation that the constants are right is subject to an OSE. Recognizing this OSE is in accordance with a *weak anthropic principle* – "what we can expect to observe must be restricted by the conditions necessary for our presence as observers" (Carter 1974: 291). The argument involves no commitment to *strong anthropic principles*. For example, there is no assertion that the correct cosmology must entail that the existence of observers such as ourselves was inevitable; nor is it claimed that our existence explains why the physical constants are right (Barrow 1988; Earman 1987; McMullin 1993).[14]

Although this point about OSEs undermines the version of the design argument that cites the fact that the physical constants are right, it does not touch other versions. For example, when Paley concludes that the vertebrate eye was produced by an intelligent designer, his argument cannot be refuted by claiming that:

(A_4) We exist, and if we exist vertebrates must have eyes with features $F_1 \ldots F_n$.

If (A_4) were true, the likelihood inequality that Paley asserted would have to be replaced with an equality, just as (1) had to be replaced by (2) and (3) had to be replaced by (4). But, fortunately for Paley, (A_4) is false. However, matters change if we think of Paley as seeking to explain the modest fact that organisms have at least one adaptive contrivance. If this were false, we would not be able to make observations; indeed, we would not exist. Paley was right to focus on the details; the more minimal description of what we observe does not sustain the argument he wanted to endorse.

The issue of OSEs can be raised in connection with other cosmic versions of the design argument. Swinburne writes that "the hypothesis of theism is that the Universe exists because there is a God who keeps it in being and that laws of nature operate because there is a God who brings it about that they do" (1990b: 191). Let us separate the *explananda*. The fact that the Universe exists does *not* favor Design over Chance; after all, if the Universe did not exist, we would not exist and so would not be able to observe that it does.[15] The same point holds with respect to the fact that the Universe is law-governed. Even supposing that lawlessness is possible, could we exist and make observations if there were no laws? If not, then the lawful character of the Universe does not discriminate between Design and Chance. Finally, we may consider the fact that our universe is governed by one set of laws, rather than another. Swinburne (1968) argues that the fact that our universe obeys *simple* laws is better explained by the hypothesis of Design than by the hypothesis of Chance. Whether this observation is also subject to an OSE depends on whether we could exist in a universe obeying alternative laws.

Before taking up an objection to this analysis of the argument from fine-tuning, I want to summarize what it has in common with the fishing example. In the fishing example, the source of the OSE is obvious – it is located in a device outside of ourselves. The net with big holes insures that the observer will make a certain observation, regardless of which of two hypotheses is true. But where is the device that induces an OSE in the fine-tuning example? There is none; rather, it is the observer's own existence that does the work. But, still, the effect is the same. Owing to the fact that we exist, we are bound to observe that the constants are right, regardless of whether our universe was produced by chance or by design.[16]

Leslie (1989: 13–4, 107–8), Swinburne (1990a: 171), and van Inwagen (1993: 135, 144) all defend the fine-tuning argument against the criticism I have just described that appeals to the idea of an OSE. Each mounts his defense by describing an analogy with a mundane example. Here is Swinburne's rendition of an analogy that Leslie presents:

> On a certain occasion the firing squad aim their rifles at the prisoner to be executed. There are twelve expert marksmen in the firing squad, and they fire twelve rounds each. However, on this occasion all 144 shots miss. The prisoner laughs and comments that the event is not something requiring any explanation because if the marksmen had not missed, he would not be here to observe them having done so. But of course, the prisoner's comment is absurd; the marksmen all having missed is indeed something requiring explanation; and so too is what goes with it – the prisoner's being alive to observe it. And the explanation will be either that it was an accident (a most unusual chance event) or that it was planned (e.g., all the marksmen had been bribed to miss). Any interpretation of the anthropic principle which suggests that the evolution of observers is something which requires no explanation in terms of boundary conditions and laws being a certain way (either inexplicably or through choice) is false.
>
> (Swinburne 1990a: 171)

First, a preliminary clarification – the issue isn't whether the prisoner's survival "requires explanation" but whether this observation provides evidence as to whether the marksmen intended to spare the prisoner or shot at random.[17]

My response to Swinburne takes the form of a dilemma. I'll argue, first, that if the firing squad example is analyzed in terms of LP, the prisoner is right and Swinburne is wrong – the prisoner's survival does not allow him to conclude that Design is more likely than Chance. However, there is a different analysis of the prisoner's situation, in terms of the *probabilities* of hypotheses, not their *likelihoods*. This second analysis concludes that the prisoner is mistaken; however, it has the consequence that the prisoner's inference differs fundamentally from the design argument that appeals to fine-tuning. Each horn of this dilemma supports the conclusion that the firing squad example does nothing to save this version of the design argument.

So let us begin. If we understand Swinburne's claim in terms of LP, we should read him as saying that

(L_1) Pr(the prisoner survived|the marksmen intended to miss) >
Pr(the prisoner survived|the marksmen fired at random).

He thinks that the anthropic principle requires us to replace this claim with the following irrelevancy:

(L_2) Pr(the prisoner survived|the marksmen intended to miss & the prisoner survived) =
Pr(the prisoner survived|the marksmen fired at random & the prisoner survived) = 1.

This equality would lead us to conclude (Swinburne thinks mistakenly) that the prisoner's survival does not discriminate between the hypotheses of Design and Chance.

To assess Swinburne's claim that the prisoner has made a mistake, it is useful to compare the prisoner's reasoning with that of a bystander who witnesses the prisoner survive the firing squad. The prisoner reasons as follows: "Given that I now am able to make observations, I must be alive, whether my survival was due to intelligent design or chance." The bystander says the following: "Given that I now am able to make observations, the fact that the prisoner is now alive is made more probable by the design hypothesis than it is by the chance hypothesis." The prisoner is claiming that he is subject to an OSE, while the bystander says that he, the bystander, is not. Both, I submit, are correct.[18]

I suggest that part of the intuitive attractiveness of Swinburne's claim that the prisoner has made a mistake derives from a shift between the prisoner's point of view to the bystander's. (L_1) is correct and involves no OSE if it expresses the bystander's judgment; however, it is flawed, and needs to be replaced by (L_2), if it expresses the prisoner's judgment. My hunch is that Swinburne thinks the prisoner errs in his assessment of likelihoods because we bystanders would be making a mistake if we reasoned as he does.[19]

The basic idea of an OSE is that we must take account of the procedures used to obtain the observations when we assess the likelihoods of hypotheses. This much was clear from the fishing example. What may seem strange about my reading of Swinburne's story is my claim that the prisoner and the bystander are in different epistemic situations, even though their observation reports differ by a mere pronoun. After the marksmen fire, the prisoner thinks "I exist" while the bystander thinks "he exists"; the bystander, but not the prisoner, is able to use his observation to say that Design is more likely than Chance, or so I say. If this seems odd, it may be useful to reflect on Sorenson's (1988) concept of *blindspots*. A proposition p is a blindspot for an individual S just in case, if p were true, S would not be able to know that p is true. Although some propositions (e.g. "nothing exists," "the constants are wrong") are blindspots for everyone, other propositions are blindspots for some people but not for others. Blindspots give rise to OSEs; if p is a blindspot for S, then if S makes an observation to determine the truth value of p, the outcome must be that not-p is observed. The prisoner, but not the bystander, has "the prisoner does not exist" as a blindspot. This is why "the prisoner exists" has an evidential significance for the bystander that it cannot have for the prisoner.[20]

I now turn to a different analysis of the prisoner's situation. The prisoner, like the rest of us, knows how firing squads work. They always or almost always follow the orders they receive, which is almost always to execute someone. Occasionally, they produce fake executions. They almost never fire at random. What is more, firing squads have firm control over outcomes; if they want to kill (or spare) someone, they always or almost always succeed. This and related items of background knowledge support the following *probability* claim:

> (P_f) Pr(the marksmen intended to spare the prisoner|the prisoner survived) > Pr(the marksmen intended to spare the prisoner).

Firing squads rarely intend to spare their victims, but the survival of the prisoner makes it very probable that his firing squad had precisely that intention. The likelihood analysis led to the conclusion that the prisoner and the bystander are in different epistemic situations; the bystander evaluates the hypotheses by using (L_1), but the prisoner is obliged to use (L_2). However, from the point of view of probabilities, the prisoner and the bystander can say the same thing; both can cite (P_f).

What does this tell us about the fine-tuning version of the design argument? I construed that argument as a claim about likelihoods. As such, it is subject to an OSE; given that we exist, the constants have to be right, regardless of whether our universe was produced by Chance or by Design. However, we now need to consider whether the fine-tuning argument can be formulated as a claim about probabilities. Can we assert that

> (P_u) Pr(the Universe was created by an intelligent designer|the constants are right) > Pr(the Universe was created by an intelligent designer)?

I don't think so. In the case of firing squads, we have frequency data and our general knowledge of human behavior on which to ground the probability statement (P_f). But we have neither data nor theory on which to ground (P_u). And we cannot defend (P_u) by saying that an intelligent designer would ensure that the constants are right, because this takes us back to the likelihood considerations we have already discussed. The prisoner's conclusion that he can say nothing about Chance and Design is mistaken if he is making a claim about probabilities. But the argument from fine-tuning can't be defended as a claim about probabilities.

The rabbit/duck quality of this problem merits review. I've discussed three examples – fishing, fine-tuning, and the firing squad. If we compare fine-tuning with fishing, they seem similar. This makes it intuitive to conclude that the design argument based on fine-tuning is wrong. However, if we compare fine-tuning with the firing squad, *they* seem similar. Since the prisoner apparently has evidence that favors Design over Chance, we are led to the conclusion that the fine-tuning argument must be right. This shifting gestalt can be stabilized by imposing a formalism. The first point is that OSEs are to be understood by comparing the *likelihoods* of hypotheses, not their *probabilities*. The second is that it is perfectly true that the prisoner can assert the *probability* claim (P_f). The question, then, is whether the design argument from fine-tuning is a likelihood argument or a probability argument. If the former, it is flawed because it fails to take account of the fact that there is an OSE. If the latter, it is flawed, but for a different reason – it makes claims about probabilities that we have no reason to accept; indeed, we cannot even *understand* them as objective claims about nature.[21]

A prediction

It was obvious to Paley and to other purveyors of the organismic design argument that, if an intelligent designer built organisms, that designer would have to be far more intelligent than any human being could ever be. This is why the organismic design argument was for them an argument for the existence of *God*. I predict that it will eventually become clear that the organismic design argument should never have been understood in this way. This is because I expect that human beings will eventually build organisms from non-living materials. This achievement will not close down the question of whether the organisms we observe were created by intelligent design or by mindless natural processes; in fact, it will give that question a practical meaning, since the organisms we will see around us will be of both kinds.[22] However, it will be abundantly clear that the fact of organismic adaptation has nothing to do with whether God exists. When the Spanish conquistadors arrived in the New World, several indigenous peoples thought these intruders were gods, so powerful was the technology that the intruders possessed. Alas, the locals were mistaken; they did not realize that these beings with guns and horses were merely *human* beings. The organismic design argument for the existence of God embodies the same mistake. Human beings in the future will be the conquistadors, and Paley will be our Montezuma.[23]

Notes

1 Does this construal of the design argument conflict with the idea that the argument is an *inference to the best explanation*? Not if one's theory of inference to the best explanation says that observations influence the assessment of explanations in this instance via the vehicle of likelihoods.

2 In light of the fact that it is possible for a hypothesis to have an objective likelihood without also having an objective probability, one should understand Bayes's theorem as specifying how the quantities it mentions are related to each other, *if all are well defined*. And just as hypotheses can have likelihoods without having (objective) probabilities, it is also possible for the reverse situation to obtain. Suppose I draw a card from a deck of unknown composition. I observe (O) that the card is the four of diamonds. I now consider the hypothesis (H) that the card is a four. The value of $Pr(H|O)$ is well defined, but the value of $Pr(O|H)$ is not.

3 Actually, Paley (1802) *does* consider a "selective retention" process, but only very briefly. In Chapter 5 (pp. 49–51) he explores the hypothesis that a random process once generated a huge range of variation, and that this variation was then culled, with only stable configurations surviving. Paley argues against this hypothesis by saying that we should see unicorns and mermaids if it were true. He also says that it mistakenly predicts that organisms should fail to form a taxonomic hierarchy. It is ironic that Darwin claimed that his own theory *predicts* hierarchy. In fact, Paley and Darwin are both right. Darwin's theory contains the idea that all living things have common ancestors, while the selection hypothesis that Paley considers does not.

4 More precisely, Fisher said that a mother should have a son with probability p and a daughter with probability $(1-p)$, where the effect of this is that the expected expenditures on the two sexes are the same; the argument is not undermined by the fact that some mothers have all sons while others have all daughters.

5 Dawkins (1986) makes the point that evolution by natural selection is not a uniform chance process by way of an analogy with a combination lock. This is discussed in Sober (1993: 36–9).

6 Dembski (1998) construes the design inference as "sweeping from the field" all possible competitors, with the effect that the design hypothesis wins by default (i.e. it never has to make successful predictions). As noted above, Paley, Arbuthnot, and other design theorists did not and could not refute all possible alternatives to Design; they were able to test only the alternatives that they were able to formulate. For other criticisms of Dembski's framework, see Fitelson *et al.* (1999).

7 Paley (1802)argues in Chapter 16 that the benevolence of the deity is demonstrated by the fact that organisms experience more pleasure than they need to (p. 295). He also argues that pain is useful (p. 320) and that few diseases are fatal; he defends the latter conclusion by citing statistics on the cure rate at a London hospital (p. 321).

8 For it to be certain that all configurations will be visited, there must be infinite time. The shorter the time frame, the lower the probability that a given configuration will occur. This means that the estimated age of the Universe may entail that it is very *im*probable that a given configuration will occur. I set this objection aside in what follows.

9 It is a standard feature of likelihood comparisons that O_s sometimes fails to discriminate between a pair of hypotheses, even though O_w is able to do so, when O_s entails O_w. You are the cook in a restaurant. The waiter brings an order into

the kitchen; someone ordered bacon and eggs. You wonder whether this information favors the hypothesis that your friend Smith ordered the meal, or that your friend Jones did. You know the eating habits of each. Table 1.1 gives the probabilities of four possible orders, conditional on the order having come from Smith and conditional on the order having come from Jones.

Table 1.1 Probabilities of four possible orders, conditional on who orders

	Smith's probabilities	*Jones's probabilities*
bacon and eggs	0.3	0.3
bacon without eggs	0.4	0.2
eggs without bacon	0.2	0.1
neither bacon nor eggs	0.1	0.4

The fact that the customer ordered bacon and eggs does not discriminate between the two hypotheses (since 0.3 = 0.3). However, the fact that the customer ordered bacon favors Smith over Jones (since 0.7 > 0.5), and so does the fact that the customer ordered eggs (since 0.5 > 0.4).

10 The example of the SETI project throws light on Paley's question as to why we think that watches must be the result of intelligent design, but don't think this when we observe a stone. It is tempting to answer this question by saying that watches are "complicated" while stones are not. However, there are many complicated natural processes (like the turbulent flow of water coming from a faucet) that don't cry out for explanation in terms of intelligent design. Similarly, narrow-band radio emissions may be physically "simple" but that doesn't mean that the SETI engineers were wrong to search for them.

11 Assessing the likelihood of a disjunction involves an additional problem. Even if the values of $Pr(O|D_1)$ and $Pr(O|D_2)$ are known, what is the value of $Pr(O|D_1 \text{ or } D_2)$? The answer is that it must be somewhere in between. But exactly where depends on further considerations, since $Pr(O|D_1 \text{ or } D_2) = [Pr(O|D_1) \bullet Pr(D_1|D_1 \text{ or } D_2)] + [Pr(O|D_2) \bullet Pr(D_2|D_1 \text{ or } D_2)]$. If either God or a super-intelligent extraterrestrial built the vertebrate eye, what is the probability that it was God who did so?

12 To isolate this new problem from the one already identified, I'll assume in what follows that the design hypothesis has built into it auxiliary assumptions that suffice for its likelihood to be well defined.

13 This general point surfaces in simple inference problems like the ravens paradox (Hempel 1965). Does the fact that the object before you is a black raven confirm the generalization that all ravens are black? That depends on how you gathered your data. Perhaps you sampled at random from the set of *ravens*; alternatively, you may have sampled at random from the set of *black ravens*. In the first case, your observation confirms the generalization, but in the second it does not. In the second case, notice that you were bound to observe that the object before you is a black raven, regardless of whether all ravens are black.

14 Although weak and strong anthropic principles differ, they have something in common. For example, the causal structure implicitly assumed in the weak anthropic principle is that of two effects of a common cause:

we exist now
↗
(WAP) origin of Universe
↘
constants now are right

In contrast, one of the strong anthropic principles assumes the following causal arrangement:

(SAP) we exist now → origin of the Universe → constants now are right.

Even though (WAP) is true and (SAP) is false, both entail a *correlation* between our existence and the constants now having the values they do. To deal with the resulting OSEs, we must decide how to take these correlations into account in assessing likelihoods.

15 Similarly, the fact that there is something rather than nothing does not discriminate between Chance and Design.

16 The fishing and fine-tuning examples involve *extreme* OSEs. More modest OSEs are possible. If C describes the circumstances in which we make our observational determination as to whether proposition O is true, and we use the outcome of this determination to decide whether H_1 or H_2 is more likely, then a *quantitative* OSE is present precisely when

$$\Pr(O|H_1 \,\&\, C) \neq \Pr(O|H_1) \text{ or}$$
$$\Pr(O|H_2 \,\&\, C) \neq \Pr(O|H_2).$$

A *qualitative* OSE occurs when taking account of C alters the likelihood ordering:

$$\Pr(O|H_1 \,\&\, C) > \Pr(O|H_2 \,\&\, C) \text{ and } \Pr(O|H_1) \leq \Pr(O|H_2) \text{ or}$$
$$\Pr(O|H_1 \,\&\, C) = \Pr(O|H_2 \,\&\, C) \text{ and } \Pr(O|H_1) \neq \Pr(O|H_2).$$

Understood in this way, an OSE is just an example of *sampling bias*.

17 There is a third possibility – that the marksmen intended to kill the prisoner – but for the sake of simplicity (and also to make the firing-squad argument more parallel with the argument from fine-tuning), I'll ignore this for most of my discussion.

18 The issue, thus, is not whether (L_1) or (L_2) are true (both are), but which one an agent should use in interpreting the bearing of observations on the likelihoods of hypotheses. In this respect the injunction of the weak anthropic principle is like the principle of total evidence – it is a pragmatic principle, concerning which statements should be used for which purposes.

19 In order to replicate in the fine-tuning argument the difference between the prisoner's and the bystander's points of view, imagine that we observe through a telescope another universe in which the constants are right. We bystanders can use this observation in a way that the inhabitants of that universe cannot.

20 Notice that "I exist," when thought by the prisoner, is *a priori*, whereas "the prisoner exists," when thought by the bystander, is *a posteriori*. Is it so surprising that an *a priori* statement should have a different evidential significance than an *a posteriori* statement? I also should note that my claim is that the proposition "I am alive" does not permit the prisoner to conclude that Design is more likely than Chance. I do not say that there is no proposition he can cite after the marksmen fire that discriminates between the two hypotheses. Consider, for

example, the observation that "no bullets hit me." This favors Design over Chance, even after the prisoner conditionalizes on the fact that he is alive. Notice also that if the prisoner were alive but riddled with bullets, this would not so clearly make Design more likely than Chance.

21 The hypothesis that our universe is one among many has been introduced as a possible explanation of the fact that the constants (in our universe) are right. A universe is here understood to be a region of space–time that is causally closed. See Leslie (1989) for discussion. If the point of the multiverse hypothesis is to challenge the design hypothesis, on the assumption that the design hypothesis has already vanquished the hypothesis of chance, then the multiverse hypothesis is not needed. Furthermore, in comparing the multiverse hypothesis and the design hypothesis, one needs to attend to the Inverse Gambler's Fallacy discussed earlier. This is not to deny that there may be other evidence for the multiverse hypothesis; however, the mere fact that the constants are right in our universe is not evidence that discriminates between the three hypotheses in contention.

22 As Dennett (1987a: 284–5) observes, human beings have been modifying the characteristics of animals and plants by *artificial selection* for thousands of years. However, the organisms thus modified were not created by human beings. Recall that I formulated the design argument as endorsing a hypothesis about how organisms were brought into being. This is why the work of plant and animal breeders, *per se*, does not show that the design argument should be stripped of its theological trappings.

23 I am grateful to Martin Barrett, Nick Bostrom, David Christensen, Ellery Eells, Branden Fitelson, Malcolm Forster, Daniel Hausman, Stephen Leeds, Lydia McGrew, Williams Mann, Roy Sorenson, and Richard Swinburne for useful comments. Thanks also to the members of the Kansas State University Philosophy Department for a very stimulating and productive discussion of this chapter.

References

Arbuthnot, J. (1710) "An argument for Divine Providence, taken from the constant regularity observ'd in the births of both sexes," *Philosophical Transactions of the Royal Society of London* 27: 186–90.

Barrow, J. (1988) *The World within the World*, Oxford: Clarendon Press.

Behe, M. (1996) *Darwin's Black Box*, New York: Free Press.

Carter, B. (1974) "Large number coincidences and the anthropic principle in cosmology," in M.S. Longair (ed.) *Confrontation of Cosmological Theories with Observational Data*, Dordrecht: Reidel, pp. 291–8.

Darwin, C. (1964 [1859]) *On the Origin of Species*, Cambridge, Massachusetts: Harvard University Press.

Dawkins, R. (1986) *The Blind Watchmaker*, New York: Norton.

Dembski, W. (1998) *The Design Inference*, Cambridge: Cambridge University Press.

Dennett, D. (1987a) "Intentional systems in cognitive ethology – the 'panglossian paradigm' defended," in *The Intentional Stance*, Cambridge, Massachusetts: MIT Press, pp. 237–86.

—— (1987b) "True believers," in *The Intentional Stance*, Cambridge, Massachusetts: MIT Press, pp. 13–42.

Dick, S. (1996) *The Biological Universe – the Twentieth-Century Extraterrestrial Life Debate and the Limits of Science*, Cambridge: Cambridge University Press.

Earman, J. (1987) "The SAP also rises – a critical examination of the anthropic principle," *American Philosophical Quarterly* 24: 307–17.

Eddington, A. (1939) *The Philosophy of Physical Science*, Cambridge: Cambridge University Press.
Edwards, A. (1972) *Likelihood*, Cambridge: Cambridge University Press.
Fisher, R.A. (1957 [1930]) *The Genetical Theory of Natural Selection*, 2nd edn, New York: Dover.
Fitelson, B., Stephens, C., and Sober, E. (1999) "How not to detect design – a review of W. Dembski's *The Design Inference*," *Philosophy of Science* 66: 472–88, available online at http://philosophy.wisc.edu/sober.
Forster, M. and Sober, E. (2003) "Why likelihood?" in M. Taper and S. Lee (eds) *The Nature of Scientific Evidence*, Chicago: University of Chicago Press, available online at http://philosophy.wisc.edu/forster.
Gould, S. J.(1980) *The Panda's Thumb*, New York: Norton.
Hacking, I. (1987) "The Inverse Gambler's Fallacy: The argument from design. The anthropic principle applied to Wheeler universes," *Mind* 96: 331–40.
—— (1965) *The Logic of Statistical Inference*, Cambridge: Cambridge University Press.
Hempel, C. (1965) "Studies in the logic of confirmation," in *Aspects of Scientific Explanation and Other Essays in the Philosophy of Science*, New York: Free Press.
Hume, D. (1990 [1779]) *Dialogues Concerning Natural Religion*, London: Penguin.
Keynes, J. (1921) *A Treatise on Probability*, London: Macmillan.
Kyburg, H. (1961) *Probability and the Logic of Rational Belief*, Middletown, Connecticut: Wesleyan University Press.
Leslie, J. (1989) *Universes*, London: Routledge.
McMullin, E. (1993) "Indifference principle and anthropic principle in cosmology," *Studies in the History and Philosophy of Science* 24: 359–89.
Paley, W. (1802) *Natural Theology, or, Evidences of the Existence and Attributes of the Deity, Collected from the Appearances of Nature*, London: Rivington.
Royall, R. (1997) *Statistical Evidence – a Likelihood Paradigm*, London: Chapman & Hall.
Sober, E. (2000a) "Evolution and the problem of other minds," *Journal of Philosophy* 97: 365–86.
—— (2000b) "Quine's two dogmas," *Proceedings of the Aristotelean Society*, supplementary volume 74: 237–80.
—— (1999) "Testability," *Proceedings and Addresses of the American Philosophical Association* 73: 47–76, available online at http://philosophy.wisc.edu/sober.
—— (1993) *Philosophy of Biology*, Boulder, Colorado: Westview Press.
Sorenson, R. (1988) *Blindspots*, Oxford: Oxford University Press.
Stigler, S. (1986) *The History of Statistics*, Cambridge, Massachusetts: Harvard University Press.
Swinburne, R. (1990a) "Argument from the fine-tuning of the Universe," in. J. Leslie (ed.) *Physical Cosmology and Philosophy*, New York: Macmillan, pp. 160–79.
—— (1990b) "The limits of explanation," in D. Knowles (ed.) *Explanation and Its Limits*, Cambridge: Cambridge University Press, pp. 177–93.
—— (1968) "The argument from design," *Philosophy* 43: 199–212.
Turing, A. (1950) "Computing machinery and intelligence," *Mind* 59: 433–60.
van Inwagen, P. (1993) *Metaphysics*, Boulder, Colorado: Westview Press.
Venn, J. (1866) *The Logic of Chance*, New York: Chelsea.
White, R. (2000) "Fine-tuning and multiple universes," *Nous* 34: 260–76.

2

THE MEANING OF DESIGN

John Leslie

"Design" means more than just order of some sort. No matter how you arrange books on a shelf, they will have some order or other, and Leibniz noted that some formula or other can always be found to fit points scattered on paper randomly. Leibniz further remarked that some kinds of order may be interesting because they have what he called "richness"; they combine obedience to fairly simple laws with results that are complex without being merely untidy. Giving a more complete account of what "Leibnizian richness" means, however, is a very hard task. You tend to end up with a collection of words such as "beauty" and "grandeur" that leaves you little the wiser. Luckily, there is no need for us to attempt the task. Instead, let us concentrate on the word "design" as it appears in the term "the argument from design" or "the design argument for God's existence."

The argument from design is really an argument *to* divine design from alleged signs of it. The idea is that the cosmos, as we can see by examining it, was selected for creation to serve a divine purpose, a purpose at least partially understandable because it is good. Leibnizian richness does enter into most people's thoughts about goodness. It is believed that a cosmos serving a divine purpose would have beauty, grandeur, etc.; but more basic, typically, is the belief that the cosmos would be good *through containing intelligent living beings*. I think that makes excellent sense. Suppose, controversially, that God is to be understood as an immensely powerful person, and that any cosmos that this person created would be entirely outside him. Now, what if he had designed the cosmos so that no life would evolve in it? What if he had designed it just for its beauty and grandeur, which he alone could appreciate since nobody else would exist? Would that not make him rather a simpleton, somebody who would have actually to *create* his cosmos before he could properly appreciate the idea of cosmic beauty and grandeur, instead of just contemplating such beauty and grandeur in his mind's eye? Would he not have created something that was worthless *in itself*, since its only value would lie in something outside it, namely, his experience of looking at its structure? While G.E. Moore wrote in his *Principia Ethica* (1903) that something could be intrinsically good merely through being beautiful, there being no need for anyone to exist to appreciate the beauty, this seems to me to be wrong, and in later writings Moore, too, came to think of it as wrong. He came to define the intrinsically good as what was worth having *in the sense in which an experience is had*. In a cosmos without living things clever enough to be worth calling *observers*, there would be nothing intrinsically good. So if a deity who is not rather a simpleton is to create a cosmos entirely outside himself, then that cosmos must contain intelligent life.

Living beings certainly *look as if* they were designed by somebody. Their parts come together to serve purposes in intricate ways. Hearts are fine mechanisms for pumping blood. Eyes are superbly constructed for collecting information. Still, we can accept this without accepting the argument from design. Darwin explained that the complex, elegant, useful arrangement of a living being's parts might well have come about without the action of a divine designer, through natural selection. When talking about hearts, a scientist of today could say "designed for pumping blood" without having to reject Darwin. All that would be meant would be that hearts were good at pumping blood, and had been produced by natural selection because of this.

It is impossible to prove firmly that Darwinian processes working on atoms that obeyed the laws of physics, and were not supplemented by any "life-forces" or miraculous acts of divine interference, would be enough to produce such structures as the human eye. Let me just say that any deity who supplemented laws of physics by life-forces and acts of interference would have produced a disappointingly untidy universe. One would wonder why he had not simply decided to run the whole thing by magic. Naturally, we must avoid being narrow-minded about what might count as laws of physics. My belief that everything obeys laws of physics is a hunch that events all conform to a fully unified set of laws, expressible by some reasonably short equation. The equation almost certainly leads to all kinds of phenomena that physicists have not yet dreamed of. The central point is merely that there are not three separate realms of matter, life, and mind, each one obeying basic laws peculiar to itself. It would, however, be absurd to try to prove this point firmly, which would involve knowing all the details of how the world works. Instead, let me try to show that anyone accepting the argument from design could have plenty to offer as evidence without needing to speak of miracles or life-forces.

We must not fancy that the only manner in which a designer could operate would be to take clay, so to speak, clay with properties beyond his control, and mould it into appropriate shapes. When the designer was God, he would have created his own clay with just the properties he wished. Divine design could be revealed by the fortunate nature of the physical laws that atoms and atomic particles obeyed. We could perhaps find evidence of design in the laws of special relativity, which permit living mechanisms to operate identically no matter how fast they move relative to one another.

There is no problem of the forces inside some system acting particularly weakly in one direction, particularly strongly in another, just because the system is in rapid motion, absolutely, in the one direction rather than the other, for special relativity recognizes no such reality as being in rapid motion *absolutely*. Again, we might detect design in the laws of quantum physics that stop atoms from collapsing and which permit seemingly dissipated wave energy to be released in concentrated bursts so that it can do useful work. Let us pay special attention, though, to the marks of design that many have seen in the apparent fine-tuning of our universe.

Recently, many physicists and cosmologists have argued that there is quite a problem in how our cosmic environment manages to be one in which Darwinian evolution can operate over long ages to produce living beings. The cosmic period known to us began with the Big Bang. It looks as if the early cosmic density, and the associated expansion speed, needed tuning with immense accuracy for there to be gas clouds able to condense into stars – tuning to perhaps one part in a trillion trillion trillion trillion trillion. Furthermore, the strength of the nuclear weak force had to fall inside narrow limits for the Big Bang to generate any hydrogen (which was needed for making water and for long-lived, stable stars like the Sun) and for the creation of all elements heavier than helium. Also, the strength ratio between electromagnetism and gravity needed extremely accurate tuning, perhaps to one part in many trillion trillion, for there to be Sun-like stars. Again, the existence of chemistry seemingly demanded very precise adjustment of the masses of the neutron, the proton, and the electron.

In a book of mine, *Universes* (1989), I made a long list of such claims about fine-tuning. No doubt some of the claims will turn out to be wrong. For instance, it might be that the early cosmic expansion speed was more or less forced to be what it was, because of a process known as "inflation," and the people who think that inflation itself needed very precise tuning could be mistaken. What is impressive, I suggest, is not any particular one of the claims about fine-tuning, but the large number of claims that seem plausible, and the consequent implausibility of thinking that every single claim is erroneous. This, then, may be our evidence of design, provided we judge that such design would have been directed towards producing living beings in a non-miraculous fashion through making the world obey physical laws that led to the existence of stable stars, planets, and an environment with a rich chemistry in which life could evolve.

The miraculous and the natural

Do not imagine that dividing divine interference from natural physical processes is an easy affair. For one thing, it is standard theology to say that the cosmos would immediately vanish if God ceased to "conserve" it in existence from moment to moment, and for the theologians to add that the laws of physics hold only because this is what God wills. Keeping everything in existence, and keeping it obedient to physical laws, is simply not counted by theologians as "interference" or "miracle." I see nothing wrong in this, but the point is a controversial one. It becomes particularly difficult to handle when you bear in mind two further points: first, that our universe can seem to obey laws of quantum physics that do not dictate precisely how events develop; and, second, that quantum randomness, perhaps together with other types of randomness, may have had major effects on the general structure of the world that we see. It is often theorized that the strengths of various forces such as electromagnetism, gravity, and the nuclear weak and strong forces, and the masses of such particles as the neutron, the proton, and the electron, could all have been settled during early instants of the Big Bang *by random processes inside an initially very tiny domain*, which later grew large enough to include everything now visible to our telescopes. Physicists speak, for example, of symmetry breaking by scalar fields whose values could have varied randomly from one very tiny domain to another. Against this background, how would things look to a theologian who believed that God *would have to choose at each instant precisely how the cosmos would be at the next instant* when he "conserved" its existence – when he preserved it, that is to say, but preserved it in a slightly changed form that led humans to speak of the action of physical forces? The laws of physics, as I said just a moment ago, could fail to dictate exactly what would have to happen in order for them to be obeyed. They could be quantum-physical laws that left this up to God, in which case God might have chosen cunningly that events would in fact develop, at early instants of the Big Bang, in such a way that there would later be the sort of world that permitted the evolution of intelligent life because the strengths of its physical forces and the masses of its particles had been settled appropriately.

Having chosen cunningly how things would happen at early instants, God might also have acted rather similarly at various crucial later moments, ensuring that events that quantum physics allowed to develop along various different paths, most of them *not* leading to the evolution of intelligent life, in fact took one or other of the few paths leading to it. We might be unjustified in calling this type of thing "divine interference," as long as it did not happen on too large a scale. The distinction between designing a world's laws in a life-encouraging fashion and then leaving them to operate, and actually designing such things as eyes by, say, putting the optic nerves in the right places, is a sufficiently clear distinction – but in between there is a fuzzy area. Here, what one person would call "messy divine interference" or "miracle" would be classified by another person as God just not choosing *perversely* to make events happen in life-excluding ways when life-encouraging ways were equally present among the possibilities allowed by physical laws, the possibilities among which God had to choose.

Another difficult point concerns whether we could say that divine design "used fine-tuning" if the fundamental laws of physics were in fact all-dictating laws: laws with no free parameters. It could at first seem that there would then be just no way in which anything could be *tuned*. The strength ratio between electromagnetism and gravity, for instance, would have to be what it was, given that the fundamental laws were what they were, and so would the masses of the neutron, the proton, and the electron. Once the laws were in place, no divine designer could have been faced with a range of possibilities among which he could have chosen cunningly. All the same, I suggest that there could be room for talk of "fine-tuning". For suppose that many slightly different systems of fundamental law, each dictating exactly how events would have developed, would all of them have led to the existence of a universe containing forces recognizable as gravity and electromagnetism, and particles recognizable as neutrons, protons, and electrons, but with the precise properties of those forces and particles differing in each case. It could then be said that a divine designer had "fine-tuned" the properties by choosing the fundamental laws appropriately.

When even the distinction between "God using physical laws" and "God operating through miraculous acts of interference" becomes fuzzy, then this field is going to supply plenty of work for philosophers like me. But unfortunately it may not be work that settles anything of much importance. It may simply amount to recommending various ways of using words, on disappointingly arbitrary grounds. What does seem to me important, however, is that we distinguish firmly between *divine selection* and *observational selection*. Many scientists who describe our cosmic situation as "fine-tuned for life" believe that observational selection is at work here. They think that a gigantic cosmos includes hugely many domains worth calling "universes." The many universes might be widely separated in space, in a cosmos that had inflated enormously; or they might be successive oscillations of an oscillating cosmos; or they might spring into existence entirely independently. Now, several mechanisms have been proposed for making the various universes differ in the strengths of their forces, in the masses of their particles, and in other respects as well. Brandon Carter's "anthropic principle" then reminds us that only life-permitting conditions give rise to beings able to observe them.

Carter's *strong anthropic principle* says this about conditions in any cosmic region you decide to call "a universe," while his *weak anthropic principle* says the same thing about conditions in anything you prefer to call a spatiotemporal locality. Inevitably, though, one speaker's "large spatiotemporal locality" is another speaker's "universe," for there are no firm rules for using these words. The point to notice is that neither Carter's weak anthropic principle nor his strong anthropic principle has anything to do with divine design. These principles concern *observational selection effects*, period. When reminding us, with his strong anthropic principle, that the universe in which we find ourselves must be (since we observers are in it, aren't we?) a universe whose properties are not totally hostile to life and to intelligence, Carter has never meant that this universe *was forced to be* of a kind that would permit intelligent life to evolve, let alone that it had been positively compelled to contain intelligent living beings. He has always accepted that a great deal of randomness might enter into whether a universe developed life-permitting properties, and, if it did, then whether living things, intelligent or otherwise, would actually evolve in it.

I ought to make clear that Carter has himself written so little about this area that he has now largely lost control of what the term "anthropic principle" means. I sometimes get the impression that most people use the phrase "believing in the anthropic principle" to mean something like "believing in divine design"; and, sure enough, when sufficiently many folk use words in a particular fashion, then that fashion can become *right*. Still, I recommend using the term "anthropic principle" in the way that Brandon Carter outlined.

Instead of confusing Carter's observational selection with divine selection, otherwise known as divine design, might we not *combine* these two things? Imagine God creating hugely many universes, the general properties of each universe being settled by random processes at early instants. Suppose that the likely outcome of such random processes would be that only a tiny proportion of the universes had properties permitting life to evolve. God could still be certain that life would arrive in many places if he created sufficiently many universes – perhaps infinitely many. And although it would now be observational selection, not divine selection, which guaranteed that intelligent beings found that their universes had properties of life-permitting kinds, God might still be counted not merely as a creator but also as a designer since he had at least ensured that the fundamental laws obeyed by all the various universes were laws leading living beings to evolve *in some of them*. Why not think along these lines?

I suspect that they would be unsatisfactory lines. Yes, a deity interested in producing good states of affairs might be expected to create infinitely many universes, for why be satisfied with creating only fifty-seven, or only 30 million? However, it might seem bizarre to imagine that this deity would create any universe *that he knew in advance would develop in a fashion totally hostile to intelligent life*. And he could of course know in advance whether a universe would become totally hostile if this depended on physical processes that were only partially controlled by fundamental laws since these laws, for instance ones of quantum physics, would fail to dictate precisely what would happen, *so that the deity himself would have to decide this* when exerting his power of conservation, of keeping things in existence while at the same time changing them slightly. Remember, divine conservation, the preservation of the existence of things, without which they would at once vanish, is very traditional theology. And theologians are not such fools as to fancy that "conservation" here means "preservation in a totally unaltered state" so that nothing ever changes.

The place of evil in a designed universe

If there exist hugely or infinitely many universes, then we could not expect our universe to be the very best of them. All the same, it may appear as though theologians face a severe difficulty in the fact that ours is a universe containing forest fires that burn animals alive, earthquakes that destroy buildings and the humans inside them, plagues, and so forth. Some people have concluded that anybody who had designed it could be interested only in producing intelligent life and not in good states of affairs. However, I suspect that a deity aiming to achieve good ends would not necessarily be in the business of saving living beings from all disasters. A universe designed by an all-powerful and benevolent being might still include many evils, for various reasons.

One possible reason is that it might, as the poet Keats suggested, be a universe designed not for its own goodness but as a "vale of soul-making." It might be a gymnasium for building up moral strength through often painful efforts. In the absence of strong moral fiber, heavenly bliss would not be deserved or perhaps could not even be *had*: the idea here is that only good souls would get pleasure from life in heaven. But a difficulty with Keats's theory is that it is not at all clear why God would not simply create souls complete with strong moral fiber. Would creating beings with pleasant personalities be dictatorial interference with freedom of the will? I cannot see that it would.

A better suggestion, I suspect, would be that any complex universe would be bound to include disasters if it obeyed causal laws. Think, here, of how it may well be impossible to create a universe in which every single coin of all the billions ever tossed was a coin that landed heads – assuming, that is to say, that the coins were governed by causal laws, not by magic. Now, it is not at all obvious that a universe designed for its own goodness would be better if it ran by magic, all such events as earthquakes being banned.

Note that a universe designed as a home for intelligent life would not necessarily include such life *from its earliest moments*. We need not picture God as forced to exist in solitary splendor until intelligent living beings had evolved in our universe. He could have created up to infinitely many earlier universes. What is more, those who agree with Einstein's views about time would say that even at our universe's earliest moments *it was true* that lives were being lived in it at later moments: moments "further along the fourth dimension." (Einstein tried to comfort the relatives of a dead friend by writing to them to say that he continued to be alive at earlier times. Many philosophers think this makes sense. They compare existing in the past or in the future to *existing on the left*, or *existing to the south*.)

Again, a universe designed as a home for intelligent life might still not be one in which any particular intelligent species, for example humankind, would be guaranteed to survive for long. Remember always that God may well have created a cosmos containing infinitely many universes, while even our own universe may, if the currently popular inflationary models are correct, stretch farther than our telescopes can probe by a factor of perhaps ten followed by a million zeros. In this connection, think of Enrico Fermi's problem of why we have detected no extraterrestrials – a possible solution being that intelligent species almost always destroy themselves soon after inventing hydrogen bombs, germ warfare, or highly polluting industrial processes. You can believe in a benevolent divine designer without rejecting this solution. In contrast, the solution that our own intelligent species happens to be *the very first of many thousands to evolve in our Galaxy* could be judged as preposterous.

If it does strike you as preposterous, then you might be interested in various themes of a book of mine published in 1996 under the potentially alarming title *The End of the World.* Let me hurry to make clear that despite the title, plus the beautiful supernova exploding on the front cover of its new paperback version, I myself think that the human race has something approaching half a chance of spreading right across its galaxy. Still, I see considerable force in a point first noticed by Brandon Carter: that just as it could appear preposterous to view our intelligent species as the very first of many thousands, so it could appear preposterous to suppose that you and I were in a human race that was fairly certain to spread right across its galaxy, which would place us *among the earliest thousandth, the earliest millionth, or even the earliest billionth of all humans who will ever have lived.* It may well seem preferable to believe that humankind will become extinct in the not-too-distant future. Believing that divine benevolence designed our universe is compatible with thinking that Carter is right.

Also, believing in divine benevolence is compatible with recognizing that the laws of physics do permit the existence of hydrogen bombs – and may actually lead to *a vacuum metastability disaster* if physicists push their experiments beyond the energies that are generally considered to be safe, the energies that have already been reached in collisions between cosmic rays. There have been some (but not nearly enough!) discussions of this last point in the physics journals. In his book *Before the Beginning* (1997), Martin Rees, who is Britain's Astronomer Royal, draws firm attention to the calamity that might lie in wait for us here. If space is filled by a scalar field in a merely metastable condition, then a sufficiently powerful collision between particles might work like a pin pricking a balloon. As S. Coleman and F. De Luccia (1980) explain, a tiny bubble of new-strength scalar field might be formed, this at once expanding at almost the speed of light and destroying first the Earth, then the Solar System, then our entire galaxy, etc. Divine design would not necessarily guarantee us against this. "Designed" need not be a word saying that the Universe is always cozy, never threatening. If that were what it said, then the design argument for God's existence would be utter rubbish.

Explaining God: the Platonic approach

When all is said and done, is belief in God really any better than belief in magic spells? I think it is. Magic cannot be understood, but if God is real then a Platonic approach might help us to understand why God is real. It could also have interesting things to suggest about the meaning of the words "God" and "divine design."

Let me first introduce the Platonism of one of the past century's finest philosophers, A.C. Ewing. In his book *Value and Reality* (1973), Ewing suggested that God exists *simply because this is good.* What sense can we make of this idea? In *Universes*, and earlier in *Value and Existence* (1979), I commented that followers of Plato think it impossible, even in theory, to get rid of all realities. Even in a blank, an absence of all existing things, it would still, be a reality, Platonists think, that two and two make four. It would still be true, in other words, that if there were ever to exist two groups of two things, then there would be four things. Similarly, it would be a reality that the blank was better than any world of people in agony that might replace it. It would be real that the absence of such a world of torment *was ethically required.* And, likewise, the presence of a good world could be ethically required despite how there would be, in the blank, nobody to have a duty to produce such a world. The Platonic suggestion is that ethical requirements can be real unconditionally, absolutely, eternally. And a further Platonic suggestion is that when it is sufficiently weighty *an ethical requirement* – such as, perhaps, the requirement that there exist a supremely good divine person – *can be directly responsible for the actual existence of whatever it is that is required.* Asking Platonists to point to some mechanism that enabled any such requirement to have this responsibility would be like asking them to point to *a mechanism that made misery an evil*, or to a mechanism that forced the experience of red to be nearer to that of orange than to that of yellow. For Platonists, these are not affairs that depend on mechanisms. Instead, they are affairs of a sort that can explain why anything at all exists, and why any mechanism ever works: why, that is to say, there is a world that obeys causal laws that mechanisms can exploit.

Much more can be said about all this, but let us simply suppose that it does make some sense, as is accepted by John Polkinghorne in his recent book *The Faith of a Physicist* (1994). Like Ewing, Polkinghorne thinks Platonism could best be used to give us insight into why there exists a benevolent divine person who selects a world among all the worlds that are possible, and who wills that it shall exist. However, Ewing and Polkinghorne are little inclined to believe that this person selects anything in quite the way you and I do, with much hard effort to reach correct evaluations, noble struggles to direct acts of will towards good results, stiffening of arm muscles, and so forth. If God is indeed a person and a designer, then we must recognize that He is at least not a person quite like you and me and a designer quite like any architect, apart, of course, from being smarter and more powerful. But many people, for instance Paul Tillich among recent theologians, have gone much further than recognizing this point. There is a long neo-Platonic tradition in which it seems to be argued (albeit obscurely) that "God" is just a name for the fact that an ethical need for the cosmos to exist is directly responsible for its existence. This tradition takes its inspiration from Plato's remark in Book VI of the *Republic* that the Form of the Good "is itself not existence but far beyond it in dignity" since it is "what bestows existence upon things."

Here the idea of an omnipotent architect is entirely abandoned. We might still speak of divine design, but only on the grounds that good things were selected for existence by virtue of being good – the word "selected" being used, evidently, in an unusual sense because nobody would be doing the selecting. As Plotinus, greatest of the neo-Platonists, expresses the matter in his *Third Ennead*, the cosmos exists "not as a result of a judgement recognizing its desirability, but by sheer necessity"; "effort and search" play no part in the creative process; yet the outcome, "even had it resulted from a considered plan, would not have disgraced its maker."

A compromise between this neo-Platonic explanation for the cosmos and belief in a divine designer can be found in Spinoza's world-picture, for which Einstein expressed admiration. On my understanding of his difficult writings, Spinoza believes that the cosmos exists because this is ethically required, which provides a reason for calling the cosmos "God." However, it is also true that God is an immensely knowledgeable mind and that there exists nothing outside this mind. How can that be so? The answer is that the divine mind contemplates everything worth knowing, including what a universe would be like if obedient to the laws that our universe obeys, and how it would feel to be each of the conscious beings in such a universe. Now, says Spinoza, the divine mind's contemplation of this *just is* the reality of our universe and of every conscious being in it. Your own knowledge of precisely what it feels like to be you is simply God's contemplating exactly how it must feel to be somebody with precisely your properties – such as, perhaps, the property of not believing a word Spinoza says.

Spinoza seems to have viewed the cosmos as being obedient throughout to a single set of laws. This strikes me as unfortunate. If the divine mind really did contemplate everything worth knowing, then presumably it would contemplate all the details of many beautiful, grand universes obeying laws that were very different from those of our universe, even to the extent of being laws incompatible with the evolution of life of any kind. Perhaps infinitely many universes would exist in the divine thought (which is, remember, where Spinoza thinks that you and I and all our surroundings exist). Yet even so, there could be limits to how far the divine thought ranged. The divine mind might not be cluttered with thoughts about absolutely all facts, including facts concerning all the messy forms that universes could take if they obeyed no laws whatsoever. We might regard all the universes that God thought about as universes *selected* for being thought about because each obeyed laws of some sort. In view of their being in this way selected, we might even speak of their law-controlled structures as "instances of divine design." It would, however, be Brandon Carter's *observational selection* which then ensured that the Universe studied by human physicists was a universe whose laws permitted the evolution of intelligent living beings. I discuss all this in detail in *Infinite Minds* (2001).

References

Coleman, S. and De Luccia, F. (1980) "Gravitational effects on and of vacuum decay," *Physical Review D* 21(12): 3,305–15.

Ewing, A.C. (1973) *Value and Reality*, London: George Allen & Unwin, Ltd.

Leslie, J. (2001) *Infinite Minds*, Oxford: Clarendon Press.

—— (1996) *The End of the World: The Science and Ethics of Human Extinction*, London and New York: Routledge.

—— (1989) *Universes*, London and New York: Routledge.

—— (1979) *Value and Existence*, Oxford: Blackwell.

Moore, G.E. (1903) *Principia Ethica*, Cambridge: Cambridge University Press.

Polkinghorne, J. (1994) *The Faith of a Physicist*, Princeton: Princeton University Press.

Rees, M. (1997) *Before the Beginning: Our Universe and Others*, Reading, Massachusetts: Addison-Wesley.

3

THE DESIGN INFERENCE

Old wine in new wineskins

Robert O'Connor

Introduction

Whatever your opinion of traditional design arguments, versions developed under the rubric of "intelligent design" (ID) are neither distinctive nor uniquely compelling. Nonetheless, proponents of intelligent design such as Michael Behe (1996), William Dembski (1998a, 1999), and Stephen Meyer (1999, 2000) represent them as substantively different from traditional design arguments and as having overcome their inherent deficiencies. Unlike traditional fare (Paley 1802; Swinburne 1979), Dembski insists that ID arguments provide "a rigorous scientific demonstration" such that "[d]emonstrating transcendent design in the universe is a scientific inference, not a philosophical pipe dream" (1999: 223). Although it is not entirely clear why an argument based on a philosophical interpretation of the ontological status of the findings of science should constitute a "pipe dream", this essay shows that ID arguments rely on philosophical premises just as much as have design inferences of the past. In this crucial feature, they are not substantively distinct.

"Intelligent design" turns on the claim that specific, scientifically determined phenomena cannot be explained by appeal to any natural processes, and their extraordinary improbability rules out appeal to chance. As Behe insists, "since intelligent agents are the only entities known to be able to construct irreducibly complex systems, the biochemical systems are better explained as the result of deliberate intelligent design" (2000a: 156). In so far as ID focuses on specific *local* phenomena (e.g. molecular machines, DNA sequencing) rather than such *global* phenomena as the very presence of life in the Universe or the natural processes by which life came to be, one might label them "local design arguments," or LDA, to distinguish them from their apparently suspect "global" counterparts (GDA). LDA claim to establish intelligent agency from *within* science. They are deemed superior to GDA in so far as they trade on the indisputable, scientifically established function of these phenomena, rather than such speculative metaphysical claims as regarding the overall purpose of the cosmos. Furthermore, because these arguments turn on such well-founded, empirically confirmed, "scientific" principles as Dembski's *Law of Conservation of Information*, they presume all the epistemic credentials of the best of contemporary science. In sum, Dembski says:

> There exists a reliable criterion for detecting design strictly from observational features of the world. This criterion belongs to probability and complexity theory, not to metaphysics and theology.... When applied to the complex information-rich structures of biology, it detects design. In particular we can say with the weight of science behind us that the complexity-specification criterion shows Michael Behe's irreducibly complex biochemical systems to be designed.
>
> (1998b: 22)

Thus, anyone subscribing to the standards of evidence adopted by science cannot, on pain of inconsistency, reject the inference to intelligent agency. LDA constitute "in-principle arguments for why undirected natural causes (i.e., chance, necessity or some combination of the two) cannot produce irreducible and specified complexity" (Dembski 1999: 276–7).

Dembski insists that, since LDA follow from the empirically confirmed, incontrovertible outcome of a well-established scientific law, their conclusions, viz. intelligent agency, constitute a proper part of *scientific* inquiry. The findings of science itself demand appeal to a designer. If so, then restrictions against appeal to a designer, especially those originating from within the domain of science, contravene the immediate and incontestable demands of its own evidence and, as such, must represent nothing less than a philosophical bias for a purely materialist, anti-theistic, philosophy. According to LDA, reason and fairness demand that scientists, theistic or otherwise, renounce commitment to any restriction in the sciences against appeal to transcendent agency, particularly "methodological naturalism's" (MN's) ban on explanatory appeal to divine agency.

In the present essay, I argue that LDA do not provide a qualitatively new or distinctive form of argument for design, much less do they constitute reason to reject MN. If this analysis is correct, LDA do not successfully avoid any objectionable features of traditional design arguments (GDA). Thus, however impressive we might find the data of contemporary molecular biology, LDA will have failed to deliver on their promise of providing scientifically compelling evidence for intelligent agency. However remarkable we might find these local phenomena to be (indeed, they are *extraordinarily* remarkable), these new arguments do not provide additional reason to suppose that they must have arisen from a non-natural agent. Thus, even if one were to agree that an intelligent agent actualized its intentions by periodically infusing history with additional order, LDA fail to present support for this belief beyond that already provided by familiar GDA. LDA can no more establish the ultimacy of mind than can GDA.

The distinctive strengths of LDA

Dembski disparages traditional philosophical arguments for design for the extent to which they rest on contentious, speculative, philosophical assumptions regarding the contingency and purpose of nature. LDA, on the other hand, rest on specific, local, empirically confirmed examples of contingent phenomena. Research shows that these local phenomena, however unlikely their origination, are necessary in order for higher-level systems to function. As such, these phenomena bear an unmistakable "mark of design," a quantitatively measurable improbability threshold that betokens intelligent agency.

The move to local, empirically confirmed phenomena is crucial since design critics typically construe such global features as simply the outcome of those natural laws and initial conditions that *happen* to hold. Rather than representing life, for example, as the end towards which the Universe aims – its intended *telos* – they regard it as the unintended result of natural processes. On this construal, the appearance of these phenomena betokens nothing more than the natural, even necessary, outcome of those natural laws and conditions that happen to characterize this universe, blindly working together in a manner that produces life. If one presumes that this material order does not require any further explanation, in the same manner that proponents of design might take mental order to be self-explanatory, then the fact that these laws and antecedent conditions produce some phenomenon, even life, carries no probative force.[1]

By focusing on particular, locally identifiable instances of functionality, LDA effectively shift attention from the contingency of the Universe as a whole, or even the contingency of the natural laws and processes, to the *scientifically established contingency* of certain specific features of that universe. LDA need not speculate as to the contingency of the universal laws, conditions or processes; rather, taking the laws of nature as given, they rest on the known contingency inherent in the outcome of these processes. This effectively undercuts the naturalist's gambit of taking these natural laws and conditions as brute features the Universe happens (always) to have had. For GDA, the contingency in question turns on the presumed improbability, from among imaginable – *metaphysically conceivable* – universes, of one having just these features. For LDA, the contingency that demands explanation arises in those phenomena that, *given what we know through careful empirical research about those laws and processes*, are exceedingly unlikely.

Furthermore, where GDA require the presumption that such remarkable features as the existence of life constitute the end or purpose of the Universe, LDA focus on the natural function of specific phenomena. There is no question as to what counts as the proper function of the systems in question, nor is there any question as to the essential role of the particular process or component in enabling it to fulfill that function. If the improbability of that process or component occurring by chance is sufficiently high, then, rather than serving as a philosophically contentious assumption in the argument, the contingency in question constitutes an empirically established scientific fact inexplicable in purely natural terms. Thus, one's philosophical or theological predilection regarding the ultimate source of the natural laws and material conditions upon which they operate does not factor into the prospects for providing a purely natural explanation. These phenomena can be explained, if at all, only by that which transcends the operational limits of the natural world.

Such are the unique strengths of LDA. In what follows, I will argue that the general form of argument is, in fact, based on more than the empirical findings of science and what we know about its laws, for LDA are at least as much indebted to philosophical assumptions as GDA. In particular, first, I'll argue that LDA trade on largely unsupported, possibly unsupportable, philosophical theses concerning the scope of scientific research. Second, they appear to presume certain *philosophical* theses concerning the presence of a mental agent, as opposed to a strictly natural cause. Of course, neither of these conclusions entails that LDA fail to provide ground for belief in transcendent design. Yet, if their success does turn on these extra-scientific assumptions, then they do not demand either the endorsement of science and its practitioners, or the radical revision of science implicated in the rejection of methodological naturalism. Neither does it follow from this analysis that the specific conclusions of LDA are mistaken, particularly the claim that there has been an infusion of information into the natural system at a specific juncture in history. Still, if I am correct in this assessment, LDA fail, not in details of probability theory or mathematical analysis, but in their broadest philosophical assumptions concerning the nature of science and the question of other minds.

Empirical evidence for design

LDA turn on the claim that the evident presence of either "irreducible complexity" or "complex specificity" (hereafter, CSI – "complex specified information") entails the presence of design. Clearly, for these arguments to work, they must have the resources to establish specific instances of CSI. However, it turns out that decisively identifying an instance of CSI requires commitment to philosophical assumptions that are not themselves concomitant with the practice of science. In particular, the empirical argument for the presence of specific tokens of CSI requires our adopting unduly optimistic assumptions about the comprehensive powers and reliability of the outcomes of scientific inquiry.[2] Thus, LDA trade on premises, implicitly forwarded without support, *about* science, rather than premises established within or *by* science itself.

In the ID literature, CSI stands for "complex specified information." In this context, "complexity" refers to the extraordinary improbability of the phenomenon in question, given the laws, processes, and conditions present.

"Specificity" refers to the very narrow range of possible configurations by which that phenomenon could fulfill its function in the broader natural system. The function of the phenomenon in question is measured in terms of its role in the operations of the organism to which it belongs. Therefore, both its function and narrow specificity are fixed by scientific knowledge of its well-understood role as a constituent of a larger system. Finally, "information" refers to the instructions required to produce this phenomenon in the specific context.[3] When cast in terms of "information theory," CSI and "irreducible complexity" come to the same thing.[4]

The general form of argument is very simple, and, if successful, a powerful apologetic. If one can establish, on the basis of the relevant empirical laws and natural conditions, that some phenomenon, *E*, is highly improbable, and if one can further establish that *E* fulfills a sufficiently narrowly constrained, yet necessary function within those natural processes, one will have successfully established the claim that *E* is the result of neither regularity, chance, nor their combination.[5] Even though *E* is as it must be in order to render a particular natural process functional, it does not result from any known natural laws and conditions, nor is it within the limits of chance. The exceedingly low probability of *E* reflects its extraordinary complexity, and the narrow range of functionality reflects it specificity.

The potential power of this form of argument rests on the assertion that these features of *E* can be established on the authority of our best scientific knowledge. Contemporary scientific findings must establish that (1) certain antecedent circumstances were satisfied, (2) given those circumstances, *E* is compatible with, but not determined by, the relevant laws of nature, and (3) *E* has actually occurred. In that case, *E* simply cannot be explained in terms of natural law.[6] Furthermore, the exceeding improbability of *E*, given those laws, rules out the chance hypothesis. So, if the best scientific understanding of the natural process requires *E*, or something very similar to *E*, for that process to function, then science cannot account for *E* by appeal to law, chance, or their combination. Therefore, if contemporary science confirms that the complexity of *E* is sufficiently great, and its specificity too narrow, one must infer design.

For instance, when Michael Behe cashes out the notion of "irreducible complexity," he speaks of a system that "cannot be produced directly by slight, successive modifications of a precursor system, since any precursor to an irreducibly complex system is by definition nonfunctional" (1996: 39). That is, the relevant natural laws (neo-Darwinian evolution) operate by means of "slight, successive modifications of a precursor system." But no such precursor system could exist, for the neo-Darwinian account requires all systems to be functional (else that precursor would not have evolved). However, the operations of the molecular machines in question are so narrowly constrained that even slight modifications to their component parts would undermine their functionality. Thus, even if the development of a system would *not specifically violate* any natural laws, nonetheless, there are simply *no laws available* by which to account for that phenomenon. Alternatively, even if the known laws could account for the phenomenon, the initial conditions necessary to do so simply did not exist.

Thus the distinctive (and most compelling) feature of LDA is their basis in the empirical evidence of phenomena that cannot be explained in terms of natural processes, and yet are too improbable to be explained by pure chance. In terms of information theory, such phenomena exhibit significantly more quantitatively measurable information, *i*, in *E* at a specific juncture in the natural process, t_2, than was available in those antecedent conditions and processes at t_1. We may symbolize this *significant quantitative increase* in *i* from t_1 to t_2 as $(i/t_2 >> i/t_1)$. LDA do not rest on mere speculation as to this relative increase in informational content. Since the probability of *E* at *t*, as determined by the natural laws and conditions, can be readily measured through scientific inquiry, this increase in informational content can be empirically quantified. Science establishes, given these conditions, the extent to which the circumstances could have differed from their actual state. Yet, in so far as they are precisely those circumstances necessary to fulfill an empirically discernable, local function, this empirical feature stands as a quantifiable mark of the explanatory failure of undirected natural causes.

The general structure of this argument suggests three strategies by which to establish the empirical contingency of *E*: (1) the violation argument, (2) the failure of imagination argument, or (3) the argument from ignorance. First, according to "the violation argument," one might hold that *E* was not the result of natural processes in so far as its occurrence explicitly contradicts the determinant results of specific and well-established empirical laws. Defying gravity violates a specific, well-established empirical law. Alternatively, the "violation" might stem from the fact that *E*'s occurrence, given those laws, would require antecedent material conditions *incompatible* with those known to have existed. Processes governing wine production require that initial conditions include ingredients other than water. There is no stronger means by which to establish that a phenomenon cannot be explained in terms of natural laws and processes than to establish its incompatibility with the material conditions known to hold. Nevertheless, proponents of LDA are decidedly resistant to establishing design on the grounds of a violation of natural laws or known conditions:

> In practice, to establish the contingency of an object, event or structure, one must establish that it is compatible with the regularities involved in its production but that these regularities [presumably given the initial conditions known to hold] also permit any number of alternatives to it....By being compatible with but not required by

> the regularities involved in its production, an object, event or structure becomes irreducible to any underlying physical necessity.
>
> (Dembski 1999: 128–9)

A distinctive strength of LDA lies in their refusal to require the belief that any particular law or known condition has been violated. LDA work *with* science, not against it, thereby allowing them to rest on the full authority of the validity of science and enable their full investigation through scientific inquiry. Thus, rather than undermining the epistemic authority of science by claiming that its laws have been violated, LDA fully endorse the methods of science as able to accurately discern the nature of causal agents. Indeed, LDA rely on the claim that the findings of science are, as far as they go, fully accurate and trustworthy.

Given this desideratum, in order to establish the empirical contingency of *E*, one might expect LDA to adopt one or the other of the two remaining alternative strategies. On the one hand, *E* might present a situation such that no *conceivable* laws and/or conditions could have produced it. Alternatively, one might argue that no *known* laws and/or conditions would have produced *E*. In the first instance, one cannot explain *E* because one cannot imagine the natural resources capable of producing it ("failure of imagination argument"). In the latter, one cannot explain *E* simply out of ignorance of such resources ("argument from ignorance").

Yet following either of these strategies for establishing the improbability of *E* requires proponents of LDA to endorse certain contestable assumptions regarding both the scope and conceptual power of scientific inquiry. In particular, the second would require the assumption that, if science does not know a specific process by which to account for these phenomena, then this provides sufficient reason to conclude that such a process does not exist. The first of these two strategies requires the assumption that the inability of scientists to even conceive of a process responsible for such phenomena means that no such process exists. However, for such claims to function as premises to these arguments would require an argument to this effect, a particularly daunting task given the extraordinary strength of each respective claim. Thus, adopting either of these strategies would render LDA vulnerable to criticism based on the presumed limitations of scientists and scientific inquiry to reveal the full complement of natural laws and processes.[7] To appreciate the strength of these assumptions, bear in mind that these arguments are primarily based on phenomena whose origins, presumably some time in the distant past, are in question. As such, the antecedent conditions are not directly available for analysis.[8]

These comments, of course, do not of themselves undermine the validity of the inference. They do, however, highlight the vulnerability of LDA to general epistemic concerns about the confidence one should invest in the ability of scientific inquiry to provide access to these otherwise unobservable conditions, or the full scope of relevant natural laws. On this construal, LDA would have to hold scientific inquiry in very high regard with respect to its capacity to accurately reveal the full scope of natural powers and processes, along with the actual antecedent conditions upon which these powers and processes work. Of course, scientific inquiry may indeed provide accurate access to the unobservable initial conditions on which the natural forces must operate. Scientific inquiry may actually have the resources to reveal the full breadth of natural powers available. That it does, however, is by no means obvious. One would have to either presume that these *epistemic* limitations (either the failure of imagination or ignorance of available means) map directly onto the *ontological* limitations of the natural world, or provide an argument to that effect. In any case, these claims extend well beyond the empirical findings of science; to pronounce as to the epistemic status and reliability of the findings of science falls to the philosophical analysis of the discipline. This would mean, then, that at their core LDA are as dependent on philosophical speculation as are GDA. This may be why proponents of LDA to be appear decidedly against any such strategy that requires defense of the claim that the limits of scientific explanation mirror limitations inherent in the causal powers of the natural world.

It would seem, then, that the fairest and, in fact, the strongest interpretation of LDA is actually a version of the "violation argument." This approach does not require anything like a comprehensive knowledge of the laws of nature, nor over-zealous confidence in the epistemic status of scientific findings. Rather, the violation argument requires only enough understanding of natural laws and processes to determine that *E* has features that nature could not have produced without having violated an empirically evident natural law or antecedent condition.

Still, there remains the overriding interest in grounding these arguments in *positive* findings of science, thereby allowing LDA to draw upon the full integrity and authority of scientific knowledge of how natural processes function. Herein lies the dilemma. If LDA rely on contentious philosophical premises regarding the scope and reliability of scientific knowledge, they stand to lose their standing as "fully scientific" arguments, and thus the presumed epistemic authority that goes along with that status. Yet, if LDA were to turn on the violation of laws regulating the behavior of natural entities and processes, for example, by positing phenomena whose occurrence would contravene a specific natural relationship, function or processes, they would undermine that very authority itself. In order to finesse these dual objectives, that is, both formulating a decisive violation argument and retaining full commitment to the scientific integrity of the project, Dembski introduces the Law of Conservation of Information, or LCI. It is this law that appears to have been violated by the appearance of natural phenomena exhibiting irreducible or specified complexity. LCI does not function as a *first-order* law or regularity by which to describe some causal power, force, liability, process, or relation. Rather, LCI functions as a *second-order* empirical principle, describing a property of the entire class of first-order empirical laws. Accordingly, LDA remain fully committed to the first-order laws of science and their ability to fully and accurately investigate and describe phenomenon *E*. They do not presume either a miraculous violation of first-order natural laws, or any sort of inexplicable "gap" in the empirical account. Nevertheless, the argument does appeal to a fully natural, empirical principle governing these natural processes. Therefore, unlike strategies that require a comprehensive and accurate knowledge of natural laws and antecedent conditions, this formulation requires only defense of this specific empirical principle, viz. LCI. As such, LDA do not appear to be vulnerable to philosophical challenges as to the scope or reliability of science from contemporary philosophy of science.

What exactly is LCI? Dembski defines the Law of Conservation of Information in the following terms:

> (1) The CSI in a closed system of natural causes remains constant or decreases. (2) CSI cannot be generated spontaneously, originate endogenously or organize itself....(3) The CSI in a closed system of natural causes either has been in the system eternally or was at some point added exogenously (implying that the system, though now closed, was not always closed). (4) In particular any closed system of natural causes that is also of finite duration received whatever CSI it contains before it became a closed system.
>
> (1999: 170)

Presumably, then, LDA hold that some phenomenon *E* violates at least one of these conditions of LCI. As we have seen, the distinctive power of LDA (*local* design arguments) stems from the claim that the violation of LCI is detectable in some specific phenomenon. This means that LDA must present evidence for (1) a specific, local increase in CSI whose source must be (2) from causes outside the system of natural causes (exogenously). The discovery of an identifiable and quantitatively measurable increase in CSI entails that (3 and 4) the natural system, if closed, has not always been so. The localized increase in CSI means that a significant quantity of information must have been infused into the system *at that specific time*. The only conclusion to draw from these scientific findings would be that at least some of the information presently empirically detectable in the system must have come from a non-natural source, that is, from a system whose functions cannot be reduced to natural causes. Thus, even if the system of natural causes were infinite in duration, having always contained a constant (or possibly decreasing) level of information, the discovery of a significant localized *increase* in CSI at a specific juncture in natural history entails the existence of a non-natural, transcendent cause.[9]

Several clarifications are in order. First, LCI speaks only to the occurrence of CSI – *complex and specified* information. This empirical principle does not prohibit every increase in information, i.e. empirical contingency, in a system. Neither does it deny the possibility that newly introduced contingencies serve a specific purpose. LDA readily acknowledge the ability of natural processes, when accompanied by random variation, to introduce contingent features that move in a specific direction; proponents of LDA show no interest in denying the general explanatory power of the Darwinian evolutionary account. Rather, LCI speaks explicitly to an increase in complexity only if accompanied by a concomitant and excessive increase in specificity; it delimits the *quantity* of specified information that can be generated by purely random variation, or a combination of random variation and deterministic causal processes in a system having a high level of contingency.[10]

Second, it is important to note that just as LCI permits a certain level of increase in information, it also allows for an accumulation of information as the result of a series of discrete and incremental steps, each one of which fall below the natural CSI threshold. As indicated above, LCI does permit the introduction of information. It even countenances the introduction of a certain level of complex and specified information. Thus, a process, over time, may continue to produce information with a modicum of specificity. Furthermore, the principle also allows for the introduction of specificity to accumulate so that the whole process, by small steps, moves in the same direction, towards a particular end. This process may easily result in a level of CSI surpassing the amount allowed under LCI for any specific process or event. Again, the ID literature does not deny the general accumulative power of, for instance, a Darwinian evolutionary process. Its focus rests on the evident increase in CSI in a particular phenomenon, occurring within a well-defined time frame ($t_1 - t_2$).[11]

Finally, the law speaks only to a significant *relative increase* in quantitatively measurable information. That is, although the law applies only to cases exhibiting a sufficiently high degree of complexity and a sufficiently narrow range of specificity, it also only applies to the origin of information within a closed system. That is, LCI applies only to situations where a phenomenon E represents the elimination of states of affairs whose possibilities are a function of the natural laws, processes, and conditions of the closed system. Natural processes that simply reorder, preserve, transfer, or even transform information are not governed by this law. Therefore, this particular law explicitly allows for CSI, *however high*, to remain constant, even (especially) eternally present in the natural system. Thus, LCI does not apply to cases where the absolute quantity of information at t is high, no matter how remarkable the absolute figure. However impressive the absolute quantity of information evident in some phenomenon E at t_2, LCI applies only where there is a sufficient increase relative to the informational content, i, at t_1, i.e. LCI is violated if and only if $(i/t_2 >> i/t_1)$. LCI applies only where there is a detectable significant increase in specified information; it speaks to the *production* of information rather than its presence, no matter how specific it might be in achieving an outcome.

It is in this respect that LDA are distinctively "scientific," and presumably more compelling, than traditional, global design arguments. LDA rest fully on empirically detectable instances of significant relative increases in quantifiably measurable CSI at some juncture *within the system* of natural processes, rather than on philosophical speculation as to the amount of information we might expect to find in the system as a whole. This means that LDA do not address the question of why there is information in the system (at all) or, even, how the bulk of information inherent to the system ultimately arises. The distinctive strength of LDA is their basis on a detectable infusion of CSI *at a particular time* during the natural history of that system. Like GDA, LDA seek to explain the origin of information in the system. Unlike GDA, LDA do not address the question of the origin of information in general, rather only the origins of those particular bits of information known to arise during the course of history that violate LCI.

According to the present strategy, if empirical investigation reveals that an increase in CSI violates the threshold implicit in LCI, then that information could not have been produced by undirected natural causes. We have championed this approach in so far as it would not require the comprehensive and accurate knowledge of natural laws and processes necessary to support the "argument from ignorance" or "lack of imagination" approach. As it turns out, however, the violation strategy, no less than these alternative formulations, requires confidence in the comprehensive scope and reliability of scientific inquiry to have provided full and accurate access to the informational content at that earlier time, t_1, as well as throughout the process from t_1 to t_2. LDA rest on the claim that the bulk of the information at t_2 was not present in any manner at t_1. Preserving, transferring, or even transforming information does not violate LCI. This means that LDA require knowledge of *all* the relevant natural laws and processes, along with detailed knowledge of the antecedent conditions at t_1, so as to establish that the CSI evident in E was in no manner present at t_1. But why suppose that one has the requisite comprehensive and accurate understanding of the natural processes and conditions at these remote times? Again, whether one can presume a sufficiently comprehensive grasp of the natural order necessary to determine that that information was not present in some form elsewhere in the system, but rather originated during that finite period of time, is a matter of philosophic argument and debate. If E involves a significant increase in CSI, then LCI has been violated. As it stands, then, establishing that it does involve that increase requires the same sort of comprehensive and accurate knowledge of the laws and conditions contributing to that phenomenon as are required in the "failure of imagination" and the "argument from ignorance" strategies. Thus, once again, even on this construal, the success of LDA crucially depends on a particular philosophical interpretation of science, rather than the findings of scientific inquiry itself.[12]

Suppose *Psci* refers to the proposition that the methods of the natural sciences provide a sufficiently comprehensive and reliable insight into the operations of natural phenomena such that scientists are justified in inferring that the inability to explain a phenomenon in terms of the known natural laws (either deterministic or stochastic), even in combination with random variation, entails the absence of any naturalistic account. According to *Psci*, scientists can reliably discern the level of information available in the system as a whole at a given time in its distant past, as well as the significant relative increase in that information in the system at some finite time later. Where the methods of natural science fail to provide an account for *E* (call this condition "*Sci*"), *Psci* holds that, in fact, no natural laws account for *E* (call this "*L*"). LDA, then, require the premise *P** that, if the methods of natural science cannot account for the increase in CSI, then there are no natural causal accounts: $P^* = \{Psci \rightarrow [(E \;\&\; \sim Sci) \rightarrow \sim L]\}$. Furthermore, LDA require the premise that, where science reveals a significant increase in CSI, this finding accurately reflects reality. Thus, LDA require the belief both that, if scientific inquiry presents a significant increase in information within the system, then there has been a significant increase in information within the system over that period of time and, if scientific inquiry cannot account for the increase in that information, then there are no naturalistic causes for the increase of information in the system. As such, LDA clearly and crucially depend on what reasons might be available either in support of *P**, or, at the least, in favor of the presumption of its truth.[13]

The philosophic assumptions of intelligent design

At this juncture one might expect the following sort of response: Surely this complex and specified information must have originated from *somewhere* outside the natural system. Even if one cannot empirically establish a significant quantitatively measurable increase in CSI from t_1 to t_2, the extraordinary absolute quantity of information present in *E* at t_2 must have come from somewhere. For even if one cannot decisively establish the absence at the earlier time, t_1, of the *CSI* evident at t_2, one might suppose that the CSI evident at that latter time must have had its origins in some source or other, presumably a source other than the natural system itself. How else could one account for the system having such a wealth of information? Indeed, this line is difficult to resist. Yet, however compelling, the intuition that the information must ultimately have come from a source *external to the natural system itself* does not have the status of empirical law. To suppose that things could have been, in terms of broad logical or metaphysical possibilities, other than they are has the status of a philosophical assumption. In particular, to suppose that the natural system might not always have contained the level of CSI detectable at any given point may constitute a reasonable conjecture; nevertheless, this supposition takes one well beyond the limits of empirical inquiry. This assumption is, I take it, central to GDA and traditional design arguments generally. Appeal to this philosophical principle betrays the logic of LDA and serves to undermine their distinctive epistemic force. It also turns "fully scientific inferences" into contentious philosophical arguments. As such, they neither challenge the propriety of the exclusivity principle of the methodological naturalist, nor suggest a distinctive research program for science, nor carry the presumed superior epistemic authority of an empirically supported inference. If, in the end, LDA turn on this gambit, then in this respect they fail to provide, as has been claimed, a distinctive, and rationally superior, form of inference.

While discussing the design argument, J.L. Mackie maintains that,

> [a]s an empirical argument, it needs not only the premise that certain objects not made by men exhibit a kind of order that is found also in the products of human design; it needs also the premise that such order is *not* found where there is no designer.

Indeed, Mackie insists that GDA actually require an "*a priori* double-barreled principle, that mental order (at least in a god) is self-explanatory, but that all material order not only is not self-explanatory, but is positively improbable and in need of further explanation" (1982: 143–4). Thus this appeal to a designing mind to explain the high-information content of the system simply mirrors the classic strategy that Dembski is at pains to repudiate.

Appeal to such *a priori* principles is evident at the following juncture in LDA. If these design arguments are not merely eliminative, they are at least that, for support for the claim that something other than natural causal processes are responsible for the local infusion of the CSI into the system rests on the argument that natural causes cannot provide an adequate account.[14] Yet LDA are not merely negative; the very same empirical evidence by which they renounce appeal to undirected natural causes also provides support for identifying a type of cause capable of accounting for this significant increase in CSI.[15] That is, the very evidence for the denial of a natural cause provides all the support necessary for appeal to a designer, that is, a discriminating, goal-directed entity. How exactly does a significant increase in CSI support this inference?

According to LDA, there must be some system having a mechanism by which to choose the outcome of its operations in such a manner that the choice is not a function of those natural laws and processes governing the operations of its several parts. The choices by which this system achieves those outcomes are not determined by any of the laws (or chance, or their combination) governing the activity of its parts. And yet achieve that end it does. Therefore, for such an entity, the goal must play an active, autonomous, direct, and essential role in determining the internal functioning of the system. The system is not merely telic, it is non-reductively telic; that end has an autonomy that allows it to *actively guide* the processes themselves. The end functions directly as a top-down determinant of the specific operations of its parts, ensuring an outcome that would have been prohibitively unlikely in its absence. It cannot be explained as the collective outcome of the individual operations of its several parts. Nothing about their individual interactions would explain the pathway taken towards this outcome.

One might suppose that a system which exhibits this kind of top-down causality *must* be a conscious, intentional, minded system, that is, a personal agent.[16] Call this "conceptual grounds" for inferring design. However, there appear to be systems exhibiting this kind of top-down causality that are not conscious, intentional agents at all. Such systems might draw upon overarching organizational principles to supervene over the lower-level causal laws that describe the processes, relations, powers, liabilities, structures, and so forth of its constituent parts. A non-reductive, global operator, functioning independently of the microlevel laws of physics, chemistry, and biology, could work directly to govern the outcome of their collective functions, utilizing those lower-level forces to effect *its* end. Thus, if denying the adequacy of "undirected natural processes" simply means conceding the absence of an explanation in terms of laws describing the activity of a system's constituent parts, it need not entail the absence of natural principles "understood only from a holistic description of the properties of the entire" set of those parts (Bak 1996: 2).[17]

Of course, appeal to the direct activity of a personal agent, one whose intended purposes are fulfilled by the choice of the overriding global principle causally responsible for governing the operations of the natural system, would also suffice to explain a significant increase in CSI. This agent need not function by violating the first-order laws so much as by guiding the outcome of their collective endeavors by a means describable in terms of a non-reductive global principle. In this case, a holistic description would indicate that its will is implemented in a law-like manner (either deterministic or stochastic). Making explicit appeal to an intentional agent would suggest that the choice of that law-like holistic principle was not determined by any feature internal to either the natural system or this transcendent agent. In this manner, appeal to personal agency would provide an account of the ultimate source of the information contained in contingent governing principles. Appeal to personal agency as a contra-causally free origin of the choice of that process would effectively transform empirically contingent phenomena into metaphysically contingent phenomena. Once again, however, this pursuit is the stuff of GDA; assumptions as to how the system as a whole came by such naturally functioning global organizational principles calls for the sort of *a priori* philosophical principles that betrays the logic of the LDA.

On the other hand, the specific local phenomena in question may simply evidence the direct causal effects of a non-reductive, goal-directing, brute feature of the natural world, working through the system of natural laws that describe the powers, processes, relations, and so forth of the constituent parts of that system. The appropriate level by which to explain this process may involve a system at a scale many times greater than the subsystem in which one describes the event itself. Nevertheless, such an account need not appeal to anything outside the system of all natural systems. This higher-level process would constitute the source of the significant net increase in local information evident in that particular phenomenon *E*. In this respect, the information present at t_2 was not present at t_1 when considered at the level of laws governing the individual behavior of its constituent parts. Nonetheless, one would have to consider it as having always been present in the processes and powers governing the whole.[18] The very possibility of such an account undermines the *conceptual* support, the "in-principle arguments," for design.

At this juncture, appeal to an *empirical* principle might serve well to forge the link between systems of this sort, viz. systems whose functions and outcomes are directed by global principles, and intentional agency. The "inductive argument" holds that, since every time CSI is traced to its source, one finds intelligent agency, one can infer that, if one were able to trace any instance of CSI to its source, one would find intelligent agency. That is, in those cases where we know the origin, the source of CSI has invariably been an intelligent agent. As a "straightforward inductive argument" (Dembski 1999: 142), however, this line of reasoning fails. This is because the evidentiary base for the inductive generalization actually contains *two* conditions, the presence of CSI plus the ability to trace the underlying causal story to its source. From that evidential base, Dembski infers that the mere presence of CSI legitimates (all but compels) the inference to intelligent design; in any other circumstance we would attribute this feature to intelligent agency.

The problem this second condition raises is obscured by the fact that, in an enumerative inductive argument, one expects to be *able* to forge a connection between the two features in question, in this case, the presence of CSI and agency. This second condition, viz. being *able* to trace the underlying causal story to its source, appears to constitute a merely formal condition of the inductive argument rather than an additional material condition. However, it actually bears on the internal structure of this inference. Strictly speaking, one can extrapolate from this experience only to cases where one *can* trace the underlying causal story to its ultimate origin, even if one fails to do so.[19] It would be illicit to extrapolate from this sample to a target population lacking any features that enable one to trace the underlying causal story to its source.

This modal difference is salient to the "straightforward inductive argument" because the sample population from which it draws, the population of instances where it is possible to trace that causal story to its source, may include factors other than CSI that are directly relevant to the inference to intelligent agency. That is, it may be these other factors that make it possible to trace the causal story; they allow one to identify that source, and so provide the support for the inference to agency. The target is crucially different from the sample population in this one important respect: although both, we will presume, involve a significant increase in CSI, the sample has some feature or other, quite possibly absent in the target, that makes it possible to detect agency. If so, then even if they share equally impressive increases in CSI, this sample does not establish the individual sufficiency of this particular feature for inferring agency.

Why not suppose that in every case it is, as LDA claim, the local increase in CSI that enables one to trace that phenomenon to an intentional agent? One reason is that a significant local increase in CSI does not represent the only grounds on which to base one's inference to agency. Since other features enable one to trace the causal story to an intelligent agent, then it may turn out that a significant local increase in CSI, however much it is commonly involved, is not directly relevant to this inference at all. If increased CSI is not necessary for the inference to an intelligent agent, then even its universal presence in such cases as constitute the sample class provides absolutely no inductive support whatsoever for its constituting sole sufficient grounds for that inference.

In LDA, the significant increase in CSI functions like a Turing Test for agency; like linguistic competency it constitutes a reliable marker for the presence of intelligence. We may call appeal to CSI the "Dembski Test." But, as John Searle persuasively argues with respect to the Turing Test (Searle 1984: 32ff.), however strongly a characteristic is correlated with a feature, it may yet fail to provide a sufficient conceptual link. Thus, even if common experience invariably involves a particular behavioral phenomenon such as linguistic competency, there are no reasons for supposing that that marker cannot be produced in the absence of an unobservable phenomenon such as intentionality.[20] Therefore, the presence of that marker, in this case a local increase in CSI, does not support the inference to agency.

But, then, neither does their invariant correlation. The inductive argument provides absolutely no support for the belief that the only system capable of producing that phenomenon has the feature in question. In fact, if a rival conceptual argument suggests that that phenomenon may be caused by something other than that characteristic, and if one does accept the legitimacy of those everyday inferences that comprise the inductive pool, then some other feature(s) of the everyday cases must actually, even if tacitly, bear the inferential burden.[21] Indeed, experience teaches that a number of features other than a local increase in CSI support the inference to human agency. Thus, even if the inference to agency were always to involve the presence of complex specificity, this would not entail that this singular feature carries the weight of the inference to intelligent agency.

Furthermore, one of these other features is *always* evident in the inference to human agency. For, in every case where a human agent is involved, the CSI must have been instantiated in a particular material, having a human generated form, or fulfilling an evident human purpose. Thus, another reason for supposing it is not a local increase in CSI that provides the support for agency stems from analysis of those cases comprising the sample class. Even in these cases where a significant increase in CSI is evident, the inference to intelligent agency may essentially rely on these other features. Although in some cases we can infer human agency by one or the other of these features *in the absence of CSI*, in no cases where we have successfully traced the underlying causal process to agency do we find CSI alone. Therefore the straight-line inductive argument cannot isolate CSI as providing sufficient support for this inference to agency.

Thus the infallibility of the inference from significant local increases in CSI to agency in cases where the ultimate source can be identified, viz. in humans or relevantly human-like entities, says nothing with respect to those circumstances where one cannot trace the source. Thus, with respect to natural phenomena, where, for the sake of this argument, one cannot say what that ultimate source might be without begging the very question at issue, in the absence of an *a priori* principle that states that the high-information content must have ultimately originated in mind, one cannot infer agency. So all that experience teaches is that, in some cases, an apparent significant increase in CSI indicates agency. In other cases, one simply cannot say whether it indicates agency. Thus the inductive argument can only support the inference that the ultimate source of that phenomenon is either an agent or something that may or may not be an agent.

This analysis suggests that, in so far as LDA rest on the evidence of a sample class – the precedence of forensics, archeology, paleontology, cryptology, etc. – it actually rests on an analogical argument. These cases involve a significant local increase of CSI, along with other features. Since local increases in CSI may have been caused by entities other than intentional agents, one cannot hold that *only* intentional agents explain these phenomena. Furthermore, since the local increase of CSI cannot be isolated from these other factors, the sample class provides no additional evidence that every such case has its origins in intelligent agency. At best, one might infer that such cases are analogous at least with respect to the local increase in CSI, and so may well have other features in common (as, for instance, the same kind of source).

Conclusion

In these remarks, I have argued that, first, the identification of a local increase in CSI requires a commitment to a particularly contentious philosophical interpretation of scientific inquiry. Second, even if one is willing to grant that if science has not yet discerned the natural pathway then it does not exist, I have argued that the mere presence of a local increase in CSI does not establish the immediate activity of an intentional, intelligent causal agent.

Dembski insists that "the principal characteristic of intelligent agency is directed contingency, what we call choice" (1998a: 62). In the end, then, even if these information-rich systems do not underwrite a "rigorous, scientifically demonstrable" argument for design, they do support explanation in terms of the intentional and transcendent choice of a personal agent. Therefore, inferring design does constitute an intelligent choice, that is, a rationally warranted, philosophically viable interpretation of certain remarkable empirical phenomena. As such, affirming design retains this central feature of intelligence: even though appeal to design is not necessary in order to account for these phenomena, it constitutes an empirically informed, discriminating choice.

Notes

1 The non-theist explains the appearance of life by appeal to the "null" hypothesis. Accordingly, life is the outcome of the operative laws working in concert with the initial conditions, either in conjunction with intrinsic randomness over time, or by necessity. Chance suggests an array of possible universes with the present universe being that which happens to exist. This strategy does not require the reification of "chance" as a competitor to mind; rather it construes the Universe as requiring *no explanation*. As with the theist's appeal to God, the material world is simply taken to be the point where explanation stops. According to Richard Dawkins, in *The Blind Watchmaker*,

> [t]o invoke a supernatural Designer is to explain precisely nothing, for it leaves unexplained the origin of the Designer. You have to say something like "God was always there" and if you allow yourself that kind of lazy way out, you might as well just say "DNA was always there," or "Life was always there," and be done with it.
>
> (1996: 28)

2 This is true particularly in the cases at hand having to do with the origins of certain kinds of natural phenomena, i.e. living organisms whose origins lie in the distant past. My comments in this section should not be read as undermining our ability to identify a miracle, that is, a specific situation where the antecedent conditions and relevant causal powers are known and well understood. Most discussions of miracles center on specific events that cannot be explained in terms of known laws and conditions; the present discussion concerns phenomena of a more general type, the sort for which a general law might be relevant.

3 These "instructions" refer to the antecedent conditions and operative laws that combine in a precise manner to produce that outcome.

4 As with the general application of "LDA," for the sake of this essay we'll adopt the nomenclature of "CSI" as referring to either Dembski's "complex specification" or Behe's "irreducible complexity." Without having to sort through the relations that hold between these two distinct notions, we'll assume that a

description, in terms of a substantial increase in empirical contingency, narrowly specified to a particular outcome, applies to each. This is what I mean by "CSI."

5 It is with respect to this negative thesis that Dembski makes the strongest claims about the design inference being a deductively valid, rigorously empirical, and fully scientific inference. It is (at least) with respect to this negative thesis that he says: "Precisely because of what they know about undirected natural causes and their limitations, science is now in a position to demonstrate design rigorously" (Dembski 1999: 107). Proponents of LDA insist upon this step as a necessary feature of their argument. LDA will not so much as get off the ground without having established that these phenomena cannot be the result of either law, chance, or their combination. It is this strictly negative sense of design that stands at the terminus of Dembski's "explanatory filter."

6 With respect to explanation by appeal to what Dembski calls "regularity," i.e. explanation by appeal to deterministic or even non-deterministic law, he comments:

> In practice, the way ~*reg*(E) gets justified is by arguing that *E* is compatible with all relevant natural laws (natural laws are the regularities that govern natural phenomena, e.g. the laws of chemistry and physics), but that these natural laws permit any number of alternatives to *E*. In this way *E* becomes irreducible to natural laws, and thus unexplainable in terms of regularities.
> (1998a: 93)

7 If by empirical contingency one means a significant quantitative increase in the informational content of *E* at t_2 relative to some earlier measure at t_1 ($i/t_2 >> i/t_1$), then establishing this leap requires a sufficiently comprehensive grasp of both the relevant laws of nature and a sufficient knowledge of the details of the antecedent conditions in order to establish both the exceeding improbability of the phenomenon and the narrow parameters of its functionality. In short, resting the argument upon an infusion of information requires confidence that that information was not already present elsewhere in the system. Accordingly, proponents must suppose that scientific inquiry is *fully* and *reliably* capable of revealing these natural laws and antecedent conditions.

8 Stuart Kauffman argues that the inability to predict subsequent states is due to quantum indeterminacy and the large-scale changes that result from very slight alterations to the initial conditions suggested by chaos theory (Kauffman 1995). If these features accurately characterize the natural domain, then the explanatory limitations are not so much a function of human cognitive fallibility and the impotence of the scientific method as they are a function of the infinitely detailed knowledge of initial conditions necessary to predict these phenomena. If a scientific explanation requires meeting a high-probability requirement (as is assumed in Dembski's discussion of the "explanatory filter"), then these features would undermine any attempt to provide a scientific explanation of a phenomenon from its initial conditions.

9 LDA focus on specific phenomena *internal* to the system, rather than raising questions about the origins of the system as a whole. Given this system, whose function is well understood, there must have been an exogenous source of *i* in order to have produced *E*. On this construal, there is simply no reason to speculate as to the origin or even duration of the system. Regardless of one's views on its origins, its history reveals that the system simply does not contain the resources to account for its own historical development.

10 Behe reiterates this point when he insists that "complexity is a quantitative property." "A system can be more or less complex, so the likelihood of coming up

with any particular interactive system by chance can be more or less probable" (Behe 2000a: 159). If so, then LDA require having both the means by which to measure relative quantities of information, and some grounds for establishing a threshold for naturally occurring information.

11 This means that in measuring the increase in CSI, one cannot simply measure the amount of information at t_1 and then also at t_2 and compare, noting that ($i/t_2 >> i/t_2$). Rather, one must know the processes well enough to discern that insufficient steps were available during that time frame to account for the CSI at t_2 relative to t_1.

12 In his discussions of design, Dembski holds that the inference to design is neutral with respect to a realist or an anti-realist interpretation of science. That is, intelligent design supports a scientific research program that seeks either the confirmation of a real designer, or, alternatively, confirmation that nature behaves "as if" it were designed. The options here, and the neutrality of the design inference to one's choice, suggest that Dembski does recognize a demarcation between what might count as belonging to the findings of science proper, and what should be regarded as providing a philosophical interpretation of those findings, with the former neutral with respect to the latter. The point at issue here is whether the twin assumptions that the findings of scientific inquiry provide (1) sufficiently reliable access into the antecedent conditions, and (2) a sufficiently comprehensive grasp of natural processes to support the claim that we have empirical evidence of a significant quantitative increase in CSI, ultimately rest on a philosophical interpretation of science. LDA are not distinctively empirical arguments.

13 Arguments against *Methodological* Naturalism (MN) constitute the crucial step in this latter strategy. If proponents of LDA can establish that this methodological commitment to naturalism, that is, taking naturalism as an operative assumption of science, is merely prejudicial, then they will have succeeded in undermining reason for resisting the claim that the inability of science to explain is a reliable indication of the lack of a naturalistic explanation. That is, arguments against MN suggest that the best strategy to adopt in the face of the failure of naturalistic accounts would be to suppose that the Universe is an open system.

14 Therefore, if the previous section undermines the inference to a designer, the design inference is a non-starter, no matter which of these separate interpretations of design is intended.

15 One reason for supposing that LDA are not intended to be purely negative is the supposition that these arguments somehow undermine commitment to MN. This follows in so far as the negative conclusion itself is *fully compatible* with MN. To hold that some natural phenomenon cannot be explained in terms of the operation of natural laws on antecedent conditions, chance, or any combination of these factors, does not of itself violate an exclusivity principle that prohibits appeal to transcendent agency. Thus, LDA must certainly take one beyond the claim that law, chance, or their combination cannot explain some phenomenon *E*, to the positing of some specific type of entity, process, system, or state of affairs, appeal to which does violate this restriction. It would appear that the support on offer for the negative thesis also serves to support, at the very least, appeal to agency.

16 Even if "[i]ntelligent agency always entails discrimination, choosing certain things and ruling out others" (Dembski 1998a: 62), directed discrimination does not entail intelligent agency. We know that, because of its capacity for goal-directed, intentional choice, intentional agency does regularly produce CSI. But since we also know that non-intentional, discriminating agents could at least in principle produce local CSI, some other consideration will have to take us this

additional step. There are simply no conceptual reasons on offer for holding that these discriminating capacities could not be built into a natural process or entity.

17 If a top-down global explanation requires specification of the mechanism of causality, then appeal to organizational principles falls flat. But so, arguably, would appeal to intelligent agency. How purpose or intentionality translate into action is notably absent from the account.

18 Neither does this conclusion conflict with MN. MN does not permit direct appeal to intentional agency, a transcendent source of governance. It does, of course, permit appeal to global-level organizational principles responsible for directing these micro-processes in a particular direction, no matter what the ultimate source of such "principles" might be. At this juncture, LDA do not require appeal to a personal agent. Unless there are other, overriding, reasons to jettison MN, as, for example, a promising breakthrough anticipated by appeal to a personal agent, one would do well to retain a principle of such long-standing merit. Nor is this line incompatible with appeal to a divine agent, one responsible for setting the initial conditions, endowing creation with the power, processes, liabilities, and relations that result in phenomena that can be described only at the holistic level of the interaction among its several parts. Science, I gather, has the task of determining whether this might be how that creative agent determined the form of creation.

19 The inductive argument also seems to beg the question, presuming that those cases comprising the sample pool, where one *can* trace the underlying causal story to a human agent, are cases where one has effectively traced the underlying causal story to its *ultimate* origin. LDA lay claim to human agency as a source for CSI, not a mere conduit by which to distribute the information already tacitly present in the natural system.

20 One might be tempted to infer rationality from bipedalism even though bipedalism is not conceptually linked with rationality. If bipedalism is not necessarily linked with rationality, then the fact that we do infer rationality from the presence of bipedalism does not provide sufficient reason to believe that it provides reliable inductive support for that conclusion. Even if in every case of bipedalism where we *can* assess the mental status of the individual we find rationality, it does not follow that, in cases where we cannot, we would also find this feature.

21 The easiest way to defeat the conceptual argument is with an alternative conceptual argument, one to the effect that directed natural processes, even random processes, could account for the increase in CSI. An interesting form of that argument holds that an intelligent being may well, with sufficient foresight, choose to actualize just that natural process known to produce the CSI in question. Notably, a divine being with sufficient foreknowledge of what would happen given the actualization of certain material conditions and natural laws could choose to actualize just those conditions that would serve its purposes. In that case, the natural processes would not have been directed towards that end; rather, the natural system was chosen for its natural outcome. If so, we could not rule out the origination of CSI by means of a natural process.

References

Bak, P. (1996) *How Nature Works: The Science of Self-organized Criticality*, New York: Copernicus.

Behe, M. (2000a) "Self-organization and irreducibly complex systems: A reply to Shanks and Joplin," *Philosophy of Science* 67(1): 155–62.

—— (2000b) "Philosophical objections to intelligent design: Response to critics," available online at http://www.discovery.org.
—— (1996) *Darwin's Black Box: The Biochemical Challenge to Evolution*, New York: The Free Press.
Dawkins, R. (1996) *The Blind Watchmaker: Why the Evidence of Evolution Reveals a Universe without Design*, New York: W.W. Norton & Co.
Dembski, W. (1999) *Intelligent Design: The Bridge between Science & Theology*, Downers Grove, Illinois: InterVarsity Press.
—— (1998a) *The Design Inference: Eliminating Chance through Small Probabilities*, Cambridge: Cambridge University Press.
—— (1998b) "Science and design," *First Things: The Journal of Religion and Public Life* 86: 21–7.
Johnson, P. (1991) *Darwin on Trial*, Downers Grove, Illinois: InterVarsity Press.
Kauffman, S. (1995) *At Home in the Universe: The Search for Laws of Self-organization and Complexity*, Oxford: Oxford University Press.
Mackie, J.L. (1982) *The Miracle of Theism: Arguments for and against the Existence of God*, Oxford: Oxford University Press.
Meyer, S. (2000) "DNA and other designs," *First Things* 102: 30–8.
—— (1999) "The return of the God hypothesis," *Journal of Interdisciplinary Studies* XI: 1–38.
Paley, W. (1802) *Natural Theology: Or Evidences of the Existence and Attributes of the Deity Collected from the Appearances of Nature*, Boston: Gould & Lincoln.
Searle, J. (1984) *Minds, Brains and Science*, Cambridge, Massachusetts: Harvard University Press.
Swinburne, R. (1979) *The Existence of God*, Oxford: Oxford University Press.

4

GOD BY DESIGN?

Jan Narveson

"Natural" theology and the design argument

The term "natural theology" has a wider and a narrower use. The wider one says, essentially, that we can arrive at some measure of understanding of the nature and existence of God on the basis of premises none of which appeal to mysticism, revelation, or other sources of belief lying outside the area of "natural reason," which we may take to include logic, mathematics, the sciences, and common-sense observation in so far as it is consistent with scientific method. This very wide use, however, would make all the familiar arguments for the existence of God into arguments from natural theology.

The narrower usage would include only those accounts of God appealing to premises that are empirical and contingent: given that the world is like *this*, rather than like *that* (as it logically could have been), we have reason to suppose it was created by a super-powerful minded being, rather than having got to be the way it is as the result of purely natural processes. This narrower view would comprehend what were classically known as arguments from design, and would exclude the ontological argument. It requires a further decision whether to include the cosmological argument, since it is unclear what to say about the empirical status of arguments whose bare premise is that there *is* a material universe, though any old material universe would do for the purpose, so long as it was laid out in time. But on the whole, I think, we should classify this as not really an argument in "natural theology" in the narrower sense. That term I will reserve for arguments to the effect that the world has some features that can only be accounted for, or at least are best accounted for, on the hypothesis of a minded super-creator. Even so, there are certain features of cosmological arguments, especially in recent treatments, which will be of interest for my further discussion, even though my main concern is the so-called argument from design. So I begin with a note about that.

The cosmological argument

According to the cosmological argument, the material world, in general, is temporal, in which respect it is a series of events, going back in time. How far, then? On the one hand, says the argument, nothing comes from nothing: events have beginnings, and in order to have them, something or other must bring it about that they do so begin. For all theoretically relevant purposes, this amounts to the claim that every event has a cause (in the usual sense of the term, which is the Aristotelian "efficient" type, as distinct from "formal," "final," and "material" causes). The argument then proceeds to invoke a premise to the effect that trains of causes cannot go on forever, and so it concludes that there must have been a "first" cause.

At this point, things get murky. On the face of it, the argument's two premises are mutually inconsistent. If events are all laid out in time – which, by hypothesis in this argument, they are – and if indeed *all* events have causes, in that usual sense of the term "cause" (what Aristotelians call "efficient" causes), then there *cannot* be a "first" cause, since the event in which that cause consisted would, contrary to that premise, have no cause.

If, on the other hand, the argument is that the whole material universe must ultimately have been caused by something *immaterial*, then the argument is no longer "natural" in the usual sense of the term. The idea that there might be "immaterial causes" of this sort is distinctly odd. Our experience of psychological causation, as we may call it, does not include creating material entities out of nothing. (It does, to be sure, include creating ideas out of what seems, from the point of view of the thinker, to be nothing – but then we are all thoroughly familiar with the fact that ideas aren't things, fantasies not accomplished real situations, and so on – and, also, that all the ideas we have occur in our heads, which are material things). But in any case, even if we were willing to allow such a strange premise into the argument, the problem it was allegedly supposed to solve seems to recur. For, after all, minded entities operate in time too, and, if so, one can obviously ask the question where that entity – say, the god whose existence the argument is an attempt to establish – is supposed to have "come from." Proponents of the argument seem to think that we aren't allowed to ask this question; somehow the immortality of the divine mind is thought to be a premise delivered on a silver platter. But if the argument was supposed to appeal only to "natural" premises, the silver platter looks tarnished.

If we are allowed to ask when the divine mind came into being, we will, of course, be precisely where we began. Either there are things that have no beginning, in which case it is unclear why we are not allowed to suppose that other things besides God are like that; or if the divine mind had a beginning too, then we are, to put it mildly, no clearer as to what brought it about than what brought the material world about. Claims that there cannot be a first event do not work any better for mental events than for physical ones.

Proponents of cosmological arguments, at this point, turn to metaphysics. Some try to argue that the idea of an infinite series of events going back into the past (or, of course, forwards into the future) is logically impossible, on the ground that there cannot be "actual infinities."

That bit of obscurantism can be rejected, however; every number in an infinite series can be finite, as we know. Dr Craig, for example, argues that time itself began, as a result of the Big Bang – not just the physical universe *in* time – and that when it did it was due to the action of a hitherto timeless God who suddenly shifts gears into timeliness at that point.[1] Those who are ready to accept this as an "explanation" of anything are not, I suggest, seriously doing science any more. We cannot help ourselves to the language of causation in the absence of the applicability of temporal notions, and so the thesis that time "began" at time t, *as the result of* an action of a deity, simply doesn't make sense. No explanation at all is surely preferable to such proposals.

Arguments from design: telling creation from non-creation

We turn now to the project of arguing for, or at least rationalizing belief in, the existence of a minded creator as having made the world we live in, on the basis of the observed characteristics of that world. It is the main purpose of this essay to cast doubt on the sense of that whole project.

The first thing we must do in order to discuss the matter at all, of course, is to contrast the hypothesis of a creator with others supposedly competing with it. In thinking about this, we at once encounter two problems. First, the events we usually call instances of "creation" are themselves natural processes, and this makes it a little difficult to get the intended contrast off the ground. Mary baking pies is a creative process, of a minor but nice sort; we don't think any magic is involved there. The pie grows by purely natural processes, unless we want to claim that Mary's thoughts as she proceeds are themselves "non"-natural. Now some may want to make that claim, but it is quite unclear what the status of the claim is, and in any case the model is quite inappropriate to the hypothesis of a minded super-creator, which would seem to have to be a pure mind, not a mind in a finite material body – which, of course, is our situation.

The second problem is more fundamental. A super-creator, clearly, could create *any* sort of universe. This is presumably true by definition, at least if we use the familiar characterization of "omnipotence" – the power, or ability, to do *anything whatever*, anything with a consistent description. If that is so, however, then we're going to need some further premises if we wish to insist that *this* universe must, uniquely or at least probably, be the work of a creator rather than having come about by natural processes. For the religious person is surely at liberty to believe, and indeed presumably must believe, that the deity also made *those very processes*.

If there are laws of nature, as exemplified by Newton's and some others, then the religious person will presumably insist that God, as it were, legislated those laws. Indeed, that is what St Thomas Aquinas, for example, appears to have thought.[2] But it is our understanding – such as it is, but fairly extensive – of natural processes, that is, our more or less intuitive grip on the "laws of nature" such as gravitation, inertial mechanics, and biological processes, which forms the background upon which we distinguish "natural" from other sorts of causes. If creation of the laws of nature themselves is in question, there is no background to fall back on, nothing to give any sense to a distinction between the natural and the supernatural. This complication is so fundamental that it would leave us hardly knowing what to say about any of these questions, and I will continue to suppose here, contrary to what we are really entitled to suppose, that we could have a workable distinction between the natural and the supposedly not natural.

Design arguments need to come down fairly heavily on the contrast between natural processes, taken independently of any idea that they are themselves due to creation anyway, and the "others"; they then argue that something about our world invites or even requires the hypothesis of creation. In view of the previous point, we will appeal to our own understanding of creation, which is the invention of things to serve human purposes. Paley's "watch" analogy will serve well to exemplify design arguments – both in what they originally claimed, and in the problem involved in claiming it. The argument was that if we came upon a watch lying on a heath, this would give us good reason to infer that there was a human somewhere or other that had invented or manufactured that watch. Watches, we reason, don't grow on trees or spring spontaneously from bogs. Given their intricacy and their evident incorporation of purpose, we take it that the watch was "created" rather than having come about by other processes.

Very well, we must ask, where do we go from there? The answer is that we are looking at arguments from design as proposing an account of the origin of the Universe, in so far as we are acquainted with it. The argument in Paley's case is that the Universe is rather like a watch, and so in like manner invites or perhaps requires the hypothesis of a creator. But the answer to the question is that the world as we know it is not very much like a watch (even though you can use it to tell time!). Rather than supposing that the planetary motions and the rest are all parts of some big mechanism with cosmic purpose, we suppose that planets do what they do because they have no purpose at all but are subject to familiar mindless physical forces, especially gravitational ones. If a thing weighs as much as a planet and gets flung out from some passing star, it just will end up going around that star, and purpose has nothing to do with it. Indeed, such motions seem to be paradigmatically, if cosmically, mechanical. And as to watches, we can see that, and why, they are produced by humans. Therein, as we will see, lies a large problem for natural theology: the world has no evident purpose, and it is impossible to ascribe such a purpose to a supposed creator on the basis of anything except wishful thinking on our part. Would the deity, after all, need to tell time? Is this why he invented the Universe? Presumably not. In the absence of any possible use for a watch, the hypothesis that humans must have invented them is not plausible. In the absence of any motive for creating a universe, we likewise have no explanation of the Universe in the hypothesis of a creator.

Clarifying "design"

Arguments "from design" go from design in one sense of that term to design in another. The conclusion of a design argument uses what we may call the "output" sense of the expression. In this sense, to have design is to have *been designed*, by somebody for some reason. But the design argument attempts to support that conclusion from premises of a different sort, using what I shall call the "input" sense. In this sense, we must confine ourselves to utilizing observable features that *look as if* or suggest that they were the product of design in the other sense. Thus regularity, or aesthetically interesting structure, or the possession of features peculiarly suited to human purposes, are all properties available at this "input" level.

The question is, when do we have evidence for divine creation in this respect? To answer this, we must attend to a distinction in the brief list just given: between (1) design in the sense of regular pattern, as with checkerboards and honeycombs, and (2) design in the sense of looking suitable for certain purposes, plausibly ascribed to proposed creators of those items.

Consider sense (2) first. Paley's watch shows some of the former and a lot of the latter. When we know how the thing works, we know it can be used to tell time, and we know that people want to know what time it is, for various reasons we also understand well. Primitive people encountering Paley's watch might well have regarded it as just some kind of oddity, having no idea what it was for, perhaps not even caring about keeping track of time to the level of precision we are used to and so never coming close to entertaining an idea that this item enables us to do so. Knowing about people independently does wonders for the plausibility of the argument. Of course, it would also make the argument from (1) to (2) virtually redundant: we know so much about time-telling, and watches, and people living in our sort of circumstances, that we scarcely recognize it as any sort of inference at all when we identify a thing on the beach that looks exactly like a watch as, indeed, a watch – that is, draw an inference from its observable features to the conclusion that it is, indeed, a watch, made by some humans and used by others to tell time.

But the other sense, involving regular patterns, is another matter, for there is no necessary connection between it and design in the "designed" sense – the sense, that is, in which design implies designer. Crystal lattices, snowflakes, and many other natural phenomena show regular patterns without having come from designers. Why they end up looking as they do is a matter of mechanics: their appearances are byproducts of the natural and mindless processes by which they come to be.

So the question now arises: Which do we discern in the Universe – "design" in the sense (2) that implies a designer, or merely design in sense (1), of pattern or regularity? Kant was impressed by the starry heavens above, though frankly they're pretty much of a mess, apart from being impressive by virtue of their vastness, and rather pretty because of the twinkle effect (which we now know, alas, to be due merely to our atmosphere, especially if it benefits from a bit of pollution, rather than to anything in the heavenly bodies themselves). In truth, the nebulae etc. look to be what they are: pretty random. Indeed, one of the merits of (and stimuli for) the "Big Bang" theory is that the galaxies etc. look as though they were just flung out, rather than neatly hung in their appropriate slots in the firmament. In any case, such regularity as they display is certainly accountable for without recourse to design. So this classic argument, at least as applied to the astronomical universe, just doesn't work. And the biological situation, attending for example to the orderliness of the parts and interrelations among the parts of the human and other animal bodies, fares, if anything, even worse, as will be seen below.

All of the foregoing reflections lean, of course, on the distinction of natural from supernatural processes – and that, unfortunately, is not clearly available at the level at which it would be really needed. For, as we saw before, we rely in all this discussion of what might have originated how on modern science and the laws of nature. Yet if those laws themselves were supposed to be created, then there isn't any background of science or common-sense causation to enable us to distinguish the natural from the supernatural. Certainly we must exit from anything that can reasonably be termed "natural" theology at this point.

The argument from design: mechanisms

All of this, however, has to be set against the backdrop of the two most spectacular deficiencies of natural theology: lack of explanatory detail about the mechanism of creation, and lack of evident purpose. Both, I shall argue, are essential to the plausibility of arguments from design to designers. Let's look at each in turn. Here we'll consider the first – the question, "Well, just how is God supposed to have done this?"

It ought to be regarded as a major embarrassment to natural theology that the very idea of something like a universe's being "created" by some minded being is sufficiently mind-boggling that any attempt to provide a detailed account of how it might be done is bound to look silly, or mythical, or a vaguely anthropomorphized version of some familiar physical process. Creation stories abound in human societies, as we know. Accounts ascribe the creation to various mythical beings, chief gods among a sizeable polytheistic committee, giant tortoises, super-mom hens, and, one is tempted to say, God-knows-what. The Judeo-Christian account does no better, and perhaps does a bit worse, in proposing a "six-day" process of creation.[3]

It is plainly no surprise that details about just *how* all this was supposed to have happened are totally lacking when they are not, as I say, silly or simply poetic. For the fundamental idea is that some infinitely powerful mind simply willed it to be thus, and, as they say, Lo!, it was so! If we aren't ready to accept that as an explanatory description – as we should not be, since it plainly doesn't *explain* anything, as distinct from merely *asserting* that it was in fact done – then where do we go from there? On all accounts, we at this point meet up with mystery. "How are we supposed to know the ways of the infinite and almighty God?" it is asked – as if that put-down made a decent substitute for an answer. But of course it doesn't. If we are serious about "natural theology," then we ought to be ready to supply content in our explication of theological hypotheses just as we do when we explicate scientific hypotheses. Such explications carry the brunt of explanation. Why does water boil when heated? The scientific story supplies an analysis of matter in its liquid state, the effects of atmospheric pressure and heat, and so on until we see, in impressive detail, just how the thing works. An explanation's right to be called "scientific" is, indeed, in considerable part earned precisely by its ability to provide such detail.

Natural theology proposes the hypothesis of creation as an explanation of how things got to be as they are. But in the absence of any remotely credible account of mechanism, in the broadest sense, it is an "explanation" in name only – a wave of the hand, or perhaps we should say a sweeping under the carpet, when scientific push comes to explanatory shove.

It has been part of the etiquette of natural theology that questions like this aren't given any attention – not even, as we might say, the time of day. Do its proponents need to do this? That is really the only serious question, for if they do, then arguments from design are hopeless from the start. But if they don't need to give attention to such questions, why is it that they don't? If the answer is that such things are beyond human understanding, then the reply is simple: Didn't you say that you were going to produce an argument appealing only to natural reason and empirically intelligible premises?

Design and cosmic purpose

When is it plausible to invoke design as an explanation? Our information and our analogies come from animal and, especially, human behavior. We can see that watches, say, are things of a sort that lie within the capability of humans, given lots of ingenious work and a certain amount of luck. But it is not the elaborateness of structure or the intricacy of mechanism that, by itself, licenses an inference to human purpose. Rather, it is that we know humans well enough to ascribe purposes to them of a kind that make sense of the production of watches, shoes, automobiles, and the rest of it. But suppose we had no source of such information? If we had no reason to suppose that the animal to which we want to ascribe watchmaking would have the slightest interest or need for keeping track of time, we would have no reason to ascribe watchmaking to it either. For that matter, there is a tradition to the effect that God is timeless. If so, one supposes that in His view, time is too crass for creators to muck around with. Why, then, produce a universe so utterly different from the ideal situation? Why would He regard it as any credit to Himself to have done so?[4]

To be sure, people do odd things, for odd reasons. Take art, for example: people have devoted lives to the production of paintings, symphonies, and the like, even though these notoriously have no "useful purpose" – apart from the supremely useful purpose of providing interesting and enjoyable visual and auditory experiences for organisms like us. These interesting activities, familiar enough to us all, might be seen as providing some grist for the budding theologian's mill. If humans are rather unpredictable, their behavior often puzzling, can't that be used to fill out a theological hypothesis to the effect that the reason why the Universe is such an odd place is that it was created by such an odd creator? But the difficulty is that with people, we have independent evidence of their existence, and can observe their behavior. That's how we know that they're pretty odd organisms in so many ways. When we explain some phenomenon by the hypothesis that some human did it for reasons unknown, we are really invoking a background of solid information: "reasons unknown," in the case of humans, is a fairly familiar category, and actually explains quite a lot, in its modest way. Of course in one sense we do not exactly have an "explanation" when we invoke reasons unknown. But in another, we do: a human did it, rather than some other sort of process independent of human contrivance, and we know humans, notably ourselves, well enough to understand that they might do things like that. And that's something, even if it doesn't also tell us why the human in question did it, or at least does not do so at a level of detail that would make it altogether clear.

What we do have, however, is recourse to chance or to submicroscopic goings-on in the brain or nervous system, or both. We can understand how some exotic process in our material bodies could affect our motivations, or our deviations from established courses of action. But those are not available in the case of gods, obviously. More generally, we have, again by hypothesis, no independent mode of verification or source of information. The inference to gods is pure, in the sense that there is no independent way of observing the entities being invoked, nor any processes by which their motives or ways of doing things may be understood. Stories in sacred books are, especially, of no avail: their authors knew a lot less than we do for one thing, and their claims to be well positioned to know the sorts of thing they ascribe to their gods are no more credible than the ascriptions themselves.

In the case of modern science, exotic entities are familiarly invoked. But exotic though they are, they are pinned down in a network of empirically supported theory. There is no reasonable doubt about the existence of molecules, atoms, or even electrons, and the still more submicroscopic entities of modern physics, while some are still conjectural, fit in fairly well-understood ways into the network of theory and observation. Moreover, it is often possible to make predictions on the basis of these hypotheses, and those predictions are confirmed or disconfirmed; either way, they count in favor of or against the hypotheses in question. In order for natural theology to be an eligible "discipline," its hypothesis would have to be in similarly decent shape. But is it?

The hypothesis is that the world as a whole was, somehow, created by a minded super-creator. The idea, improving on the supposedly more primitive religious notions of ancient Babylonians, aborigines, and many others, is that instead of some mythical man-like being, we have an individual of *unlimited* powers. Being unlimited, no wonder he can make this entire universe!

But in order for the explanation to have any content, we need to know something that is not often addressed: *Why* is this being supposed to have done this? Consider that a being of this type already knows everything there is to know, so He can hardly have created the world to satisfy His curiosity. And since He has no body, no senses, and no needs in any usual sense of the word, where are we to get the psychological premises we would require in order to make an inference to His creative activity plausible?

Note my natural use of the pronoun 'His' in the preceding sentence. Feminists may object that it is a sexist characterization of the Deity, and I cheerfully accept the charge. Were we to try to correct this by saying 'She', a counter-charge would be equally in order. But of course both would be wrong, and for the same reason: having no body, and no reproductive system, and all of that familiar stuff that immerses you and me in the world we live in, there is no reason to attribute to a super-creator *any* properties of the kind. Unfortunately, that serves to point us in the direction of the basic problem: the situation is quite a lot worse than that, for there is in turn no reason to attribute *any* motives of *any* kind to a being so described.

Below we shall consider the hypothesis that the Creator has one other essential attribute: moral goodness. But even that attribution is in for heavy weather when purely spiritual super-creators are in question, for it is hard to see why gods should have any "sense of obligation" whatsoever. Who's to complain, and why should they listen to any complaints? Indeed, since God is alleged to have created people along with the rest of the world, the whole game would seem to be quite thoroughly rigged anyway.

But first, bracketing that discussion for the moment, consider such ideas as that God would at any rate be supremely "rational" – at least consistent, say. But unfortunately, when carefully considered, this is of no help. Consistency is primarily defined with reference to propositions, and we may agree that you won't catch a deity believing inconsistent pairs of propositions; and of course we must not describe the Deity in ways that are mutually incompatible (which is not a virtue of the Deity but a needed virtue in human accounts of the Deity). But why won't we catch Him, contra Einstein, playing dice, for example? People, after all, sometimes do, so it's not as if it can't be done. And if the point is that God would always know how the throws come out, fine: He would always know how *everything* comes out, so why bother to do any of them? Indeed, what was supposed to have been the point of creating universes in the first place? What's He going to *do* with the darned things, anyway? We are, obviously, at a loss for reasonable answers to any such questions.

But since we are, we are also unable to get anywhere at all with the project of natural theology. No matter what the Universe is like, it could have been created by a super-creator who, for some utterly unknowable reason, just wanted to create one of those, precisely the way it is. There is, for instance, no reason to suppose that such a being would necessarily like a nice orderly universe (rather than the cluttered, messy one we actually have). That He would "prefer" a "consistent" one is obvious, but only because it is a misunderstanding to suppose that consistency is a property of universes. It is, rather, a property of what purport to be descriptions of them, and of course if the description is inconsistent then it does not describe a possible universe and that's that. But that's consistent with universes being just incredibly messy, or incredibly simple and boring, or whatever. This being so, there is obviously no sense in saying, "Well, it's *this* way rather than *that* way, and so, you see, that makes it more likely that it was created by a deity." Alas, no such specification of a "way" makes it more likely. Not only doesn't it make it "necessary," as overly enthusiastic formulations of the argument from design might have it, but it makes it utterly arbitrary – just like any other hypothesis in this particular field.

Three examples

In light of this, consider three recent ideas – the major impetus, indeed, behind the conference that was the occasion for writing this paper.[5] They are, respectively, the "Big Bang" hypothesis in physics, singularities in evolutionary biology, and the "fine-tuning" thesis. To this we can add one very old idea: appeals to miracles. Each deserves a fairly brief comment – that is, given the level of interest and attention they've aroused, they deserve comment, and, given the actual merits of the proposals, the comment need hardly be long.

The Big Bang

According to recent science, it begins to look as though the whole known universe emerged from a quite fantastic explosion some 15 billion years ago, give or take a few billion. Some are seizing on this theory to proclaim that this, indeed, was that very first creation that the Deity indulged in. Mind you, the story is a bit different from what we read in the Book of Genesis, where it took God – for some reason – six days to create various categories of things, animals, and people. But what's a few billion years among friends? The real problem with the Big Bang idea as it relates to theology is that there simply isn't any reason to see why a deity would do things that way rather than any other way. Indeed, it is not clear what a "way" is for a being that is supposed to be able to will things into existence, effortlessly. Bertrand Russell used to point out that, logically, the entire world could have come into existence five minutes ago, complete with all the features that make us think it has been around a lot longer than that. Might not the Deity have done it that way instead? Assuredly He might. (We need hardly add that there is no known reason why the Big Bang, if it occurred, should not have been just one in an unending and unbeginning series of Big Bangs.) So why didn't He? Those who think they can give a reasonable answer to that question have not, I think, considered the nature of their hypothesis with sufficient care. Being unlimited is a major hazard in this business.

Evolutionary biology

There has been, especially on the North American side of the Atlantic, considerable public kerfuffle over the supposed issue of "creationism" versus "Darwinism" in biological theory. According to Darwinians, the various species we have are here because they were, in the circumstances, equipped to survive. Over millions of years, such factors as mutation, changing gene frequencies, random splicing, and other matters too subtle for ordinary folk like the writer (and his audience) to have much of a grip on (but clearly relevant for the purpose of explaining these things) bring about alterations in organisms. Some of these result in extinction while others result in organisms that survive long enough to reproduce in their turn. Where we are now is simply the latest in this prolonged and increasingly elaborate show. The supposedly alternative view is that the Deity created the different species – but all at once, contra the evidence, complete with leftover fossils to make evil scientists think somebody else did it? Whatever. The basic point remains clear: since the deity could have done it either way, what's to argue about? Believers can believe that God started, or continues to support, an evolutionary process just as well as any other way of doing such things. And of course the epistemic situation remains the same: there is no credible reason why He would have done it one way, or another, or for that matter – worse yet – at all. As for the idea that there are mechanisms in some or all of our species that "cannot" arise by "natural processes," it is a bit late in the game to entertain notions like that, is it not? At the beginning of the twenty-first century, when we have insight into computers, atomic fission, voice recognition, and so on, it takes a good deal more than rashness to insist that there are structures that "cannot be explained" on the basis of recognizable, law-like processes, especially when we include chaos theory.[6]

Fine-tuning

The "fine-tuning" thesis has it that, for example, the human species requires a combination of conditions whose antecedent probability (however you compute *that*!) is astronomically small, making it a cosmic accident that there are people. From this it is inferred that the existence of people must be due to divine intervention, divine fine-tuning, after all. This is perhaps the most remarkable of all of these arguments, for it evidently implies that the Deity prefers vastly improbable ways of bringing about intended results (the existence of people) to the much more plausible ones that presumably He could also have done at the drop of the divine hat. Why on Earth would He behave like that?

And that's just the trouble. The lack of answers to this question is matched by the lack of good answers to any other question of this general type that you can ask. Bodiless minded super-creators are a category that is way, way out of control. To all questions, there is but one general answer: "Who are we to understand the mysterious ways of the Deity?" A good point, in its way – but one that utterly undermines the project of design arguments, since we no longer know what is to count as "evidence" for or against any such hypothesis.

A note on miracles

It was once popular to suppose that the various stories in the Bible – but not, of course, the various comparable stories in innumerable other sacred texts of the world's religions – constitute *prima facie* evidence for the existence of God. The intermediate premise, of course, is that the stories are true, and analysts from David Hume onwards have rightly pointed out that the stories in question are short on credibility, to put it mildly. But it seems to me that another question must be asked about them: namely, why on Earth (or in heaven) would the Deity engage in such shenanigans anyway? Here, we are told, we have a Deity who goes way out of His way to subject everything in nature to laws; then He proposes to induce people to believe in His existence (again, for reasons unknown) by speaking to them out of burning bushes, or curing lepers on the spot, or what-have-you. Well, what are the more rationalistically minded among His flock supposed to make of this odd stuff? Why miracles on Monday but not on Tuesday, to this lot of simple fishermen or shepherds rather than that? The God this would induce people to believe in must, evidently, be a remarkably arbitrary one – contrary to initial billing. Again, the theological story tends to unravel before our eyes. In the end, of course, the only conclusion about this that commends itself to reason is that the stories were invented or embroidered by believers, and in particular by believers not inclined to ask too many embarrassing questions.

Thus far, I have complained about the project of shoring up natural theology by arguments from design on two counts: first, that the hypothesis is devoid, at the crucial points, of explanatory power because mysteries must always be invoked when it comes to embarrassing details such as how the Deity did it; and, second, that the hypothesis is of such a sort as to deprive us of an essential premise, namely, a clear insight into the motivations of the supposed Creator, beginning with the question of why He would have any "motives" at all. But it behooves us to consider the remarkably popular thesis that the Deity is not only super-powerful but, somehow, *good.*

The goodness of God and the badness of theological explanations

People are moved to worship the divinity not only because of its alleged omnipotence, but also because of its moral perfection. We must admit that a tendency to fall on one's knees before potentates good, bad, and indifferent has been a prominent feature on the human scene, but the more thoughtful among believers will readily agree that the worship-worthiness of a divinity is crucially a function of its moral character and not just its impressive assets for big-time coercion.

Bringing morality into the picture complicates matters for the would-be natural theologian. On the one hand, it adds – or at least, should add – real content to the story, for the hypothesis that the world exists because of the good taste of a super-creator certainly suggests that we are no longer left wide open in our choice of worlds that might have been created by it. Good worlds, one assumes, are a subset (and perhaps a very small subset) of all the logically possible worlds. Leibniz, indeed, seems only consistent in affirming that God would, of course, create the "best" of all possible worlds.

Well – best *how*? If the world attests to the goodness of God, we need relevant criteria of assessment. The claim that the world was created by a *morally* good being should, one hopes, be quite different from the claim that it was created by somebody who was really good at creating worlds, for instance; or good at sowing confusion, or playing dice, or whatever. Moral goodness is a narrow notion on any tolerable account.

We should attend to one important point about this at the beginning: there are some, still, who affect to believe that goodness is whatever the Creator wants to say it is – no controls! If that were so, of course, the subject would be otiose, and any appeal to argument pointless. If whatever the

Deity wants is *ipso facto* good, then we can't appeal from the goodness of the world to the conclusion that it must have been made by God, since the claim amounts to no more than that it is the way He made it, which could be just any way and therefore provides no information whatsoever that could contribute credibility to an argument of this type.

But these same people rarely claim that it is actually a *good thing* to stuff several million Jews in ovens or exterminate tens of millions of people in Siberian labor camps or inflict cancer on a few hundred million randomly selected humans, or unleash *E. coli* on us unsuspecting humans. Not, that is, a good thing in itself.

The idea that God could somehow invent morality while He was at it is a nonsense claim, taken at face value. When people believe that God is good, they attribute some non-arbitrary properties to the personage in question. Most especially, given the requirement that God is to be *maximally* good, and not just ordinarily or somewhat good, they presumably mean to claim that God is ultra-benevolent as well as ultra-fair, and other things that we expect of persons claimed to be outstanding in moral respects.

Notoriously, this is not going to look too plausible in the face of the world we actually have, in which all of the preceding examples and countless more are standard fare. The serious proponent of an ultra-moral God needs, then, to do something about this. One option, of course, it to try to blame it all on *us*. Humans didn't cause cancer, tuberculosis, typhus, malaria, and the rest of it, to be sure, but they do cause wars, great and small, and maybe the blame can be fobbed off on us and diverted from the Divinity via the hypothesis that we are being punished, or our souls tried, or whatever. Such maneuvers do have a problem, though: Why did this supposedly super-good being who is also super-powerful *let* all these bullies get away with it? The standard reply is the "free-will defense," but it is an odd one. Here comes the assassin with his dagger or his .45. If we manage to get him first, we will certainly short-circuit his free will, which in fact is just what, at this point, we certainly want to, ought to, and surely have a right to be doing, considering what the miscreant is bent on doing to us. The fact that the murderer was acting "of his own free will" isn't an excuse, either for him or for whoever is supposed to be defending us from him, and it certainly isn't the remotest shadow of a reason why the victims of his ferocity should put up with it – yet this shabby story is supposed to be good enough to exculpate the Creator from what, on any usual view of the matter, would be regarded as horrendous crimes.

Now, the wily theologian will invoke assorted other hypotheses to square the facts with the apparent aspirations of the being whose existence they hope to infer from the facts about the world we live in. But all of them amount to the same thing: there is no doubt some justification for all this, somewhere down the pike, but it is beyond our understanding, or words to that effect. And so we are back where we were at the end of the preceding section: with a "hypothesis" that can be "squared" with any facts you like. And that has the usual result – that *nothing whatsoever* can be inferred from it, putting it beyond refutation and, consequently, beyond the reach of science. And, to remind the reader of our project here, putting us beyond the pale of good arguments with plausible premises and valid reasoning.

According to the ancient astronomer Ptolemy, the heavenly bodies must have circular orbits. On the face of it, this implies a lot of predictions that, alas, don't accord with the facts as reported by Lydian shepherds and other worthies. Never you mind, says Ptolemy: for they don't just go in plain ordinary big circles, they also go in little circles (epicycles) that operate in the big orbits, thus accounting for deviations from the big circles. Moreover, you get to add littler circles onto the epicycles too. Indeed, I am told by persons with much more mathematical prowess than I that if you add enough epicycles, you can account for any sort of orbit you like, including of course the elliptical ones that Newtonian physics tells us are what you'd expect of heavenly bodies with gravitational forces acting on them. But then we pay the price: once the number of epicycles is unlimited, you no longer have a hypothesis with predictive power, but rather one that can be squared, *ex post facto*, with the facts, *whatever* they may be.

Exactly the same is true of natural theology with morality added on. You may add what you like to your characterization of the Deity – and noting the floridity of religious stories around the world and back in time in assorted cultures, not to mention contemporary variants such as feminist theology, gay theology, and so on, it would seem that the sky is the limit on this; so it is clear enough that adding what you like is pretty much the name of the game, what with several thousand distinct religions flourishing in the USA alone at present and no doubt many more to come. But the more you add, the more your hypothesis looks as though it predicts, and therefore the more amenable to refutation at the hands of the facts. No religion can afford to let that happen, of course, and they don't. All you need is that old reliable, supreme epicyclical gambit: that "the Lord works in mysterious ways His wonders to perform," and you needn't worry about mere facts.

But of course you do need to worry about "mere facts" if you're a scientist, and that is what the natural theologian purports to be. The need for the sort of maneuvers that are commonplace in all religions is clear enough, but resorting to them takes the hypothesis out of the realm of science. And that is my point. People have religious beliefs for emotional reasons, not as genuine explanatory hypotheses, and once you adopt a "hypothesis" because you like the idea rather than because it is genuinely helpful in explaining phenomena, then you aren't doing science any more. On the whole, then, natural theology is not a genuine enterprise. Its apparent insight, its air of explaining things, is all smoke and no genuine content; the talk of "evidence" is really beside the point – window dressing rather than the real thing.

A note on religion as a social phenomenon

Religions have been extremely popular through the ages, and indeed some want to take that very fact as evidence for their truth – ignoring the complication that there are an enormous number of religions, each specifically distinct from and *prima facie* incompatible with all the others, so that if the fact of belief in one of them is to be taken as evidence for that religion, it must also count as evidence against all the others. But plausible explanations for the phenomenon of religion itself are not so difficult to come by. They also suggest that religions have the potential to be very serious problems for mankind, as they certainly have been. It can hardly escape notice that religions, especially the Western monotheistic ones, are models of absolute despotism: here is the mighty ruler, whose word must therefore be taken as law by all, and put beyond question, even to the point that daring to question at all is often regarded as a deadly sin. What better for aspiring earthly authorities than to present themselves as the indispensable intermediaries between ordinary people and the fearful but ultimately (the emphasis must, of course, be on the "ultimacy" of this aspect) benevolent God who rules over us all? Cushy jobs in the church hierarchy, the respect of one's "flock," and of course their malleability before the seats of power are all grist to the aspiring politician/prelate's mill. A brief excerpt from a recent Internet publication, though it was about democracy rather than theology, makes the point well enough:

> Make as many rules as possible. Leave the reasons for them obscure. Enforce them arbitrarily. Accuse your child of breaking rules you have never told him about and carefully explain that ignorance of your rules is not an excuse for breaking them. Keep him anxious that he may be violating commands you haven't yet issued. Instil in him the feeling that rules are utterly irrational. This will prepare him for living under a democratic government.[7]

Substitute any kind of government, or the governance of churches themselves, as well as of unruly flocks of human sheep by divine absolute monarchs, for "democratic" and the point holds even better.

Of course, this all leads to problems, not only with internal rebels, who may require the Inquisition or other familiar modes of control, but also with rival tribes and their different but equally inscrutable religions. If the god of people H calls for stern measures against those miscreants, the Ps, and vice versa, we may be sure that the proposal to assemble an academic conference to explore the merits of the rival hypotheses and settle on the best one isn't going to cut it. Instead, we will have the Thirty Years' War, the Palestinians versus the Israelis, Shiites versus Sunnis, and the like, and it will take (has taken) many centuries to see the futility of all that, and the absolute necessity of the principle of freedom of religion (including irreligion) if we are to enjoy peace among men (and women). Religion as a human institution has, indeed, a great deal to answer for. But the point is, its general properties are quite sufficient to account for its considerable prominence on the human scene – though not in a way that attests to its truth, or even its plausibility.

This is not meant to be a general condemnation of religion, though it may sound like it. For one thing, we owe much of the world's great art to religion, and it may be that we also owe some of the Western world's considerable measure of civility to its influence, though I think that in this respect it's influence is overrated. The point, rather, is that one has no difficulty in understanding why a human group might soon equip itself with the general sort of mythology that religions abound in. In particular, its epistemically refractory features are just what the leaders of an ignorant multitude need: intelligible, testable stories are bound to suffer at the hands of the facts, and simple ones won't do because they might lead people to think that they don't need human religious leaders to expound them. That the ideology of religion should be subservient to human purposes – some of them, unfortunately, not very nice ones – makes ample sense on reflection. But it also further undermines the suggestion that theology should be regarded as a respectable entry on the ledger books of science, via versions of the argument from design. That it is not.[8]

Notes

1 I refer to William Lane Craig's address to the Gifford Bequest International Conference, "Natural Theology: Problems and Prospects," in Aberdeen, Scotland, 26 May 2000.

2 I am not an Aquinas scholar, and accept correction on all interpretive points from my betters at such things, who are numerous. When I say that Aquinas "believed p," understand this to mean that there is a moderately popular understanding – or misunderstanding, as may be – of his work according to which he believed p.

3 The account, as scholars have noted, includes two slightly differing and specifically incompatible accounts within a few lines of each other (Genesis 1:27 and 2:22 *et seq.*) of the creation of woman.

4 My thanks to Neil A. Manson for noting this point.

5 The Gifford Bequest International Conference, "Natural Theology: Problems and Prospects," in Aberdeen, Scotland, 25–8 May 2000.

6 I note, recently, a book by the eminent geneticist Richard Dawkins entitled *The Blind Watchmaker*, which has the subtitle, *Why the Evidence of Evolution Reveals a Universe without Design* (New York: W.W. Norton & Company, 1986).

7 "Offshore & privacy secrets," 29 May 2000. Published by OPC International, available online at http://permanenttourist.com.

8 My thanks to discussants too numerous to list at presentations in Waterloo, at the Gifford Conference in Aberdeen, and at the University of Glasgow. Thanks also to several e-mail discussants. I am especially grateful to George Mavrodes for calling my attention to a major problem in the version read at Aberdeen. My thanks also to Neil A. Manson for encouragement in this project.

5

THE ARGUMENT TO GOD FROM FINE-TUNING REASSESSED[1]

Richard Swinburne

A posteriori arguments for the existence of God can be arranged in an order by the generality of their premises. The cosmological argument argues from the fact that there is a universe at all; one form of argument from design argues from the operation of laws of nature (i.e. that all the constituents of the Universe behave in a law-like way), and another form of argument from design argues from the laws and boundary conditions of the Universe being such as to lead to the evolution of humans, claiming that rather special laws and boundary conditions are required if the Universe is to be human-life-evolving. The normal way in which this latter is expressed is to claim that the constants and variables of those laws and boundary conditions have to lie within very narrow limits in order to be human-life-evolving. This argument is therefore called the argument from fine-tuning. There are, then, many other arguments that begin from narrower premises. The arguments are, I believe, cumulative. That is, the existence of a universe raises the probability of the existence of God above its intrinsic probability, its probability on zero contingent evidence. The operation of laws of nature raises it a bit more, and so on. Counter-evidence, e.g. from the existence of evil, might lower that probability. I have argued elsewhere that the total evidence (i.e. everything we – theists and atheists – agree that we know about the Universe) makes the existence of God more probable than not.[2] My concern in this chapter is solely with the force of the argument from fine-tuning: how much more probable the human-life-producing character of the laws and boundary conditions makes it that there is a God than does the fact that there is a law-governed universe.

By the "boundary conditions" I mean, if the Universe began a finite time ago, its initial conditions, such as the density of mass-energy and the initial velocity of its expansion at the instant of the Big Bang. If the Universe has lasted for an infinite time, I mean those overall features of the Universe not determined by the laws that characterize it at all periods of time – e.g. perhaps the total quantity of its matter-energy. But having made the point that the argument need not depend on the Universe having lasted only a finite time, I shall assume – for the sake of simplicity of exposition – that it has only lasted a finite time, and that it began with the Big Bang. The argument might prove to be somewhat weaker if the Universe has lasted for an infinite time (because the range of boundary conditions conducive to human evolution would then be wider), but my guess is that it would not be very much weaker. By "the Universe" I mean our universe; and by that I mean the system of physical objects spatiotemporally related to us. (Two things are spatially related if they are at some distance in some direction from each other. Two things are temporally related if they are before, after, or simultaneous with each other. I shall assume that both the relation of being spatially related and the relation of being temporally related are reflexive, symmetric, and transitive.) Any other actual systems of spatially and/or temporally related objects I shall classify as another universe.

I shall understand by a "person" a being with a capacity to have sensations and thoughts, desires, beliefs, and purposes (of a certain degree of sophistication). I shall understand by a "human being" a special kind of person – one with a capacity to learn about the world through perception, and to make a difference (on his own or through co-operation with others) to all aspects of his own life, that of others, and the world, with free will to choose which differences, good or bad, to make. Such a person will have good and bad desires (inclinations) – good desires to enable him to recognize the good, and bad desires in order to have a choice between good and evil. (To be able to choose the good, you need to be able to recognize it, and, if you can, that will give you a minimum inclination to pursue it. But without any desire for the bad, a creature will inevitably pursue the good.)[3] I include also in my concept of a human being a capacity to reason, including to reason (in at least a primitive way) about metaphysics, and to have the concept of God. I emphasize that this sense of "human being" as that of a person with all these capacities is not the ordinary sense, but one stipulated for the purposes of this chapter.

We, I and my readers, are essentially persons (if we didn't have a capacity to have desires, beliefs, etc., we wouldn't exist), but not essentially humans (we could continue to exist, for example, even if we ceased to have bad desires). I shall assume, however, that we are humans in my sense.[4] We are not merely humans but humans with bodies, although on my definition embodiedness is not an essential attribute of humanity. My body is that public object, a chunk of matter through natural processes in which I learn about the world and retain beliefs about it, through natural processes in which I make differences to the world, and natural processes in which cause me pleasant or unpleasant sensations. I learn about the world through light, sound, etc., impinging on my body; I make differences to the world by moving my arms, legs, mouth, and so on. I have no other means of learning about and influencing the world except by using my body, and it is the detailed processes within my body that enable me to perceive and act. Nerves translate the pattern of light impinging on my retina into a pattern of neural firings, which interacts with the neural networks established in the brain, through previous perceptions and genetically, to yield a new pattern of neural firings, which causes me to have the perceptions I do. Events in my body cause me pleasure or pain. And the purposes that I seek to execute cause brain states that interact with brain states resulting from beliefs about which actions will realize my purposes, to cause the motions of my limbs. A human body is a functioning public object of this kind suitable as a vehicle for human perception and action. A human being embodied involves there being such a public object through which alone that human perceives and acts, and in which alone that human feels. I shall assume that it is logically possible that humans could exist without bodies; and that human bodies could exist and behave as ours do without being the vehicles of human perception and action.[5] We can now characterize the "argument from fine-tuning" more precisely than I did in my opening paragraph, as the argument from the world being such as to permit the existence of human bodies; and so – if the world contains only one universe – from that universe's laws and boundary conditions being such as to permit the evolution of human bodies, public vehicles which make possible human perception and action.

An argument from fine-tuning will be a strong argument to the extent to which it is not too improbable that there should be such fine-tuning if there is a God, but very improbable that it should exist if there is no God. In attempting to compare these probabilities, I shall, for the sake of simplicity of exposition, assume that the only God up for consideration is the traditional theistic one. I shall not consider the possibility of evil gods or lesser gods, my reason being one for which I have argued elsewhere – that hypotheses that such beings exist are more complicated hypotheses than the hypothesis of the existence of the God of traditional theism, and so have lower prior probabilities than the latter.[6] The God of traditional theism, as I construe it, is a being essentially eternal, omnipotent (in the sense that He can do anything logically possible), omniscient, perfectly free, and perfectly good.[7]

Why a world with human bodies is likely if God exists

So what sort of a world will God's perfect goodness lead him to make? An omnipotent God can only do what is logically possible; one thing, for example, which it is not logically possible for God to do is both to create creatures with a libertarian freedom to choose between two alternatives, and at the same time to determine how they will choose. So our question must be – in so far as it is logically possible for God to determine what sort of a world there shall be, what sort of a world will He bring about?[8] A perfectly good being will try to realize goodness as much as He can. So in so far as there is a unique best possible world, God will surely make it. If there is no one best of all possible worlds but a number of incompatible equal best worlds, He will surely make one of them. But if every possible world is less good than some other incompatible possible world, all that He can do in virtue of His perfect goodness is to create a very good world. In any of these cases, the goodness of a world may be greater for including some bad aspects or the possibility unprevented (by God) of some bad aspects. God will therefore necessarily create any state of affairs that belongs to any best of all possible worlds, or to all the equal best possible worlds, or to all the good possible worlds. But what can we say about the certainty or probability of God bringing about some state of affairs that belongs only to some of the equal best possible worlds, or to some members of the series of ever better worlds? I suggest that it follows from His perfect goodness (as expilcated above) that if it is better that a state of a certain kind exist than that it should not (whatever else is the case), then God will bring about a state of that kind; and if it is as good that a state of a certain kind should exist as that it should not (whatever else is the case) then there is a probability of 0.5 that God will bring about a state of that kind. For states of a kind that belong to a series, each less good than the next, where their relative goodness can be measured, it will be enormously probable that God will bring about a state greater than any one you care to name. That is because there will be an infinitely larger range of states above that state than below it. Perfect goodness that cannot produce the best will very probably be very generous.

Now God, being essentially perfectly good, cannot but choose the good; He has no free choice between good and evil. But it is plausibly a good thing that there shall be beings that have this great choice, and the responsibility significantly to benefit or harm themselves, their fellows, and the world. We recognize this as a good when we ourselves have children and seek to make them free and responsible; and there would seem to be good in God creating free creatures with a finite limit to the amount of harm they can do to each other. Yet this good carries with it a risk of much evil. Any significant degree of freedom and responsibility will involve a significant risk of much harm being done; and God must – I suggest – impose some limits on the possible harm that creatures can do to each other (e.g. a limit constituted by creatures having a short finite life). Whether a perfectly good God will create such creatures (even within the limits of the harm they can do to each other) must depend on the extent of responsibility to be possessed by the creatures and the degree of risk of their misusing it; and the exact weighing-up of the moral worth of the different states that God must do is not easy for us to do. But, to oversimplify vastly, I suggest that, because of the risk of the evil that might result from significant freedom, any world in which creatures have significant freedom (within certain limits) would be as good as the same world without a state of this kind, whatever else might be the case – in which case, there would be a probability of 0.5 that God would create such a state. But the complexities are such that perhaps all that one can reasonably say is that since freedom and responsibility are such good things, then there is a significant (say between 0.2 and 0.8) probability that God will create a world containing such creatures. Such creatures I have called human beings.

If creatures are to have significant responsibility for themselves and for others, they must be able to affect their own and each other's mental lives of sensation and belief. They need to be able to cause in themselves and others pleasant or unpleasant sensations, investigate the world and acquire true beliefs (which I shall call knowledge), and tell others about it. But significant responsibility involves also a capacity for long-term influence over those capacities themselves. They must be able through choice to influence the capacities of themselves and others to acquire these beliefs and cause sensations, to influence what they find pleasant or unpleasant, and to influence the ways (for good or evil) in which they are naturally inclined to use their powers. They must thus be able to help each other to grow – in knowledge, factual and moral; in the capacity to influence things; and in the desire to use their powers and knowledge for good. And they must also, in order to have significant responsibility, be able – if they so choose – to restrict their own and each other's knowledge, capacities, and desire for good. So creatures must start life with (or acquire by natural processes) limited unchosen power and knowledge and desires for good and bad, and the choice of whether to extend that power and knowledge and improve those desires, or not to bother. And if that choice is to be a serious one it must involve some difficulty; time, effort, and no guarantee of success must be involved in the search for new knowledge, power, and improved desires. So creatures need an initial range of basic actions. (Basic actions are intentional actions that we just do – that we do not do by doing any other action. I may kill you by shooting, shoot by pulling the trigger, and pull the trigger by squeezing my finger. But if I don't squeeze my finger by doing any other intentional action – whether or not things of which I may have no knowledge have to happen in my body if I am to perform that intentional action – squeezing my finger is a basic action.) We may call the kinds of effects that a creature can (at some time) intentionally bring about by his basic actions his region of basic control. Creatures need an initial region of basic control, and creatures need, too, as we have noted, an initial range within which they can acquire largely true beliefs about what is the case. Let us call the kinds of such beliefs that a creature can acquire his region of basic perception. Creatures need an initial region of basic perception. The region of basic perception will have to include the region of basic control. For we cannot bring about effects intentionally unless we know which effects we are bringing about.

Extending our region of control beyond the basic region will involve discovering (that is, acquiring true beliefs about) which of our basic actions will have further effects. For the possibility of a large extension of our region of control, it needs to be the case that our basic actions will have different effects beyond the basic region that vary with the circumstances in which they are done. What these circumstances are must themselves be alterable by our basic actions; and if we are to affect the region of control of others, we must be able to alter the circumstances in which those others are to be found. Effects "beyond" the basic region mean in some sense effects more "distant" than it; and altering "the circumstances" involves in some sense "movement." We can learn what effects we have when we change circumstances if our region of basic perception moves with our region of basic control – though that may not always be necessary if the former region is much larger than the latter region. We can learn how to produce some effect in another room by moving into the room, and when we are there (but not here) we can see the effects of our actions there – our region of basic perception has moved with our region of basic control. But we can learn how to hit some distant person with a stone without altering our region of basic perception, for it is large enough for us to discover without moving the effects of throwing stones in different ways. The region of our control may be increased not merely by movement at a time, but also by discovering by previous movements what distant effects some kind of basic action normally has. By going to see where our bullet lands when we fire our gun at different angles, we can learn the distant effects of firing a gun at different angles and in this way, again, extend our region of control. And the region of perception may be increased by discovering (through previous movement) which basic perceptions are normally evidence of more distant phenomena. We can learn to see things far away through a telescope where we have discovered (through going to see) the correlation of things a little way away with their images in the telescope, and extrapolating from that to a similar connection between their images in the telescope and things at a great distance. Control may be widened so as to include events well in the future; and perception may be widened so as to include events well in the past.

So, in order to have significant freedom and responsibility, humans need at any time to be situated in a "space" in which there is a region of basic control and perception, and a wider region into which we can extend our perception and control by learning which of our basic actions and perceptions have which more distant effects and causes when we are stationary, and by learning which of our basic actions cause movement into which part of the wider region. If we are to learn which of our basic actions done where have which more distant effects (including which ones move us into which parts of the wider region), and which distant events will have which basically perceptible effects, the spatial world must be governed by laws of nature. For only if there are such regularities will there be recipes for changing things and recipes for extending knowledge that creatures can learn and utilize. So humans need a spatial location in a law-governed universe in which to exercise their capacities, and so there is an argument from our being thus situated to God.

Now, if humans are not merely to find themselves with beliefs about each other's beliefs and purposes (which they will need to do if they are to be able to influence them), but are also to be able to choose to learn about each other's beliefs and purposes and to communicate with them in the public way needed for co-operative action and co-operative rational discussion (which will involve language), then they need to be able to re-identify humans. That means that there need to be public objects – human bodies – which they can re-identify and the behavior of which manifests their beliefs and purposes. Those bodies need to behave in such a way that the simplest explanation of their behavior is often in terms of some combination of belief-and-purpose. In consequence, for example, we must be able to attribute to each other (on the grounds of being the simplest explanation) beliefs sensitive to input, e.g. beliefs that some object is present when light comes from that object on to their eyes, and purposes that – although not fully determined by brain states – do show some constancy. We can, for example, only come to understand the language of another human if we assume that he normally seeks to tell the truth, that he has some language constant over time by which he expresses his beliefs, and that his beliefs are often sensitive to incoming stimuli in the ways our own are. We then notice that he says "*il pleût*" when input to his eyes and ears is caused by rain, and so infer that he means by "*il pleût*" "it is raining."

This public communication of a kind that can be learned and refined can be achieved, as it is achieved in our world, by our having spatially extended bodies formed of constituents, some of which are stable (and so permit a continuing organized body) and some of which are metastable (i.e. change their states quickly in response to new input, e.g. of sensory stimuli) and so store new memories. Given such constituents, there can be machines sensitive to input that produce an output (out of a large variety of possible outputs), which is such as – given the input – will more probably attain some goal than will any other output. That allows us, if we thought that the machine was conscious, to attribute to it the belief that the means being used would attain the goal, and to attribute to it the purpose of attaining the goal. Given stable bits and metastable bits such machines can be constructed, and can – plausibly – occasionally arise, through endless reassembling of the bits. All this, though, does not ensure that there are humans embodied in these bodies, but only if humans have extended bodies of this kind rather than any other will they be able to have public knowledge about each other and public communication with each other of a kind that they can learn and refine.

But if humans had only spatial location and not extension, they would have just "particle-bodies." Some of the constituents of the physical world would be then "particle-bodies," and it is they (not combinations of them) that would need to exhibit the requisite input–output behavior that would be required to understand other humans.

It would be good that we should have the power not merely to extend the region of our control and perceptions beyond the basic, but also that we should have the power to extend or restrict (or prevent being restricted by others or by natural processes) the region of basic perception and control (including the ability to move) itself of ourselves and others, and the pleasant or unpleasant sensations that we have. There need to be basic actions that we can do, or non-basic actions that we can learn to do, which under various circumstances will make differences to our capacities for basic action and perception, and to our sensations. That involves there being natural processes that we can discover and so affect, which enable us to perform our basic actions, to acquire and retain in memory basic perceptions, and to diminish or increase pain or pleasure. And if these processes are to be manipulable not merely by the human whose they are, but by other humans as well, they must be public processes. The obvious way in which our capacities for basic action and perception can depend on public processes is again by our having a spatially extended body.

We actual humans do have a range of basic control: it is what we can do with our limbs, mouths, and tongues – "just like that," not by doing anything else. The region of basic control varies with the age of a human; it increases and then decreases again with time, even unhelped by other humans (how fast we can move our arms and legs does not depend too much on learning or help from others). But we can discover or be taught how to increase that region of control in many respects – above all, to influence others by uttering sentences of a language. And we have a range of basic perception – increasing or decreasing with the age of a human independently of intentional action. Recognizing inanimate objects is a perceptual capacity that develops without much help; learning to understand people's words needs more by way of help from others. We learn by our basic actions to hurt or benefit others, to use tools, build houses, or cut down trees. We utilize principles of what is evidence for what to detect the previous presence of others from footprints and remains of fires, and to detect the passage of elementary particles from tracks in cloud chambers. Through our growth of knowledge and control, we learn how to cause pleasure and pain, to give knowledge and control to others or to refuse to do so. We can allow ourselves to get into situations where it is difficult to do good, and so fall into bad habits – or, alternatively, prevent this happening. And, through learning, we can acquire the ability to influence the ways in which others are naturally inclined to use their powers – we can educate them morally or immorally.

But, as well as learning how to extend the region of control and perception (of ourselves and others) beyond the basic, we can also learn how to extend or restrict (or prevent being restricted) the region of basic control and perception itself. By starving ourselves or others, we can restrict basic abilities and perceptual capacities; as we can by cutting off arms or tongues or eyes. In ways unintended by us or others, our powers may diminish through disease, and we can learn to prevent the effects of disease by using medicine and surgery; or we can not bother to take the trouble to discover how to do so. And our present capacities to affect our basic capacities and perceptual powers seem fairly small compared with what medical science will surely provide for us in the course of the present millennium. Medical intervention will surely enable us even within the next century to grow new limbs and sense organs, and to slow down memory decay. Having a body thus involves the surface of mind–brain interaction lying within the body (i.e. in the brain); it also involves events elsewhere within the body affecting what we basically perceive and how we can act basically. Having a body thus allows us to diminish or extend the basic capacities and perceptual powers of each other, or to prevent such diminution being produced by natural processes. With such powers we have very much more substantial power over each other than we would have otherwise.

It would also be possible for our basic capacities for perception and action to depend on public processes if we had a "particle-body."[9] The public processes would then need to consist of temporally extended input to a spatially unextended object. The latter would be like a totally impenetrable black box. We would discover how to improve or damage our sight, or weaken or strengthen our memories, by giving a certain input to the box over a long period of time. But the box could not be opened; indeed, it would have no spatial extension. Our memories would not then depend on a brain; they would depend on input over time that affected memory by action at a temporal distance (as far as the physical was concerned – there could be processes within the "mental" realm, the "soul," which the physical input affected and on which the mental life more directly depended). This would provide a less immediate kind of embodiment – for the dependence of the mental on the physical would not be instantaneous. But it might seem, nevertheless, to be an alternative way in which humans could have the ability to affect their and each other's basic capacities without the normal kind of embodiment.

So if humans are to have the great goods of being able to learn to communicate with each other and of being able to extend or restrict the range of their basic perception and control,[10] human embodiment needs more than spatial location in an orderly world. Humans need to have extended bodies, made of stable and metastable constituents, or alternatively particle-bodies of special kinds.

I suggested earlier that if God brings about the world there is a significant probability (say between 0.2 and 0.8) that He will bring about an orderly, spatially extended world in which humans have a location. This is because He has a reason for giving to humans responsibility for themselves and each other. I have now argued that if humans are to be able publicly to learn about each other (or choose not to do so), and co-operatively to affect their and each other's basic capacities for action and perception, the embodiment will have to be of a more specific kind; and this means that some universe must have laws and conditions allowing the existence of constituents, either of the stable and metastable kind permitting the occurrence of (spatially extended) human bodies, or particle-like ones of a kind that would evince patterns of stimulus and response interpretable in terms of purpose and belief. Either way, the bodies must not be totally deterministic in their behavior, and probably there would be some scope for reproduction. Perhaps it is somewhat more risky for God to give to humans the more significant kind of responsibility involved in embodiment. Yet the harm they are likely to do to each other does not seem to be greatly increased. (It is less in the respect that they have to choose to learn how to harm rather than being born with the knowledge; but greater in the respect that, if they acquire the knowledge, they can harm each other's basic capacities.) And, above all, surely a God who created creatures capable of choosing freely to love each other (as these with their good and evil desires etc. would be) would make them capable of entering into a loving relation with Himself – a perfectly good creator would surely do that. And so they need to be able to grasp the concept of God. So I'm inclined to suggest a number for the probability that, if there is a God, He will create embodied human beings which is similar to the probability that He will give them merely a location. Artificial though my numbers ("between 0.2 and 0.8") are, the crucial point is that the probability is a significant one. A good God will want to bring about finite free beings with considerable responsibility for the well-being of themselves and each other. The vagueness of the probability values takes seriously the goodness of there being such bodies, without exaggerating our ability to calculate just how good a thing this is. Our situation with regard to predicting what a good God will do is not dissimilar in principle to that of predicting the kind of things that any human person with a postulated character is likely to bring about – we know the kinds of things they might be expected to bring about, but cannot be certain exactly what they will bring about. There would be no content to the supposition that God was perfectly good unless we supposed – barring the disadvantages of so doing – that bringing about creatures who could understand each other, and freely choose whether to grow in understanding and power so as to mould themselves (including in respect of their basic capacities) and other things, so as to come to a knowledge of God Himself, was one of the things He would naturally choose to do. If God does so choose, He may bring about this result in one of two ways. One way is by acting at the beginning of a universe (i.e. the Universe, if there is only one; or some one universe, if there is more than one) by creating laws and initial conditions such that conditions hospitable to embodied human life will evolve at some stage. The other way is by acting at each moment (of finite or infinite time) so as to conserve laws of the right kind to bring about boundary conditions and conditions hospitable to embodied human life at some time or other.

Why a world with human bodies is unlikely if there is no God

Now what is the probability of there being human bodies, or human "particle-bodies" with the required properties, if there is no God? While science is as yet in no position to discuss the probability of the occurrence of some of the features of extended human bodies (such as the ability to exhibit the physical correlate of moral awareness), it can discuss and has discussed the necessary conditions for the existence of bodies with most of the features required for human bodies. We saw earlier that extended bodies require stable and metastable constituents. It seems to be generally agreed that given the kinds of law presently believed to be operative in our Universe (the laws of quantum theory, the four forces, etc.), the constants of laws of nature and the values of the variables of the initial conditions needed to lie within very narrow limits if such constituents were to evolve. If the initial velocity of the Big Bang had been slightly greater than the actual velocity, stars and thus the heavier elements would not have formed; if it had been slightly less, the Universe would have collapsed before it was cool enough for the elements to form. And there had to have been a slight excess of baryons over anti-baryons. If the proportion had been slightly less, there would not have been enough matter for galaxies or stars to form; and if it had been greater, there would have been too much radiation to allow planets to form.[11] Similar constraints would apply to a universe that had a beginning of a less concentrated and violent kind than our universe seems to have had. An everlasting universe would also have to have features additional to the values of the physical constants mentioned, if planets and the heavy elements were to be formed at any time at all – although these constraints would be less than the constraints on a universe with a beginning.

It may be that the kinds of postulated laws for which the fine-tuning of initial conditions at the time of the Big Bang would be required will prove mere approximations to the true laws for which far less fine-tuning of their conditions would be required. One could postulate laws that would yield values for the expansion of the Universe after a few seconds, and thus to the evolution of planets and the heavy elements, starting from more or less any initial conditions. But for those laws to fit all our other data, they would probably have to contain some very fine-tuned constants, even more fine-tuned than if you suppose rather special initial conditions. The "inflation" hypothesis, in its many variants, looks as though it might be successful in removing the need for fine-tuning from the initial conditions only by putting it into the laws.[12] But maybe it won't turn out that way, and the inflation hypothesis will reduce somewhat the need for fine-tuning.

More deeply, it may prove that those laws that we presently believe to be fundamental are derivative from more fundamental laws which have the consequence that the values of one or two of the physical constants of the former laws uniquely constrain the values of all the others; and that the true fundamental laws permit only a restricted set of boundary conditions. In that case, the need for the "fine-tuning" of values of constants and variables in a literal sense would be much diminished.[13] But it would remain the case that the Universe needed to be "fine-tuned" in the very wide sense that rather special laws and boundary conditions were required if the Universe was to be life-evolving. It will be evident that many possible universes with laws of a kind different from our own would not be hospitable to embodied creatures, whatever the constants of their laws – for example, a universe in which all atoms lasted for ever and the only forces were forces of repulsion between them. Other universes (say, universes with seven kinds of force instead of four) could still be hospitable, but only given certain values of the constants of their laws.

So what are the principles for determining the prior probability, that is, the probability on solely *a priori* grounds (which I call the "intrinsic probability"), of a universe governed by laws having laws and boundary conditions that are life-evolving? Laws and boundary conditions of universes have intrinsic probabilities that vary with their simplicity, and so do ranges of laws. That must be the case, because if they did not then, any hypothesis (however complex and *ad hoc*) about the nature of our universe that predicted what we have observed so far would be equally probable on the evidence of observation. That is clearly not so, and hence *a priori* factors enter into the assessment of the probability of hypotheses on evidence. These factors, I have argued, are scope and simplicity. The "scope" of a hypothesis is a function of how much it tells you – how detailed are its claims about how many objects; but, as all hypotheses about the laws and boundary conditions of universes will have the same scope, we can ignore this factor. Simplicity alone will determine the intrinsic probabilities of universe-explaining hypotheses. A full study of the criteria which are used, and we think it right to use, to judge the relative probabilities of hypotheses of equal scope that have had equal success in their predictions so far (when there is no other empirical evidence, or "background evidence") should make it possible to develop a set of criteria for how simplicity determines intrinsic probability.[14]

Note that the same laws of nature can be expressed in innumerable logically equivalent forms. The criteria for determining the simplicity of a hypothesis are criteria for determining its simplicity by means of the simplicity of its simplest formulation – that is, the one in which the variables designate properties as close as can be to being observable, and the equations connect these by fewer laws involving fewer terms in mathematically simpler, i.e. more primitive, ways. It will in general be the case that these criteria have the consequence that a hypothesis so formulated in which the constants and variables lie within a certain range will be as probable intrinsically as one in which the constants and variables lie within a different range of equal length. That is, the density of the intrinsic probability of values of physical constants and variables of boundary conditions is constant for hypotheses of a given kind (i.e. which vary only in respect of these constants and variables).[15]

It is beyond my ability to calculate, using these criteria, what is the intrinsic probability that a universe belongs to the set of possible universes hospitable to embodied humans, and – I suspect – it is beyond the ability of any present-day mathematician to calculate this. But the problem seems well defined, and so hopefully is one that can be solved by some mathematician of the future. Given that solution, we would then have a precise proven answer to the question of the prior probability of a sole universe being fine-tuned (in the wide sense). In the absence of a proven solution to the problem of what is the intrinsic probability that a universe belongs to the set of possible universes hospitable to embodied humans, we must conjecture. I suggest that there is no reason to suppose that universes with our kinds of laws and boundary conditions are untypical in the respect that only a minuscule range of them, given probabilities in virtue of their intrinsic natures that I explicate as the relative simplicity of their laws and initial conditions, are fine-tuned.

But, atheists have suggested, perhaps there are very many actual universes; in which case, it would not be surprising if at least one of them was fine-tuned. But to postulate a large number of such universes as a brute uncaused fact merely in order to explain why there is a fine-tuned universe would seem the height of irrationality.[16] Rational inference requires postulating one simple entity to explain why there are many complex entities. But to postulate many complex entities to explain why there is one no less complex entity is crazy. In terms of probability, this is because the intrinsic probability of there existing a large number of universes uncaused is vastly less than the intrinsic probability of there existing one universe uncaused. If the atheist is to claim that a fine-tuned universe exists because there are innumerable universes of different kinds, what he needs to do to begin to make his claim plausible is postulate a mechanism producing universes of all kinds, including the occasional fine-tuned one.

Let us look at this hypothesis of a "mechanism" in a little more detail. It could be the suggestion of a law operating by itself, dictating the continued coming-into-being of new universes of different kinds. But this is not a kind of "explanation" that we can recognize as such. Scientific explanations by means of laws require states of affairs on which the laws operate in order to produce new states of affairs (or to prevent the occurrence of certain kinds of states). "All copper expands when heated" has no effect on the world unless there is any heated copper. Apparent exceptions, like conservation principles, are really limitations on how states can evolve; they have no consequences by themselves for what exists. (Indeed, in my view, laws of nature are simply generalizations about the powers and liabilities of existing objects. But I don't press that point here.) There are, it seems to me, two alternative ways to make the hypothesis of a universe-generating mechanism intelligible. One is to suppose that there is a master-universe governed by the law that it generates daughter-universes with innumerable different laws and initial conditions, either at a first instant or continually. The other is to suppose that there is a law governing all universes, that each old universe generates many new universes all with different laws and initial conditions (but including the law of generation, in most cases). In each case the new universe would be related temporally, though not spatially, to its parent. These hypotheses seem to me coherent. They also seem to me far less simple than the rival theistic hypothesis, which explains the existence of our universe by the action of God. An atheistic hypothesis needs to have a very detailed law ensuring the diversity of universes that result; it would need to have a certain form rather than innumerable possible other forms, and probably constants too that need fine-tuning in the narrow sense (and maybe universes with initial conditions of certain kinds on which to operate) if that diversity of universes is to result. That the law was just like this would be the atheist's brute fact. Theism simply postulates infinite degrees of the four properties (power, knowledge, freedom, and temporal extension), some amounts of which are essential for persons to be persons, and all else then follows (with significant probability). The detailed atheistic law will have to state that while matter on a universe scale produces other universes, matter within each universe never produces more matter (governed by different laws) – for we do not find any such process of matter generation within our universe. There is no process at work in our universe throwing up little regions not governed by quantum theory. The law has to postulate kinds of processes at work on a large scale that do not operate on a small scale, and is thus complicated in confining the range of its processes. Theism, by contrast, postulates the same kind of causality at work in creating the Universe as we find on a very small scale within the Universe – intentional causality, by agents seeking to bring about their purposes seen in some way as a good thing. And if the universe-generating hypothesis took the form that all universes generate new universes, we'd need an explanation of why the beginnings of such a process have not so far been observed within our universe.

However, if we do postulate a universe-generating mechanism, it is to be expected that there will be at least one fine-tuned universe, and so a universe containing human bodies. But there is no particular reason why we, persons, even if we are human beings in the sense defined at the beginning of the chapter, should find ourselves in such a universe. Human beings could exist in any orderly universe at all, although they could only have bodies in a fine-tuned universe. In other universes we could have a location on a particle (though not a particle-body in the sense defined earlier) and have a range of control and perception that we could increase through learning, though our doing so would not be dependent on the operation on any process within the particle. The particle itself would be the locus of mind–body interaction. We could not learn about the purposes and beliefs of others by studying their public behavior, but we could perhaps find ourselves believing that some particles were controlled by other humans; or perhaps we would be solipsists. Furthermore, even if we persons had bodies, we need not be humans in the sense defined at the beginning of this chapter and thus find ourselves in a fine-tuned universe. For being humans, as I defined it, includes having moral beliefs and being able to exert significant influence on ourselves and each other for good or ill (and also having a capacity for conceiving God). We might be encased in hard shells and be unable to cause pain or pleasure to each other; there might be an abundant supply of food and everything else we wanted and so no possibility of depriving others of anything; we might have fixed non-moral characters; and we might not be able to produce descendants – let alone influence them for good or ill. Our universe has all the features God might be interested in giving to some conscious beings, and it has them in a very big way. In our universe the scope for growth of knowledge and control is enormous compared with many a fine-tuned universe. A universe is still fine-tuned, for example, even if it gives rise to only one generation of humans with no power to influence future generations. In our universe humans can influence their children, grandchildren, and many future generations (e.g. by affecting the climate in which the latter live, and the availability to them of raw materials). Our universe is also untypical in that the kind of orderliness that makes possible the evolution of human-like beings is characteristic not merely of a small spatial region and of a small temporal region, but also of every observable part of an enormous universe over a very long period of time. If it is good that God gives us some freedom and responsibility, and degrees of freedom and amounts of responsibility can be measured, and the goodness of God's gift is proportional to the measure, it follows from an earlier result that even if there is no limit to the amount which He could give us and that the more, the better, He will very probably give us a lot (within limits of the harm we can do to particular individuals, such as the limits provided by the length of a human life). But on the atheist hypothesis, we are very lucky to have any descendants at all. So the atheistic generating hypothesis, as well as not being very simple, is pretty poor in predicting the particular features of our universe. Given all these other possibilities, the probability, bared on the hypothesis of a universe-generating mechanism, that a given human being would find itself in a fine-tuned universe would be small. So if there is a better explanation of why the Universe in which we are situated is fine-tuned, we should prefer that. Theism provides that in a way sketched in the first part of this chapter, for it gives a reason why God should put us in a fine-tuned universe (and is not in any way committed to there being any other universes).

Similar results follow, I suggest, with respect to the possibility of humans being embodied in "particle-bodies." Special kinds of law would be required for these bodies to be such that other humans could infer from their behavior what are the purposes and beliefs of the embodied humans, and to be able to influence their basic capacities. As far as I know, no detailed mathematical work has been done on what, by way of fundamental laws and initial conditions, would be required for this; and so the great improbability of such a state of affairs is a conjecture. But it is a reasonable conjecture, if only for the reason that no possible universe currently investigated by physicists could contain any such states. In none of them are particles sensitive to streams of stimuli over time in such a way that we can attribute beliefs and purposes to them. By arguments similar to those given earlier with respect to ordinary bodies, a universe-generating mechanism would need to be of a certain complex and so somewhat improbable kind to throw up a universe containing particle-bodies of the kind discussed. And, again, even if it did, there is not much probability that we humans would find ourselves in such a universe – since, whether as humans or not, we could exist in many other universes.

I conclude that while it is significantly probable that there would be a universe fine-tuned for the occurrence of human bodies or "particle-bodies" if there is a God, it is not at all probable that there would be such a universe if there is not a God. Hence "fine-tuning" (in the sense in which I have defined it) contributes significantly to a cumulative case for the existence of God.

Notes

1 This chapter meets various deficiencies in my previous account of this matter – "Argument from the fine-tuning of the universe," first published in Leslie (1989) and (in large part) republished as Appendix B to Swinburne (1991). I am most grateful to Dr Pedro Ferreira for guidance on the physical theories discussed in this chapter.
2 See Swinburne (1991) and Swinburne (1996).
3 For argument in defense of this claim, see (e.g.) Swinburne (1994: 65–71).
4 The only controversial element in this assumption is that we have libertarian freedom – that is, the uncaused freedom to choose between alternative actions, given the state of the world (and in particular of our brains) in all its detail at the time of our choice. I give a (probabilistic) argument in favor of this in Swinburne (1997: Ch. 13).
5 It seems fairly obvious to many people that there is no logical inconsistency in supposing that they could exist without their bodies, or that their bodies could exist as robots unconnected with any conscious life. For detailed defense of this claim and generally of my substance dualist view of the nature of humans, see Swinburne (1997).
6 For my claim that polytheism is a more complicated hypothesis than traditional theism, see Swinburne (1991: 141). My grounds for holding the hypothesis of an omnipotent evil deity to be more complicated than traditional theism are that (see note 7) perfect goodness follows from omniscience and perfect freedom, and so one or other of knowledge or freedom would need to be limited in an omnipotent evil deity – which would make the hypothesis more complicated than traditional theism.
7 For analysis of what it means to say that God has such properties, and a demonstration of how the divine properties fit together, see Swinburne (1993) and Swinburne (1994: Chs 6 and 7). For the argument that perfect goodness follows from omniscience and perfect freedom, see Swinburne (1994: 65–71, 134–6).

8 In Plantinga's terminology, the question is what kind of a world God will "strongly actualize" (1974: 173). I use the word "world" to include all that exists apart from God, and its way of behaving – whether (in part or totally) indeterministic or determined by its intrinsic powers and liabilities to act codified in natural laws. A world may or may not include many universes. Possible worlds are, however, to be individuated, as stated in the text, only by those features that it is logically possible for an omnipotent being to bring about. In this terminology (which is not standard) a world counts as a possible world if God given only His omnipotence and not His other properties could bring it about. There are, therefore, possible worlds that God could not bring about in virtue of His other properties – e.g. in virtue of His essential moral goodness.

9 I owe this suggestion to Joseph Jedwab and Tim Mawson.

10 And – I should add – if humans are to reproduce in such ways that their characteristics are in part inherited (through some DNA-like constituents).

11 For full details of constraints on physical constants and initial conditions, see Barrow and Tipler (1986), especially Chapters 5 and 6.

12 For the suggestion that the inflation hypothesis does not solve the alleged problems that it was devised to solve, without it becoming quite unnaturally complex, see Earman and Mosterin (1999).

13 It is possible that the derivation of the fundamental laws of nature from string theory would greatly reduce the need for fine-tuning. This has been argued by Kane *et al.* (2000). They suggest that all string theories are equivalent, and different possible "vacua" uniquely determine all the constants and initial values of the variables of laws of nature. They acknowledge that much work needs to be done before (if ever) string theory is established and their result can be demonstrated. But, even granted all this tentative speculation, they acknowledge that "there will be a large number of possible vacua"; and that means both having string theory rather than any other fundamental laws and requiring special variables of initial conditions.

14 For an attempt at an analysis of the various facets that determine the relative simplicity of hypotheses, see Swinburne (2001: Ch. 4).

15 My insistence that the probability of a constant or the value of a variable lying within any interval of the same length of possible values thereof be the same (and thus there be a constant probability density distribution) and be determined in respect of the laws in their simplest and so most fundamental form avoids versions of Bertrand's paradox. To take a very simple example of the problems that might otherwise arise – Newton's law of gravitational attraction $F = G (mm'/r^2)$ could be expressed as $F = mm'/d^3r^2$, where d is defined as $G^{-1/3}$. A constant probability density distribution for d will not yield a constant probability density distribution for G, and conversely. Expressing the laws of nature in very complicated forms, logically equivalent to their simplest forms, and assuming a constant probability density for the constants and variables of these forms, could have the consequence that much greater variations of these (much less "fine-tuning") would be required for the Universe to be human-life producing. But laws are judged simpler and so to have greater prior probability in virtue of the features of their simplest forms. Since a constant is simpler than a constant to the power $(-1/3)$, the traditional form of Newton's law is the simplest and so most fundamental form. And, more generally, insistence on the simplest form of a law should yield a unique probability density distribution for the constants and variables of laws of that kind (or, at most, if there are a number of equally simple forms of a law, a few different probability density distributions that should not make much difference to the extent of the need for fine-tuning).

16 However, some writers have proposed just this. For example, Tegmark says:

> Our TOE [Theory of Everything]...postulates that all structures that exist in the mathematical sense...exist in the physical sense as well. The elegance of this theory lies in its extreme simplicity, since it contains neither any free parameters nor any arbitrary assumptions about which of all mathematical equations are assumed to be "the real ones".
>
> (1998: 38)

He explicitly assumes an account of simplicity according to which a theory is simpler the fewer the number of computational symbols needed to express that theory (Tegmark 1998: 44). This "algorithmic" account has the consequence that, for example:

> the set of all perfect fluid solutions to the Einstein field equations has a smaller algorithmic complexity than a generic particular solution, since the former is specified simply by giving a few equations and the latter requires the specification of vast amounts of initial data on some hypersurface.
>
> (Tegmark 1998: 44)

So it is simplest of all to postulate that every possible universe exists, since that needs very few computational symbols indeed to state!

This seems to me a bizarre account of simplicity, totally out of line with our inductive practice. If we are postulating entities to explain phenomena, we postulate the fewest number of entities (possessing some causal interconnectedness) needed for the job. (For a more detailed criticism of the "computational" account of simplicity that Tegmark is using, see Additional Note F in Swinburne (2001). And just how seriously is Tegmark taking "every possible universe"? The only ones he discusses are governed by natural laws, and he assumes that persons are embodied. But there is a vast infinity of possible universes in which neither of these conditions is satisfied. And the possibility of our being disembodied and/or being non-human persons has the consequence that we (particular individuals – you, me, and Tegmark) could have existed in innumerable universes that are not in my sense "fine-tuned." But it is not logically possible for an actually embodied individual to exist in more than one universe at a given time? So why do we exist in a fine-tuned universe? For Tegmark, that must be just something very improbable. The theist can explain this in terms of the goodness of our existing as embodied humans along the lines developed earlier in this chapter.

References

Barrow, J.D. and Tipler, F.J. (1986) *The Anthropic Cosmological Principle*, Oxford: Clarendon Press.

Earman, J. and Mosterin, J. (1999) "A critical look at inflationary cosmology," *Philosophy of Science* 66: 1–49.

Kane, G.L., Perry, M.J., and Zytkow, A.N. (2000) "The beginning of the end of the anthropic principle," available online at www.lanl.gov/abs/astro-ph/0001197.

Leslie, J. (ed) (1989) *Physical Cosmology and Philosophy*, New York: Macmillan.

Plantinga, A. (1974) *The Nature of Necessity*, Oxford: Clarendon Press.

Swinburne, R. (2001) *Epistemic Justification*, Oxford: Oxford University Press.
—— (1997) *The Evolution of the Soul*, Oxford: Clarendon Press.
—— (1996) *Is There a God?*, Oxford: Oxford University Press.
—— (1994) *The Christian God*, Oxford: Clarendon Press.
—— (1993) *Coherence of Theism*, revised edn, Oxford: Clarendon Press.
—— (1991) *The Existence of God*, Oxford: Clarendon Press.
Tegmark, M. (1998) "Is 'the theory of everything' merely the ultimate ensemble theory?," *Annals of Physics* 270: 1–51.

6

PERCEIVING DESIGN

Del Ratzsch

In 1885, the Duke of Argyll recounted a conversation he had had with Charles Darwin the year before Darwin's death:

> [I]n the course of that conversation I said to Mr. Darwin, with reference to some of his own remarkable works on the *Fertilisation of Orchids*, and upon *The Earthworms*, and various other observations he made of the wonderful contrivances for certain purposes in nature – I said it was impossible to look at these without seeing that they were the effect and the expression of Mind. I shall never forget Mr. Darwin's answer. He looked at me very hard and said, "Well, that often comes over me with overwhelming force; but at other times," and he shook his head vaguely, adding, "it seems to go away."
>
> (Argyll 1885: 244)[1]

It is interesting – even surprising – that in the last year of his life, over twenty years *after* publication of the *Origin of Species*, Darwin could still say that he sometimes found the idea of deliberate design in nature to have "overwhelming force," and sometimes found himself in the grip of an impression of designedness. Of course, sometimes it all evaporated.

But equally interesting here is what Darwin does *not* say. There are no mentioned inferential or evidential processes driving those swings. There are no mentioned intellectual exertions, decisions, acts of will, etc. Darwin here merely *reports* these alternations in passive, experiential, phenomenological terms, portraying himself as a *spectator* – not an actor – in and to this ebb and flow within his cognitive landscape. This belief – or absence – is something that "comes over" him, something that *happens to* him. And he seems to have little say in the matter – *overwhelming force* is involved. Indeed, *both* parties to the conversation implicitly presuppose a passive, perceptual view. Note that Argyll asserts the *impossibility* (no choice, no decision) of not *seeing* (perceptual, non-inferential) that mind was involved in the contrivances in question.

Although the familiar examples of design are generally read as inferential, contrary suggestions were not completely absent historically. For instance, William Whewell, in 1834, said:

> When we collect design and purpose from the arrangements of the universe, we do not arrive at our conclusion by a train of deductive reasoning, but by the conviction which such combinations as we perceive, immediately and directly impress upon the mind.
>
> (Whewell 1834: 344)

And among views noted by Hume is the following expressed by Cleanthes:

> Consider, anatomize the eye: Survey its structure and contrivance; and tell me, *from your own feeling*, if the idea of a contriver does not immediately flow in upon you *with a force like that of sensation* [emphasis in original].
>
> (Hume 1947: 104)[2]

In what follows, I shall explore a picture of design perception and recognition as non-inferential – as passive and experiential. Current attempts by advocates of intelligent design follow the presumed historical precedent in focusing nearly exclusively upon *inferences* to design conclusions.[3] Perhaps that is the correct approach. But even if so, it is, I will suggest, incomplete in a crucial respect.

Background: Reid

In order to get an initial fix on what a non-inferential picture of design recognition might look like, I shall begin with a brief exploratory sketch of a non-inferential treatment of everyday cases of belief and recognition in familiar domains. The most plausible such case historically comes from Thomas Reid (1710–96), so let us begin with his treatment of ordinary perceptual judgments:

> When I grasp an ivory ball in my hand, I feel a certain sensation of touch. In the sensation there is nothing external, nothing corporeal. The sensation is neither round nor hard; it is an act of feeling of the mind, from which I cannot, by reasoning, infer the existence of any body. But, by the constitution of my nature, the sensation carries along with it the conception and belief of a round hard body really existing in my hand.
>
> (Reid 1872b: 450)

Reid's position is that in certain experiential situations, specific sensory, phenomenological content triggers particular cognitive states – *de re* beliefs, conceptions, etc. – which do not *follow* inferentially from that content. (Nor does that content *resemble* in any relevant sense the character attributed to the object of the belief.) The resultant cognitive state is a causal consequence of the triggering sensory experience given our constitution (and relevant antecedent states), and does not result from any inference, decision, choice, or other volition. In the above case, we simply *find* ourselves in the grip of a conviction of the cognitively apprehended object both really existing and really having the cognitively apprehended properties, and we are in general powerless to resist. Acquiring the belief in question is something that *happens* to us. Sensation accompanied by involuntary conviction is Reid's basic picture of *perception*.

Of course, some beliefs are based on inferences. But, quite obviously, constituents of such inferences must *ultimately* track back to beginning points independent of further inferences, on pain of infinite regress. It is Reid's position that such beginning points must be consequences of the constitution of our nature. The innate processing structures from which such beliefs emerge can be articulated in a variety of "first principles" that, according to Reid, shape our conceptual systems. These first principles are accepted involuntarily, characterize all mature minds, and are both unproven and unprovable. For instance, in sensory perception our mind is presented with a certain complex apprehension (e.g. the round hard body above), and is carried involuntarily to a belief in the objective reality of that object by a cognitive structure articulated as a first principle that Reid states thus:

> [T]hose things do really exist which we distinctly perceive by our senses, and are what we perceive them to be.
>
> (1872b: 445)[4]

Suppose for the moment that Reid is basically right. The fact, then, that our common beliefs concerning objects in the external world in no way rest upon an inference of any sort would rather neatly explain why it is that, despite the efforts of many of the best thinkers historically, attempts to construct (or reconstruct) satisfactorily powerful inferences from, for example, sense data to physical objects have been hard to come by – or, perhaps more accurately, have been without exception abysmal failures (much to the detriment of classical foundationalism, for instance).[5] It would equally well explain why the formal failure of such arguments has never made the slightest substantive difference to anyone. It would furthermore explain why the seeming irrefutability of skeptical arguments has also not made the slightest practical difference to nearly anyone – even including proponents of such arguments themselves.[6]

This picture fits nicely with human personal experience as well. We simply do not find ourselves engaging in inferences (whether from sense data, incorrigible beliefs about our own inner states, self-evident propositions, or whatever) out of which emerge our convictions concerning the existence of the objects of everyday experience. Indeed (as Reid points out), each of us firmly held such beliefs well before constructing inferences was, so far as we know, within our capabilities. Nor do we find ourselves deciding whether or not to hold such beliefs, or able to discard such beliefs at will. We simply have the convictions under the right experiential circumstances, with few options to do much about it.

It might be argued that although we do not in practice *acquire* the beliefs in question on the basis of inference, the *rational justification* of such claims requires that appropriate inferences at least be in principle *constructible*. But that is at least problematic. Indeed, any such arguments may be irremediably invidiously circular (Alston 1991: Ch. 3). In any case, Reid's position is that starting points that are consequences of the constitution of our nature are thereby, other things being equal, rationally justified for us as are other beliefs properly derived from them.

As Reid argued, this general schema fits not only ordinary perceptual beliefs, but a wide range of other essential human beliefs. Our memory beliefs about the past, our beliefs concerning the existence of other minds, even our beliefs about fundamental axioms of reason – all arise not from specifiable inferences, but in specific circumstances as nearly ineluctable causal consequences of our nature.

Reid and design

According to the usual (non-Reidian) account of design recognition, we observe (and participate in) the coming into existence of humanly designed artifacts, and by some type of *abstraction* we notice certain commonalities among them. We *infer* that those constitute generally reliable marks of design, and we then attempt to *inductively extend* this generality to things in nature, thereby identifying relevant things as also designed.

Reid, however, suggests a different story. To begin with, for Reid the primary sense of *design* applies not to designed objects, but to *minds*. For Reid, objects have design only in a derivative sense of carrying marks of design, or being effects of design in a mind.[7] (Such derivative senses are common, as when we speak of "intelligent solutions" to problems.) The underlying question for Reid, then, is how we recognize relevant (logically prior) qualities of minds.

Reid contends that this recognition cannot be based on past experiences, as the usual account has it. First, we never directly experience the minds of others, and thus could never experience a single positive instance of principles linking other people's behavior or other observable evidences to their mental qualities.[8] Nor could we always generalize even from direct awareness of our own mental qualities, since, Reid claims, we know many relevant qualities *even in our own case* only via their signs and effects.[9] (That is not as peculiar as it might initially sound. We do not, for instance, discover via sheer introspection that we have a talent for solving crossword puzzles – we do that *empirically* by seeing how we do on them.)

But Reid believes that we do learn about the mental states and characteristics of other beings via marks, signs, and the effects of such states and characteristics.[10] If experiencing signs does underlie our perception of relevant mental properties in others, and if that process rests neither upon inference, nor prior experience, nor anything of that sort, then the connection linking signs with states and characteristics must be simply built into our cognitive nature.

That is Reid's direction in the case of design. Among our inbuilt cognitive structures, he claims, is one articulated as follows:

> [D]esign and intelligence in the cause may be inferred with certainty, from marks or signs of it in the effect.
>
> (1872b: 457ff.)[11]

Reid explicitly compares this general principle to the earlier quoted first principle concerning external physical objects.[12] According to Reid, this design first principle is not only crucial to everyday life, underlying many of our beliefs concerning those around us, but it also applies to natural phenomena:[13]

> [T]here are in fact the clearest marks of design and wisdom in the works of nature;

From that we can conclude that

> the works of nature are the effects of a wise and intelligent Cause.
>
> (1872b: 461)[14]

A crucial question, of course, is: what *are* those "clearest marks" and signs? Reid unfortunately does not address that question systematically, but the marks referenced in scattered passages include contrivance, order, organization, intent, purpose, usefulness, adaptation, aptness/fitness of means to ends, regularity, and beauty.[15] Of those, contrivance generally seems to be the default value.[16]

But surely, it might be objected, from such marks we cannot infer "with certainty" the existence of a designing mind. We now have alternative (e.g. Darwinian) possible explanations for such marks as function, adaptation, etc. That claim may indeed be plausible, but I think it slightly misses Reid's mark. Reid placed the design first principle among *necessary* truths, and had a correspondingly strong conception of marks. And for some marks of design, the principle as stated is plausible – e.g. genuine intent does indeed require an intender, and from intent we can infer an intender with certainty. Reid may have been led by the science of his day into thinking that *adaptation* shared relevant properties with *intent*, and thus fell under the same principle. Perhaps he was mistaken about that. But that does raise another, even more crucial question: How are true marks of design reliably identified *as* marks of design?

Unfortunately, Reid does not seem particularly concerned over that issue, and (so far as I can find) does not address it systematically. One reason might have been his belief that, in many cases, we simply do not know what the cues are from which our beliefs – even about the mental states of others – arise. For instance, infants can recognize moods of adults from facial expressions, but neither they nor most adults could produce a defensible catalogue of the visual cues (Reid's example). But one's beliefs concerning the mental states of others are not rationally any the worse for that. Similarly, recognition of the relevant marks *as* design marks is presumably not essential to recognition of design (or, by extension, of designedness), given that design in its primary sense refers to a mental characteristic.[17]

But I think that Reid must be read as holding the view that, when it does occur, recognition of a mark *as* a mark of design is perceptual. For one thing, the claim that one needs an inference *to* a mark of design and then another from that mark to design itself (in accord with the design first principle) begins to look suspiciously regressive – something to which Reid was relatively sensitive – and runs counter to Reid's overall general approach. And although I will not go into detail here, Reid was more explicit in one area that he himself closely linked to design – aesthetics – and his views there very strongly (if circumstantially) reinforce the key points concerning design – that recognition of designedness (and often of design itself) is not inferential.[18] For instance, in discussing mental properties associated with beauty, Reid says:

> Other minds we perceive only through the medium of material objects, on which their signatures are impressed. It is through this medium that we perceive...wisdom, and every...intellectual quality in other beings. The signs of those qualities are immediately perceived by the senses; by them the qualities themselves are reflected to our understanding.
>
> (1872b: 503)

Here, not only are the mental qualities in question perceived (in Reid's sense) but the *signs themselves* are "immediately perceived." Since for Reid every perception embodies a judgment, perception of the signs would involve a conviction (at least *de re*) about those signs. Note also that resultant convictions about the qualities of the mind in question are results neither of prior experience nor of arcane inference from the signs: the qualities of other minds are *perceived* – indeed, they are "*reflected to* our understanding" (my emphasis) by the marks or signs.

Finally, Reid's remarks involving apprehending marks of design are almost always perceptual. Suggestive examples are numerous, of which a small sample follows (all emphases mine):

> [T]hus then we *see evident* marks of design
>
> (1973: 25)

> [I]n our own planetary system we *perceive*...marks of wisdom and design.
>
> (1973: 19)

> Everyone *sees* that [plant] roots are *designed*.
>
> (1981: 30)

These are very much in line with other remarks on closely allied topics.[19] Reid at one point explicitly mentions *purpose* – one mark of design – as something that, in other humans at least, is

> discovered to us by a natural principle, without reasoning or experience.
>
> (1872a: 122)

In some cases, judgments of design proper may, of course, rest upon inferences. But, in the primary cases, the requisite conceptual, cognitive, and epistemic transitions are simply functions of faculties that are constitutive parts of our natures. So I think that we can take Reid as holding that recognizing various characteristics *as* marks or signs of design is a *perceptual* process upon which in appropriate circumstances we simply discover in ourselves the requisite recognitions as one component of our sensory contact with the world.[20]

Reid further holds that design recognition, like any other type of perception, is not subject to our will – choice, decision, and anything else voluntary is typically absent. Concerning judgments of mental properties (as design fundamentally is) Reid says:

> Every man of common understanding forms such judgments of those he converses with, he can no more avoid it, than he can seeing objects that are placed before his eyes.
>
> (1981: 51–2)[21]

In the case of perception more generally, resisting, says Reid, is often

> not in my power.... My belief is carried along by perception, as irresistibly as my body by the earth.
>
> (1872a: 183–4)

In a related aesthetic case involving the experience of *grandeur*, Reid turns the dial up even further, saying that one's belief is

> carrie[d]...along...involuntarily and by a kind of violence rather than by cool conviction.
>
> (1872b: 496)

Reid's fuller picture, then, is that we perceive in objects in the world qualities that we often involuntarily and non-inferentially recognize *as* marks of design. From them, via a constitutionally constitutive processing structure (articulatable in terms of the various first principles), we acquire the belief of a power of design in the mind of the ultimate cause of these marks.[22] That sequence involves *at most* one trivial inferential step riding the rails of the relevant "self-evident" (and, according to Reid, necessary) first principle.[23] Even if an inference is involved, the real work is done at the initial design-recognition stage.

Whether or not inferences are ultimately unavoidable in identifying design in nature (and I think Reid is not completely clear here), Reid does push the case that *basic* design recognition is perceptual, and that, consequently, even if inductive design inferences are required, those arguments must themselves ultimately rest on a foundation of base cases of perceptually, non-inferentially recognized designedness.

Assessment

Prima facie *plausibility*

Reid's position has some initial plausibility. He at least *seems* to be right that in our ordinary, everyday recognition and identification of designedness we do not engage in inferences, calculate probabilities, or anything of the sort.[24] (Truth to tell, we likely have almost no clue as to what the relevant probabilities even are.) Nor do we typically choose whether or not to believe that selected things around us are or are not designed. We seem to be very much in a Darwinesque position – when we see anything, from a muffin to a space shuttle, we simply *find* that a belief in its designedness *happens* to us.

There are, of course, examples of more specialized cases where we do examine evidence etc. in attempting to come to a reasoned conclusion concerning design – when trying to distinguish, for example, extremely primitive hand axes from naturally chipped stones. We also engage in such processes when attempting to distinguish very subtle codes from sheer noise. But, in these cases, a Reidian might claim with some plausibility that we were really trying to identify properties that, once discovered, would be directly seen as marks of design from which – once found – full recognition would spring.

To take a less familiar example, when a string of prime numbers in binary was received in the movie *Contact*, not only was the string instantly perceived as designed (once its character was identified), but the content of the aliens' thinking was also instantly discerned as well – they had been thinking of the produced string of primes in binary. Again, if there was an inference, it was a monumentally trivial one.[25]

Counter-suggestions and responses

Tacit arguments

Of course, it can be maintained that complex inferences do underlie design identification – the inferences being so familiar and reflexive that, they go unnoticed. As Sherlock Holmes once remarked concerning an inference of the sort for which he is famous:

> From long habit the train of thoughts ran so swiftly through my mind that I arrived at the conclusion without being conscious of intermediate steps.
>
> (Doyle 1905: 24)

That may be right. It is often difficult in specific types of cases to determine whether or not unconscious tacit inferences are involved. For instance, philosophers for centuries sought tacit inferences to physical objects, other minds, etc. – inferences that, if Reid is correct, simply did not exist, but which many philosophers were firmly and mistakenly convinced were tacitly employed.

But even if we do sometimes employ tacit inferences, as it turns out that cannot be the whole story. One of the crucial components of ordinary design identification is recognition of *artifactuality*. Typically, it is the recognition that some phenomenon is not (or cannot be) a product of nature that turns our thoughts onto a design track. Recognition of artifactuality has profound effects concerning design judgments. We might disagree over whether the eye is a product of design or of fundamentally chance processes, but, *were we to agree that the eye was an artifact*, we would not for a moment consider the possibility that it was not deliberately designed. Whatever chance may or may not produce, the evidential status of properties like those possessed by the eye change drastically in the context of artifactuality.[26]

But nature is typically identified *as* nature precisely because of its intuitively non-artifactual character.[27] Consequently, the very foundation of design recognition procedures in ordinary cases – artifactuality – is systematically missing from cases in nature. It might be possible, of course, to construct inductive inferences concerning objects in nature, using for known base cases artifacts recognized as designed in the ordinary non-inferential way. But the evident artifactuality on which the base case recognition rested would be systematically absent in the natural cases, so the sample property for the induction would have to be some other property that was consistently present in the base cases and the proposed natural cases, and to which ordinary design recognition was *not* connected.

Here is a simple analogy. We recognize blue things directly, experientially, and non-inferentially. But suppose that we needed to identify blue things in total darkness. If we found that in our experience all and only blue things tend to have temperatures two degrees above their surroundings, we could then inductively identify blue things in total darkness by that temperature difference. The induction could not be based on the actual ordinary identifying process – seeing the color – but would depend upon something totally irrelevant to normal identification processes – temperature variation – but which the two classes of objects (observed blue and unobserved test cases) shared. Identification of cases in the dark would be inferential, and would not represent a simple extension of ordinary recognition processes. That would potentially affect the evidential force of the induction, as would the fact that in the design issue all the base cases fell within one category – artifacts – whereas every relevant attempted conclusion involved phenomena in what appears to be an importantly distinct category – natural phenomena.[28]

In any case, the key implication above is that even *if* inductive inferences figure in important cases of design identification, the *base* cases for those inductions must involve a qualitatively different – evidently non-inferential – recognition process. Inference, then, even *if* essential, cannot be the entire story.

Inference to the best explanation

Some proponents of inferential design pictures see design arguments not as strictly formal, but as looser "inferences to the best explanation." Although there is some plausibility to this approach, not only is this argument–genre problematic in some respects in itself, but it, too, cannot be the whole story. Taking design to be the best explanation of some phenomenon requires recognition of specific properties of the phenomenon as *design-relevant*. How exactly is that supposed to be done?

That issue is closely tied to the one discussed in the previous section. Significantly, exactly the same question arises concerning other design inference theories as well. But even the most rigorous current attempt to formalize design inferences – William Dembski's *The Design Inference* – does not address *in inferential terms* the core process of design recognition.[29] Without going into detail, running Dembski's formal inference structure requires first identifying what he terms "side information," which functions as a key component in more than one aspect of the formalism.[30] Very intuitively, side information identifies patterns that would be reasonable candidates for being deliberately designed. This side information is thoroughly mind-related and, it seems to me, constitutes the deep core of design identification even within Dembski's inferential system. How do we identify this key information? Dembski says:

> [I]dentifying suitable patterns and side information for eliminating chance requires of [the inferer] S insight....What's needed is insight, and insight admits no hard and fast rules....We have such insights all the time, and use them to identify patterns and side information that eliminate chance. But the logic of discovery by which we identify such patterns and side information is largely a mystery.
>
> (1998: 148)

However, a Reidian approach might actually solve that mystery. Even *if* inferences are ineliminable in identifying and attributing design in cases involving nature, the foundations for such inferences might still, perhaps unavoidably, be Reidian. Reid might suggest here that there simply is no such *logic* of discovery – such discovery being perceptual – and that it is to Dembski's credit that he sees that an inferential approach can penetrate only so far before leaving one on the edge of "mystery." Perhaps the mystery is that we simply *see* patterns that speak of design, that we simply find that certain things speak design to us perceptually. If so, that aspect of that particular mystery evaporates.[31]

So even *if* design advocates are correct in taking broader design cases to be fundamentally inferential, non-inferential design recognition may still be essential to the foundations of their case. But Reid's reach may well be greater even than that.

Additional support for Reid

The strength of the convictions of design advocates might suggest that such convictions are underpinned by more than just an induction with possible confirmation problems. Critics, of course, will (and do) immediately claim that religious commitments are the source of and provide the driving force for such convictions. Perhaps so. But that cannot be the whole story.

Design resistance

If design beliefs were founded on mere problematic inferences, or if their force arose merely from religious convictions, then we might not expect beliefs concerning design in nature to exert a very substantial tug on, for example, biologists – especially those not sympathetic to religion. That is not quite what we find. Even professional biologists seem to have an almost innate tendency to see biological systems in design terms. Thus, Francis Crick, co-discoverer of the structure of DNA (and no great fan of religion):

> Biologists must constantly keep in mind that what they see was not designed, but rather evolved.
>
> (1988: 138)

(And recall Darwin himself – also by the 1880s no great fan of religion.) But, according to historian of science Timothy Lenoir, commenting on a related area, maintaining rigorous vigilance and successfully *resisting* teleological inclinations are two quite different matters – even for professional biologists:

> Teleological thinking has been steadfastly resisted by modern biology. And yet, in nearly every area of research biologists are hard pressed to find language that does not impute purposiveness to living forms.
>
> (1992: ix)

That we have a deeply embedded conceptual tilt of this sort against which resistance is futile would come as no surprise to any Reidian.

Missing analyses

Historically there have been almost no attempts to construct rigorous analyses of the very concept of design. That is rather striking – bordering on astonishing – given that arguments from design have been discussed in great detail for centuries, that design evidences were the focus of enormous intellectual attention during the early nineteenth-century natural theology movement, and that the presence or absence of design in nature was one of the more volatile issues surrounding Darwinism. It was – and is – just assumed that everyone was on the same conceptual page concerning design. This otherwise puzzling absence is reasonably explainable, however, if a Reidian view is correct. It is not unusual where familiar *experiential* matters are involved.[32]

Missing arguments

One final consideration tells in Reid's favor. Convictions concerning design in nature have a long, influential, sometimes intellectually dominant history. But it is worth noting that most historically proposed attempts to reconstruct design inferences have been no more successful than historically proposed attempts to reconstruct physical object inferences. Perhaps the same explanation – the non-existence of the sought arguments – holds in both cases. If so, then the widespread recognition (or at least conviction) of design in nature must have some non-inferential source. Reid's proposal, then, might apply much more broadly than merely to some set of base cases.

Implications

The various foregoing considerations do not, of course, establish that a Reidian, non-inferential view is correct. But suppose for the moment that it was. Then attempts to co-opt familiar design-recognition procedures for application in nature may be problematic, for several reasons.

First, if basic design recognition is perceptual and not inferential, it cannot be claimed that *inference*-based searches for design in nature involve simply applying to objects in nature the identical procedures and criteria we employ in familiar situations. This could put attempts to identify design inferentially into roughly the same category as attempts to inferentially establish the existence of physical objects – efforts that, as noted earlier, have a track record which is not only abysmal but which may be inevitably so. In addition, if ordinary design recognition is experiential and not inferential, then constructed inferential design procedures and criteria cannot claim to automatically be the heirs of the legitimacy and justification of those familiar processes.

Second, Reid believed that some perceptual processes were susceptible to interference. Perception of beauty (which Reid tied closely to design) could be skewed by, among other things, fashion, habit, opinion, custom, fancy, casual association, education, and perhaps individual constitution, as well as being relativized to varying ends and purposes.[33] Thus it might be that disputes over design in nature arise out of the variable clarity or variable ability to perceive such design. If design recognition is perceptual, and if that perception is sensitive to subjective influences above some threshold level in cases of natural objects, the prospects for design playing a substantive role in *natural* science seem problematic – at least, that is, given the usual scientific preferences for objectivity and commonality of observation. On the other hand, if Reid is right, there are objective properties in objects that trigger the experience in question. But if the properties that trigger design experiences are objective, measurable properties in their own right, design recognition might be scientifically superfluous, the underpinning properties themselves being able to do any requisite scientific work. It may not be just coincidence that in the most influential specifically empirical piece of current design advocacy – Michael Behe's *Darwin's Black Box* – the scientific work is done not by the concept of design itself, but by a specific complex property (irreducible complexity) for which, it is claimed, Darwinian processes are unable to account. The subsidiary claim that this sort of complexity is a sign of design carries none of the scientific load.

Finally, if design recognition is ultimately experiential there will be strategy implications for current design efforts as well. Presenting arguments in issues where positions are experience- or perception-based does not seem particularly effective. One cannot be argued into having an experience of blueness. One cannot be argued into having a genuine aesthetic experience. Similarly, it may be that one cannot be argued into recognition of design. The most effective strategy may be (as in the case of blueness) simply to situate a person in experientially favorable circumstances, and hope that any scales will fall from his or her eyes.

Some prospects

Both historically and presently, most discussions of design in nature have been read as attempts to inductively extend inferential structures of design recognition from the artifactual realm into the natural realm.[34] If Reid is correct – and I have suggested some reasons for thinking that he may be – then both the foundation of that approach (inferentially identified base cases) and attempts to build cases for design upon it (inductive extensions to nature) may be seriously problematic. But even if inferences necessarily play some role in recognizing design in nature (and, again, it may be possible to read Reid either way here), the primary design recognition upon which inductive extensions would be based may demand a Reidian analysis.

One could construe this as solving a current "mystery" for design advocates. And a Reidian analysis might also offer them additional potential attractions. If basic design recognition is perceptual, then many of the historically popular criticisms of design cases will be irrelevant. Furthermore, if design recognition *in nature* is non-inferential, then standard criticisms of the inferential moves in design cases will have no bearing on properly constructed design positions. And such objections as that we have only one observed universe, and thus cannot attribute design to it – having no comparison cases – would be equally irrelevant.[35] After all, *perception* and *experience* function in single cases perfectly unproblematically. And if design recognition is fundamentally perceptual, charges that design is not an empirical matter – and thus scientifically illegitimate – will at least require modification.

So, if a thoroughgoing Reidian view is correct, current design advocates may be on the wrong evidential track and may have to redirect their efforts. But it is not clear that they must simply give up.[36]

Notes

1 This conversation is also referenced in Darwin (1958: 68n). With respect to specifically religious belief, in 1879, only three years prior to his death, Darwin wrote:

> [M]y judgment often fluctuates....In my most extreme fluctuations I have never been an Atheist in the sense of denying the existence of a God. I think that generally (and more and more as I grow older), *but not always*, that an Agnostic would be the more correct description of my state of mind [emphasis mine].
>
> (1958: 55)

2 Cleanthes may not speak for Hume in the dialogues, but Philo – who is generally thought to represent Hume – says concerning "the inexplicable contrivance and artifice of nature":

> A purpose, an intention, or design strikes everywhere the most careless, the most stupid thinker; and no man can be so hardened in absurd systems, as at all times to reject it.
>
> (Hume 1947: 214)

3 Indeed, one of the most influential attempts is titled simply *The Design Inference* (Dembski 1998).
4 This is the fifth of twelve contingent first principles Reid outlines.
5 Any such inferences would have to be new *discoveries* since they have no connection to the actual acquisition of our beliefs.
6 Reid himself mentions Zeno's arguments (unanswerable for centuries, but yet not "moving" anyone).
7 As Reid says:

> The works of men in science, in the arts of taste, and in the mechanical arts, bear the signatures of those qualities of mind which were employed in their production.
>
> (1872b: 503)

8 This is frequently stressed in Reid. See, for example, Reid (1973: 30–1; 1872b: 449–50, 460, 461; 1981: 53–4).
9 Reid says:

> [I]ntelligence, design, and skill, are not objects of the external senses, nor can we be conscious of them in any person but ourselves. Even in ourselves, we cannot, with propriety, be said to be conscious of the natural or acquired talents we possess. We are conscious only of the operations of mind in which they are exerted. Indeed, a man comes to know his own mental abilities, just as he knows another man's, by the effects they produce, when there is occasion to put them to exercise.
>
> (1872b: 458)

A virtually identical passage is contained in Reid (1981: 51–2). See also Reid (1973: 30–1; 1872b: 449–50, 460; 1981: 53–4).
10 As he puts it:

> How do I know that any man of my acquaintance has understanding?...I see only certain effects, which my judgment leads me to conclude to be marks and tokens of it.
>
> (Reid 1872b: 461)

A slightly different version appears at Reid (1981: 56).

11 It is not clear that Reid is correct concerning the modal status of that principle, but, even if he is not, the substance of his position concerning design need not be affected.

12 He says:

> [I]t is no less a part of the human constitution, to judge of men's characters, and of their intellectual powers, from the signs of them in their actions and discourse, than to judge of corporeal objects by our senses.
>
> (Reid 1872b: 458)

See also Reid (1981: 52).

13 We do, says Reid, "conclude...[human wisdom] from tokens that are visible" and (in a slightly different context) Reid affirms that:

> The very same argument applied to the works of nature, leads us to conclude that there is an intelligent Author of nature, and appears equally strong and obvious in the last case as in the first.
>
> (1872b: 449)

Thus, from signs and marks of design in nature, we move via a first principle built into our cognitive constitution to a belief in the existence of a wise, intelligent Author of nature. In a slightly more extended comment, Reid says:

> [F]rom the marks of wisdom and design to be met with in the Universe we infer it is the work of a wise and intelligent cause....[I]ntelligence, wisdom and skill are not objects of our external senses....A man's wisdom can be known only by its effects, by the signs of it....Yet it may be observed that we judge of these talents with as little hesitation as if they were objects of our senses....Every man of common understanding forms such judgments of those he converses with, he can no more avoid it, than he can seeing objects that are placed before his eyes. Yet in all these the talent is not immediately perceived, it is discerned only by the effects it produces. From this it is evident it is no less a part of the human constitution to judge of powers by their effects than of corporeal objects by the senses....[N]ow every judgment of this kind is only an application of that general rule, that from marks of intelligence and wisdom in effects, a wise and intelligent cause may be inferred....[This] is...to be received as a first principle. Some however have thought that we learn this by reasoning or by experience. I apprehend that it can be got from neither of them.
>
> (1981: 51–2)

14 A nearly identical passage occurs at Reid (1981: 54). See also Reid (1981: 15).

15 Why he did not is not totally clear. See, for example, Diamond (1998: 226–7). Even the status of some of the relevant materials is unclear. The *Lectures on Natural Theology* (Reid 1981) are lecture notes, but it is unsettled whether they are Reid's own notes or student notes, and, if the latter, how reliable they are. Some lecturers of the time read the same lectures over a course of years, and student notes were passed along, corrected, and came close to being lecture (and lecturer's notes) transcripts. I take the present notes (whatever their nature) to be quite reliable. In fact, in places they seem to be a virtually verbatim copy of passages Reid published elsewhere. I shall take the content of Reid (1981) to be Reid's own. Some even of what he does say is ambiguous. Complicating matters is the fact that Reid also sometimes uses key terms in

potentially misleading ways. See, for example, Somervill (1989: 259). See also Diamond (1998: 226–7).

16 This is especially evident in Reid (1981) – see, for example, pages 49 and 74.

17 For instance, Reid notes that we can often read mental states off subtle features of facial expressions without having any clue specifically as to what triggers that recognition. See also Reid (1973: 31–2, 37).

18 In fact, Reid may have thought of design as a subspecies or a subcomponent of beauty:

> I come now to consider what this beauty is or in what it consists.
> It consists then, I apprehend, in those actions and qualities of mind which command our admiration and esteem.…Beauty in material objects arises from those actions and qualities of mind which excite our esteem, in a secondary manner, as signs.…[B]eauty in figures, theorems, &c., arises from a consideration of some excellence in them or in some quality of mind which excites our esteem, either *as marks of design* or excellence or some other qualities [my emphasis].
> (1973: 41)

See also Reid (1872b: 503).

19 Reid says the following [all emphases mine]:

> [W]e *evidently see* the intentions of nature
> (1973: 27)

> When we consider attentively the works of nature we *see clear indications* of power, wisdom, and goodness.
> (1973: 61)

> [T]hose who have the least discernment will *observe* that it is *intended.*
> (1981: 40)

> [God's wisdom is] *conspicuous*
> (1981: 113)

> The invisible creator…hath stamped upon all his works *signatures* of his divine wisdom…which are *visible* to all men.
> (1872b: 503)

> [Y]et it is *manifest* that we were *designed* for this.
> (1981: 47)

There are also numerous instances throughout Reid (1981). However, Reid sounds a possibly contrary note:

> The ignorance of true philosophy which leads men to discern marks of wisdom and design in the formation and government of things may be considered then one cause of Speculative Atheism.
>
> (1981: 3)

20 S.A. Grave refers to "*self-identifying* marks of intelligence and will" (my emphasis) (1967: 121). In a related area, Roger Gallie comes to a similar reading of Reid:

> The beauty of the virtues lies in their real excellence which is, it seems, *immediately recognised* [his emphasis].
>
> (1998: 152)

21 He also makes more informal comments, e.g. "It is impossible not to see that man was intended to take care of his own preservation" (Reid 1981: 35).

22 That there must be a mind follows from design being a quality of mind. Reid also makes the connections more explicit in some limited contexts. For instance:

> Regularity and uniformity are the marks of design; nothing produced by chance can possibly be regular. Hence it is evident that regularity must be the sign of intelligence and of mind as well as of design.
>
> (Reid 1973: 42)

23 Reid's sense of "self-evident" does not require analyticity, but is a bit weaker. See, for example, page 6 of D.D. Todd's Introduction to Reid (1989).

24 If there are inferences here, they are often tacit ones of which we are unaware and which we might not even be able to reconstruct. It might be held that usual design conclusions implicitly rest upon a deep familiarity with human intents, purposes, etc. However, we do recognize design even in cases where the intent, purpose, design specifics, source, means of production, and all such other matters are utter mysteries to us. I have discussed relevant cases in Ratzsch (2001).

25 Were we to land on Mars, climb out of our vehicle, and see in front of us a diesel bulldozer, we would not begin a search for unusual Martian natural laws or unusual Martian chemistry. We wouldn't begin some complicated probability calculations. Nor would we construct some inductive inference based upon prior experiences with diesel bulldozers back on Earth. Indeed, we might do nothing remotely like that even if confronted with some alien construct whose purpose, intent, mode of construction, etc. was not only unknown to us but completely beyond us, involving concepts that we were unable to fit into any human conceptual categories at all. But we might still simply find ourselves with the belief that we were dealing with something designed, even were we unable to fathom specifically what the content of the design was. But, whatever it was, we'd attribute it to a mind having the requisite design capacities.

Of course, we might, upon further study, come to understand what the thing did, and then take ourselves to also understand what it was intended for – what the content of the design was, what the intent of the designing mind was. And although inferences, calculations, and the like might be involved in the determination of the function, the transition from conviction concerning function to conviction concerning intent would require either no inferential steps or a spectacularly short one.

26 I have discussed this effect in more detail in Ratzsch (2001).

27 I have developed this distinction in detail in Ratzsch (2001). Reid suggests it also (Reid 1981: 58).

28 Beyond the systemic weakness just indicated, inductive cases for design in nature may also have attenuated confirmability. An induction concerning the sun rising tomorrow can eventually be tested and confirmed independently of the induction itself – we can gain independent access to the truth of that matter. But with inductive cases for design in nature, there is no obvious independent access to the truth of the matter. This is connected to Reid's contention that we can not learn first principles connecting signs and others' inner states by experience, because we never experience those inner states and thus can never experience positive instances of the connections. See Reid (1973: 30–1; 1872b: 449–50, 460, 461 1981: 53–4). In the absence of any non-inferential means, how would we (aside from the initial inference itself) establish whether or not the propulsion system of *E. coli* is in fact designed? Of course, science deals routinely with a multitude of things to which it has no inference-independent or theory-independent access – quarks, the past, etc. In such cases, however, there are numerous independent but converging lines of evidence. Are such "consilience" cases available here? I do not know if anyone has attempted to make that case.

29 Again, Reid does speak in some places of inferences involved in design cases. (See the earlier quotation concerning the argument involving final causes, for instance.) Any inferences in such cases, however, would be inferences to conclusions concerning the mind of a designer – not to the existence of design or to evidence of design in natural phenomena. Inferences of this kind would leave intact the core Reidian contentions that both (1) recognition of marks and signs of design *as* such, and (2) acceptance of the relevant first principle connecting marks and signs to the mind of a designer, are non-inferential and involuntary.

30 I have discussed this and related matters in detail in the appendix of Ratzsch (2001).

31 It could then be argued that the remaining substantive difference between Dembski and Reid involves the modal status of the link between perceptually identified marks of design on the one hand, and the existence and qualities of a designing mind on the other – Reid taking that link to be both necessary and a constitutive part of our cognitive natures, Dembski taking it to be much weaker, and requiring serious probabilistic analysis. Whether this hybrid position would be remotely tolerable to either is, of course, another question.

32 There are some other interesting matters which *may* tell in Reid's favor, but which are, I think, more difficult to assess. For instance, *when* we have fully grasped an inference, or have seen our way along a trail of reason, the conviction of the cogency of the inference seems relatively stable – we don't usually just *find* at some point that the inference no longer seems cogent. Yet Darwin reported a marked variability in the conviction that nature exhibited design. That sort of variability seems more in keeping with our experiences of perception than with those of inference.

33 Reid (1973: 48, 36, 42–3) and Reid (1872b: 490, 491, 492, 501, 506). Natural taste can also, Reid claims, be corrupted; see Reid (1973: 36) and Reid (1872b: 491–2).

34 Although I will not pursue the issue here, contrary to the nearly universal reading I am not convinced that Paley intended to present an inductive *argument* for design. I think that Paley can be read as presenting cases of human design as *examples* of nearly direct design recognition, calling our attention to the relevant recognition processes, then showing us those *same* processes operating in our

interactions with natural phenomena. The uncritically received presumption that design cases are inferential has masked intriguing contrary hints in Paley – a masking that sensitivity to non-inferential possibilities could remove.

35 According to Reid, Hume has apparently woefully understated the problem. It isn't merely that we have not experienced multiple worlds – some designed with wisdom, some not – but that we have never directly experienced wisdom – *even our own* – at all. If, then, we do recognize wisdom – which we surely can – it cannot rest on prior direct experience, either internal or external. Recognition of signs *as* signs must evidently arise elsewhere. See again the earlier quoted passage from "Of judgment" in Reid (1872b: 458).

36 My thanks to Nicholas Wolterstorff, Neil A. Manson, C. Stephan Evans, and the members of the Philosophy Department of Calvin College.

References

Alston, W. (1991) *Perceiving God*, Ithaca, New York: Cornell University Press.

Argyll, George Douglas Campbell, 8th Duke of (1885) "What is science?" in *Good Words* April: 236–45.

Behe, M. (1996) *Darwin's Black Box*, New York: Free Press.

Crick, F. (1988) *What Mad Pursuit*, London: Penguin.

Darwin, F. (ed) (1958) *Autobiography of Charles Darwin and Selected Letters*, New York: Dover.

Dembski, W. (1998) *The Design Inference*, Cambridge: Cambridge University Press.

Diamond, P.J. (1998) *Common Sense and Improvement: Thomas Reid as Social Theorist*, New York: Peter Lang.

Doyle, A.C. (1905) *A Study in Scarlet*, in *The Complete Sherlock Holmes*, Garden City, New Jersey: Doubleday.

Gallie, R. (1998) *Thomas Reid: Ethics, Aesthetics and the Anatomy of the Self*, Dordrecht: Kluwer Academic Publishing.

Grave, S.A. (1967) "Thomas Reid," in P. Edwards (ed.) *Encyclopedia of Philosophy*, New York: Macmillan, pp. 118–21.

Hamilton, W. (1872) *The Works of Thomas Reid*, vol. I, 7th edn, Edinburgh: Maclachlan & Stewart.

Hume, D. (1947) *Dialogues Concerning Natural Religion*, ed. N.K. Smith, Indianapolis: Bobbs-Merrill.

Lenoir, T. (1992) *The Strategy of Life*, Chicago: University of Chicago Press.

Ratzsch, D. (2001) *Nature, Design, and Science*, Albany, New York: SUNY Press.

Reid, T. (1989) *The Philosophical Orations of Thomas Reid*, ed. D.D. Todd, trans. S.D. Sullivan, Carbondale, Illinois: Southern Illinois University Press.

—— (1981) *Thomas Reid's Lectures on Natural Theology (1780)*, ed. E.H. Duncan, Washington, DC: University Press of America.

—— (1973) *Thomas Reid's Lectures on the Fine Arts: Transcribed from the Original Manuscript, with an Introduction and Notes*, ed. Peter Kivy, The Hague: Martinus Nijhoff.

—— (1872a) *Inquiry into the Human Mind*, in W. Hamilton *The Works of Thomas Reid*, vol. I, 7th edn, Edinburgh: Maclachlan & Stewart.

—— (1872b) *Essays on the Intellectual Powers of Man*, in W. Hamilton *The Works of Thomas Reid*, vol. I, 7th edn, Edinburgh: Maclachlan & Stewart.

Somervill, J. (1989) "Making out the signatures: Reid's account of the knowledge of other minds," in M. Dalgarno and E. Matthews (eds) *The Philosophy of Thomas Reid*, Boston: Kluwer Academic Publishing, pp. 249–73.

Whewell, W. (1834) *Astronomy and General Physics Considered with Reference to Natural Theology*, London: William Pickering.

Part II

PHYSICAL COSMOLOGY

7

THE APPEARANCE OF DESIGN IN PHYSICS AND COSMOLOGY[1]

Paul Davies

I learned science as a set of procedures that would reveal how nature works, but I never questioned *why* we were able to do this thing called science so successfully. It was only after a long career of research and scholarship that I began to appreciate just how deep scientific knowledge is, and how incredibly privileged we human beings are to be able to unlock the secrets of nature in such a powerful way.

Of course, science didn't spring ready-made into the minds of Newton and his colleagues. They were strongly influenced by two longstanding traditions that pervaded European thought. The first was Greek philosophy. Most ancient cultures were aware that the Universe is not completely chaotic and capricious; there is a definite order in nature. The Greeks believed that this order could be understood, at least in part, by the application of human reasoning. They maintained that physical existence was not absurd, but rational and logical, and therefore in principle intelligible to us. They discovered that some physical processes had a hidden mathematical basis, and they sought to build a model of reality based on arithmetical and geometrical principles.

The second great tradition was the Judaic worldview, according to which the Universe was created by God at some definite moment in the past and ordered according to a fixed set of laws. The Jews taught that the Universe unfolds in a unidirectional sequence – what we now call linear time – according to a definite historical process: creation, evolution, and dissolution. This notion of linear time – in which the story of the Universe has a beginning, a middle, and an end – stands in marked contrast to the concept of cosmic cyclicity, the pervading mythology of almost all ancient cultures. Cyclic time – the myth of the eternal return – springs from mankind's close association with the cycles and rhythms of nature, and remains a key component in the belief systems of many cultures today. It also lurks just beneath the surface of the Western mind, erupting occasionally to infuse our art, our folklore, and our literature.

A world freely created by God, and ordered in a particular, felicitous way at the origin of a linear time, constitutes a powerful set of beliefs, and was taken up by both Christianity and Islam. An essential element of this belief system is that the Universe does not *have* to be as it is; it could have been otherwise. Einstein once said that the thing that most interested him is whether God had any choice in his creation. According to the Judeo-Islamic-Christian tradition, the answer is a resounding "yes."

Although not conventionally religious, Einstein often spoke of God, and expressed a sentiment shared, I believe, by many scientists, including professed atheists. It is a sentiment best described as a reverence for nature and a deep fascination for the natural order of the cosmos. If the Universe did not have to be as it is, of necessity – if, to paraphrase Einstein, God did have a choice – then the fact that nature is so fruitful, that the Universe is so full of richness, diversity, and novelty, is profoundly significant.

Some scientists have tried to argue that if only we knew enough about the laws of physics, if we were to discover a final theory that united all the fundamental forces and particles of nature into a single mathematical scheme, then we would find that this superlaw, or theory of everything, would describe the only logically consistent world. In other words, the nature of the physical world would be entirely a consequence of logical and mathematical necessity. There would be no choice about it. I think this is demonstrably wrong. There is not a shred of evidence that the Universe is logically necessary. Indeed, as a theoretical physicist I find it rather easy to imagine alternative universes that are logically consistent, and therefore equal contenders for reality.

It was from the intellectual ferment brought about by the merging of Greek philosophy and Judeo-Islamic-Christian thought that modern science emerged, with its unidirectional linear time, its insistence on nature's rationality, and its emphasis on mathematical principles. All the early scientists such as Newton were religious in one way or another. They saw their science as a means of uncovering traces of God's handiwork in the Universe. What we now call the laws of physics they regarded as God's abstract creation: thoughts, so to speak, in the mind of God. So in doing science, they supposed, one might be able to glimpse the mind of God. What an exhilarating and audacious claim!

In the ensuing 300 years, the theological dimension of science has faded. People take it for granted that the physical world is both ordered and intelligible. The underlying order in nature – the laws of physics – is simply accepted as given, as brute fact. Nobody asks where the laws come from – at least they don't in polite company. However, even the most atheistic scientist accepts as an act of faith the existence of a law-like order in nature that is at least in part comprehensible to us. So science can proceed only if the scientist adopts an essentially theological worldview.

It has become fashionable in some circles to argue that science is ultimately a sham, that we scientists read order into nature, not out of nature, and that the laws of physics are our laws, not nature's. I believe this is arrant nonsense. You'd be hard-pressed to convince a physicist that Newton's inverse square law of gravitation is a purely cultural concoction. The laws of physics, I submit, *really exist* in the world out there, and the job of the scientist is to uncover them, not invent them. True, at any given time, the laws you find in the textbooks are tentative and approximate, but they mirror, albeit imperfectly, a really existing order in the physical world. Of course, many scientists don't recognize that in accepting the reality of an order in nature – the existence of laws "out there" – they are adopting a theological worldview. Ironically, one of the staunchest defenders of the reality of the laws of physics is the US physicist Steven Weinberg, a sort of apologetic atheist who, though able to wax lyrical about the mathematical elegance of nature, nevertheless felt compelled to pen the notorious words: "The more the universe seems comprehensible, the more it also seems pointless."

Let us accept, then, that nature really is ordered in a mathematical way – that "the book of nature," to quote Galileo, "is written in mathematical language." Even so, it is easy to imagine an ordered universe that nevertheless remains utterly beyond human comprehension, due to its complexity and subtlety. For me, the magic of science is that we *can* understand at least part of nature – perhaps in principle all of it – using the scientific method of inquiry. How utterly astonishing that we human beings can do this! Why should the rules on which the Universe runs be accessible to humans?

The mystery is all the greater when one takes into account the cryptic character of the laws of nature. When Newton saw the apple fall, he saw a falling apple. He didn't see a set of differential equations that link the motion of the apple to the motion of the Moon. The mathematical laws that underlie physical phenomena are not apparent to us through direct observation; they have to be painstakingly extracted from nature using arcane procedures of laboratory experiment and mathematical theory. The laws of nature are hidden from us and are revealed only after much labor. The late Heinz Pagels – another atheistic physicist – described this by saying that the laws of nature are written in a sort of cosmic code, and that the job of the scientist is to crack the code and reveal the message – nature's message, God's message, take your choice, but not *our* message. The extraordinary thing is that human beings have evolved such a fantastic code-breaking talent. This is the wonder and the magnificence of science; we can use it to decode nature and discover the secret laws that make the Universe tick!

Many people want to find God in the *creation* of the Universe, in the Big Bang that started it all off. They imagine a superbeing who deliberates for all eternity, then presses a metaphysical button, and produces a huge explosion. I believe this image is entirely misconceived. Einstein showed us that space and time are *part of* the physical Universe, not a pre-existing arena in which the Universe happens. Cosmologists are convinced that the Big Bang was the coming-into-being, not just of matter and energy, but also of space and time as well. Time itself began with the Big Bang. If this sounds baffling, it is by no means new. Already in the fifth century St Augustine proclaimed that "the world was made with time, not in time." According to James Hartle and Stephen Hawking, this coming-into-being of the Universe need not be a supernatural process, but could occur entirely naturally, in accordance with the laws of quantum physics, which permit the occurrence of genuinely spontaneous events.

The origin of the Universe, however, is hardly the end of the story. The evidence suggests that in its primordial phase the Universe was in a highly simple, almost featureless state: perhaps a uniform soup of subatomic particles, or even just expanding empty space. All the richness and diversity of matter and energy we observe today has emerged since the beginning in a long and complicated sequence of self-organizing physical processes. What an incredible thing these laws of physics are! Not only do they permit a universe to originate spontaneously; they also encourage it to self-organize and self-complexify to the point where conscious beings emerge, and can look back on the great cosmic drama and reflect on what it all means.

Now you may think I have written God entirely out of the picture. Who needs a God when the laws of physics can do such a splendid job? But we are bound to return to that burning question: Where do the laws of physics come from? And why *those* laws rather than some other set? Most especially: Why a set of laws that drives the searing, featureless gases coughed out of the Big Bang towards life and consciousness and intelligence and cultural activities such as religion, art, mathematics, and science?

If there is a meaning or purpose to existence, as I believe there is, we are wrong to dwell too much on the originating event. The Big Bang is sometimes referred to as "the creation," but in truth nature has never *ceased* to be creative. This ongoing creativity, which manifests itself in the spontaneous emergence of novelty and complexity, and in the organization of physical systems, is permitted through, or guided by, the underlying mathematical laws that scientists are so busy discovering.

Now the laws of which I speak have the status of timeless eternal truths, in contrast to the physical states of the Universe that change with time, and bring forth the genuinely new. So we here confront in physics a re-emergence of the oldest of all philosophical and theological debates: the paradoxical conjunction of the eternal and the temporal. Early Christian thinkers wrestled with the problem of time: Is God within the stream of time, or outside of it? How can a truly timeless God relate in any way to temporal beings such as us? But how can a God who relates to a changing universe be considered eternal and unchangingly perfect?

Well, physics has its own variations on this theme. In the last century, Einstein showed us that time is not simply "there" as a universal and absolute backdrop to existence; rather, it is intimately interwoven with space and matter. As I have mentioned, time is revealed to be an integral part of the physical universe; indeed, it can be warped by motion and gravitation. Clearly something that can be changed in this manner is not absolute, but a contingent part of the physical world.

In my own field of research – called quantum gravity – a lot of attention has been devoted to understanding how time itself could have come into existence in the Big Bang. We know that matter can be created by quantum processes. There is now a general acceptance among physicists and cosmologists that space–time can also originate in a quantum process. According to the latest thinking, time might not be a primitive concept at all, but something that has "congealed" from the fuzzy quantum ferment of the Big Bang, a relic, so to speak, of a particular state that froze out of the fiery cosmic birth.

If it is the case that time is a contingent property of the physical world rather than a necessary consequence of existence, then any attempt to trace the ultimate purpose or design of nature to a *temporal* Being or Principle seems doomed to failure. While I do not wish to claim that physics has solved the riddle of time – far from it – I do believe that our advancing scientific understanding of time has illuminated the ancient theological debate in important ways. I cite this topic as just one example of the lively dialogue that is continuing between science and theology.

A lot of people are hostile to science because it demystifies nature. They prefer the mystery. They would rather live in ignorance of the way the world works and our place within it. For me, the beauty of science is *precisely* the demystification, because it reveals just how truly wonderful the physical universe really is. It is impossible to be a scientist working at the frontier without being awed by the elegance, ingenuity, and harmony of the law-like order in nature. In my attempts to popularize science, I'm driven by the desire to share my own sense of excitement and awe with the wider community; I want to tell people the good news. The fact that we are able to do science, that we can comprehend the hidden laws of nature, I regard as a gift of immense significance. Science, properly conducted, is a wonderfully enriching and humanizing enterprise. I cannot believe that using this gift called science – using it wisely, of course – is wrong. It is good that we should know.

So where is God in this story? Not especially in the Big Bang that starts the Universe off, nor meddling fitfully in the physical processes that generate life and consciousness. I would rather that nature can take care of itself. The idea of a God who is just another force or agency at work in nature, moving atoms here and there in competition with physical forces, is profoundly uninspiring. To me, the true miracle of nature is to be found in the ingenious and unswerving lawfulness of the cosmos, a lawfulness that permits complex order to emerge from chaos, life to emerge from inanimate matter, and consciousness to emerge from life, without the need for the occasional supernatural prod – a lawfulness that produces beings who not only ask great questions of existence, but who also, through science and other methods of inquiry, are even beginning to find answers.

You might be tempted to suppose that any old rag-bag of laws would produce a complex universe of some sort, with attendant inhabitants convinced of their own specialness. Not so. It turns out that randomly selected laws lead almost inevitably either to unrelieved chaos or boring and uneventful simplicity. Our own universe is poised exquisitely between these unpalatable alternatives, offering a potent mix of freedom and discipline, a sort of restrained creativity. The laws do not tie down physical systems so rigidly that they can accomplish little, nor are they a recipe for cosmic anarchy. Instead, they encourage matter and energy to develop along pathways of evolution that lead to novel variety, what Freeman Dyson has called the principle of maximum diversity: that in some sense we live in the most interesting possible universe.

Scientists have recently identified a regime dubbed "the edge of chaos," a description that certainly characterizes living organisms, where innovation and novelty combine with coherence and co-operation. The edge of chaos seems to imply the sort of lawful freedom I have just described. Mathematical studies suggest that to engineer such a state of affairs requires laws of a very special form. If we could twiddle a knob and change the existing laws, even very slightly, the chances are that the Universe as we know it would fall apart, descending into chaos. Certainly the existence of life as we know it, and even of less elaborate systems such as stable stars, would be threatened by just the tiniest change in the strengths of the fundamental forces, for example. The laws that characterize our actual universe, as opposed to an infinite number of alternative possible universes, seem almost contrived – fine-tuned some commentators have claimed – so that life and consciousness may emerge. To quote Dyson again: it is almost as if "the universe knew we were coming." I can't prove to you that that is design, but whatever it is it is certainly very clever!

Now some of my colleagues embrace the same scientific facts as I, but deny any deeper significance. They shrug aside the breathtaking ingenuity of the laws of physics, the extraordinary felicity of nature, and the surprising intelligibility of the physical world, accepting these things as a package of marvels that just happens to be. But I cannot do this. To me, the contrived nature of physical existence is just too fantastic for me to take on board as simply "given." It points forcefully to a deeper underlying meaning to existence. Some call it purpose, some design. These loaded words, which derive from human categories, capture only imperfectly what it is that the Universe is *about*. But that it is about something, I have absolutely no doubt.

Where do we human beings fit into this great cosmic scheme? Can we gaze out into the cosmos, as did our remote ancestors, and declare: "God made all this for us!" Well, I think not. Are we then but an accident of nature, the freakish outcome of blind and purposeless forces, an incidental byproduct of a mindless, mechanistic universe? I reject that too. The emergence of life and consciousness, I maintain, are written into the laws of the Universe in a very basic way. True, the actual physical form and general mental make-up of *homo sapiens* contains many accidental features of no particular significance. If the Universe were rerun a second time, there would be no Solar System, no Earth, and no people. But the emergence of life and consciousness somewhere and somewhen in the cosmos is, I believe, assured by the underlying laws of nature. The origins of life and consciousness were not interventionist miracles, but nor were they stupendously improbable accidents. They were, I believe, part of the natural outworking of the laws of nature, and as such our existence as conscious inquiring beings springs ultimately from the bedrock of physical existence – those ingenious, felicitous laws. This is the sense in which I have written in my book *The Mind of God* (1992), "We are truly meant to be here." I mean "we" in the sense of conscious beings, not *homo sapiens* specifically. Thus although we are not at the center of the Universe, human existence *does* have a powerful wider significance. Whatever the Universe as a whole may be about, the scientific evidence suggests that we, in some limited yet ultimately still profound way, are an integral part of its purpose.

How can we test these ideas scientifically? One of the great challenges for science is to understand the nature of consciousness in general and human consciousness in particular. We still haven't a clue how mind and matter are related, nor what process led to the emergence of mind from matter in the first place. This is an area of research that is attracting considerable attention at present, and for my part I intend to pursue my own research in this field. I expect that when we do come to understand how consciousness fits into the physical universe, my contention that mind is an emergent and in principle predictable product of the laws of the Universe will be borne out.

Second, if I am right that the Universe is fundamentally creative in a pervasive and continuing manner, and that the laws of nature encourage matter and energy to self-organize and self-complexify to the point that life and consciousness emerge naturally, then there will be a universal trend or directionality towards the emergence of greater complexity and diversity. We might then expect life and consciousness to exist throughout the Universe. That is why I attach such importance to the search for extraterrestrial organisms, be they bacteria on Mars or advanced technological communities on the other side of the Galaxy. The search may prove hopeless – the distances and numbers are certainly daunting – but it is a glorious quest. If we *are* alone in the Universe, if the Earth is the only life-bearing planet among countless trillions, then the choice is stark. Either we are the product of a unique supernatural event in a universe of profligate overprovision, or else an accident of mind-numbing improbability and irrelevance. On the other hand, if life and mind are universal phenomena, if they are written into nature at its deepest level, then the case for an ultimate purpose to existence would be compelling.

Notes

1 The following excerpt is taken from my acceptance address given at the 1995 Templeton Prize for Progress in Religion award ceremony.

References

Davies, Paul (1992) *The Mind of God*, New York: Simon & Schuster.

8

DESIGN AND THE ANTHROPIC FINE-TUNING OF THE UNIVERSE

William Lane Craig

Introduction

Widely thought to have been demolished by Hume and Darwin, the teleological argument for God's existence came roaring back into prominence during the latter half of the last century. Defenders of the argument earlier in the same century appealed to what F.R. Tennant called "wider teleology," which emphasizes the necessary conditions for the existence and evolution of intelligent life, rather than specific instances of purposive design. Unfortunately, they could speak of this wider teleology for the most part only in generalities, for example, "the thickness of the earth's crust, the quantity of water, the amount of carbon dioxide," and so forth, but could furnish few details to describe this alleged teleology (Tennant 1935, vol. 2: 87).

In recent years, however, the scientific community has been stunned by its discovery of how complex and sensitive a nexus of conditions must be given in order for the Universe to permit the origin and evolution of intelligent life.[1] The Universe appears, in fact, to have been incredibly fine-tuned from the moment of its inception for the production of intelligent life. In the various fields of physics and astrophysics, classical cosmology, quantum mechanics, and biochemistry various discoveries have repeatedly disclosed that the existence of intelligent, carbon-based life depends upon a delicate balance of physical and cosmological quantities, such that were any one of these quantities to be slightly altered, the balance would be destroyed and life would not exist.

Examples of wider teleology

For example, the values of the various forces of nature appear to be fine-tuned for the existence of intelligent life. The world is conditioned principally by the values of the fundamental constants α (the fine structure constant, or electromagnetic interaction), m_p/m_e (proton to electron mass ratio), α_G (gravitation), α_w (the weak force), and α_s (the strong force). When one assigns different values to these constants or forces, one discovers that the number of observable universes, that is to say, universes capable of supporting intelligent life, is very small. Just a slight variation in any one of these values would render life impossible.

For example, if α_s were increased by as much as 1 percent, nuclear resonance levels would be so altered that almost all carbon would be burned into oxygen; an increase of 2 percent would preclude formation of protons out of quarks, preventing the existence of atoms. Furthermore, weakening α_s by as much as 5 percent would unbind deuteron, which is essential to stellar nucleo-synthesis, leading to a universe composed only of hydrogen. It has been estimated that α_s must be within 0.8 and 1.2 times its actual strength or all elements of atomic weight greater than 4 would not have formed. Or again, if α_w had been appreciably stronger, then the Big Bang's nuclear burning would have proceeded past helium to iron, making fusion-powered stars impossible. But if it had been much weaker, then we would have had a universe entirely of helium. Or again, if α_G had been a little greater, all stars would have been red dwarfs, which are too cold to support life-bearing planets. If it had been a little smaller, the universe would have been composed exclusively of blue giants, which burn too briefly for life to develop. According to Davies, changes in either α_G or electromagnetism by only one part in 10^{40} would have spelled disaster for stars like the Sun. Moreover, the fact that life can develop on a planet orbiting a star at the right distance depends on the close proximity of the spectral temperature of starlight to the molecular binding energy. Were it greatly to exceed this value, living organisms would be sterilized or destroyed; but, were it far below this value, then the photochemical reactions necessary to life would proceed too slowly for life to exist. Or again, atmospheric composition, upon which life depends, is constrained by planetary mass. But planetary mass is the inevitable consequence of electromagnetic and gravitational interactions. And there simply is no physical theory that can explain the numerical values of α and m_p/m_e that determine electromagnetic interaction.

Several of these same constants play a crucial role in determining the temporal phases of the development of the Universe and thus control features of the Universe essential to life. For example, α_G and m_p/m_e constrain (1) the main-sequence stellar lifetime, (2) the time before which the expansion dynamics of the expanding Universe are determined by radiation rather than matter, (3) the time after which the Universe is cool enough for atoms and molecules to form, (4) the time necessary for protons to decay, and (5) the Planck time.

Furthermore, a fine balance must exist between the gravitational and weak interactions. If the balance were upset in one direction, the Universe would have been 100 percent helium in its early phase, which would have made it impossible for life to exist now. If the balance were tipped in the other direction, then it would not have been possible for neutrinos to blast the envelopes of supernovae into space and so distribute the heavy elements essential to life.

Moreover, the difference between the masses of the neutron and the proton is also part of a very delicate coincidence that is crucial to a life-supporting environment. This difference prevents protons from decaying into neutrons, which, if it were to happen, would make life impossible. This ratio is also balanced with the electron mass, for, if the neutron mass failed to exceed the proton mass by a little more than the electron mass, then atoms would simply collapse.

Considerations of classical cosmology also serve to highlight a new parameter, S, the entropy per baryon in the Universe. The total observed entropy of the Universe is 10^{88}. Since there are around 10^{80} baryons in the Universe, the observed entropy per baryon, which is about 10^9, must be regarded as extremely small. Unless S were $< 10^{11}$, galaxies would not have been able to form, making planetary life impossible. In a collapsing Universe the total entropy would be 10^{123} near the end. Comparison of the total observed entropy of 10^{88} with a possible 10^{123} reveals how incredibly small 10^{88} is compared to what it might have been. Thus, the structure of the Big Bang must have been severely constrained in order for thermodynamics as we know it to have arisen. Not only so, but S is itself a consequence of the baryon asymmetry in the Universe, which arises from the inexplicable, built-in asymmetry of quarks over anti-quarks prior to 10^{-6} seconds after the Big Bang. Penrose calculates that the odds of the special low-entropy condition having arisen sheerly by chance in the absence of any constraining principles is at least as small as about one part in $10^{1,000B(3/2)}$ where B is the present baryon number of the Universe $\sim 10^{80}$. Thus, aiming at a manifold whose points represent the various possible initial configurations of the Universe, "the accuracy of the Creator's aim" would have to have been one part in $10^{10(123)}$ in order for our universe to exist. Penrose comments, "I cannot even recall seeing anything else in physics whose accuracy is known to approach, even remotely, a figure like one part in $10^{10(123)}$" (1981: 249).

In investigating the initial conditions of the Big Bang, one also confronts two arbitrary parameters governing the expansion of the Universe: Ω_0, related to the density of the Universe, and H_0, related to the speed of the expansion. Observations indicate that at 10^{-43} seconds after the Big Bang the Universe was expanding at a fantastically special rate of speed with a total density close to the critical value on the borderline between recollapse and everlasting expansion. Hawking (1988:123) estimates that even a decrease of one part in a million million when the temperature of the Universe was 10^{10} degrees would have resulted in the Universe's recollapse long ago; a similar increase would have precluded the galaxies from condensing out of the expanding matter. At the Planck time, 10^{-43} seconds after the Big Bang, the density of the Universe must have apparently been within about one part in 10^{60} of the critical density at which space is flat. This results in the so-called "flatness problem": Why is the Universe expanding at just such a rate that space is Euclidean rather than curved? A second problem that arises is the "homogeneity problem." There is a very narrow range of initial conditions that must obtain if galaxies are to form later. If the initial inhomogeneity ratio were $> 10^{-2}$, then non-uniformities would condense prematurely into black holes before the stars form. But if the ratio were $< 10^{-5}$, inhomogeneities would be insufficient to condense into galaxies. Because matter in the Universe is clumped into galaxies, which is a necessary condition of life, the initial inhomogeneity ratio appears to be incredibly fine-tuned. Third, there is the "isotropy problem." The temperature of the Universe is amazing in its isotropy: it varies by only about one part in 100,000 over the whole of the sky. But, at very early stages of the Universe, the different regions of the Universe were causally disjointed, since light beams could not travel fast enough to connect the rapidly receding regions. How then did these unconnected regions all happen to possess the same temperature and radiation density?

Contemporary cosmologists believe that they have found an answer to these three problems – or at least seem certain that they are on the right track – in inflationary models of the early universe. According to this proposed adjustment to the standard Big Bang cosmology, the very early universe briefly underwent an exponentially rapid inflation of space faster than the speed of light. This inflationary epoch resulted in the nearly flat curvature of space, pushed inhomogeneities beyond our horizon, and served to bury us far within a single region of space–time whose parts were causally connected at preinflationary times.

Inflationary scenarios – by 1997 Alan Guth could count over fifty versions (Guth, 1997) – have, however, been plagued by difficulties. The original "old inflationary model" and its successor the "new inflationary model" are now dead. As Earman and Mosterin have shown in their recent survey of inflationary cosmology, even the newest inflationary scenarios like Linde's do not "overcome the glaring deficiencies of the original versions of inflationary cosmology" (Earman and Mosterin 1999: 36). They write:

> Proponents of inflationary cosmology originally charged that the standard big bang model was beset by problems which inflation could cure in a natural and straightforward way. But (a) results showing that inflation is likely to occur under generic conditions in the Universe were not forthcoming, (b) cosmic no-hair theorems showing that inflation is effective in ironing out generic non-uniformities were not forthcoming (and by our reckoning are probably not true), and (c) in the straightforward version of inflationary cosmology where an inflationary era is inserted into a hot big bang model, the presence of enough inflation to solve the monopole, horizon, and uniformity problems in an open FRW [Friedmann-Robertson-Walker] universe ($k = -1$, $\Omega < 1$) and to explain the

> origin of density perturbations is difficult to reconcile with a low value of Ω_0....
>
> In sum, inflationary cosmologists have never delivered on their original promises. The newer models to which they have been driven depart radically from the original goal of improving the standard big bang model by means of a straightforward modification. And the link to concrete theories of elementary particle physics that initially made inflationary cosmology so exciting has been severed. The idea that was "too good to be wrong" has led to models that an impartial observer might well find contrived or fanciful or both.
>
> (Earman and Mosterin 1999: 36, 38)

In addition they note that inflationary cosmology has not enjoyed any successful empirical predictions: either the predictions tend to be falsified by the evidence or else new models are constructed to be compatible with the evidence, so that the predictions become non-predictions.

Inflationary cosmology is, in a way, eloquent testimony to the fact that the fine-tuning of the Universe for intelligent life does cry out for explanation and cannot remain a brute fact. Earman and Mosterin contend that the motivation for inflationary theory lies not in alleged empirical inadequacies of standard Big Bang cosmology, but rather in dissatisfaction with the style of explanation in the standard model: "the explanation given by the standard big bang model is found wanting because it must rely on special initial conditions" (1999: 23). The standard model's explanation of the flatness, homogeneity, and isotropy of the Universe is rejected by inflationary theorists as "not a good or satisfying explanation because it must rely on highly special initial conditions" (Earman and Mosterin 1999: 19). Given this dissatisfaction, interest in inflationary cosmology will continue unabated. They conclude:

> Despite the lack of empirical successes, the unkept promises, and the increasingly contrived and speculative character of the models, the inflationary juggernaut has not lost steam. The reasons are complicated, but we suspect that the main one is simply the sense among theorists that inflationary cosmology, if not the only game in town, is the only one around to provide computationally tractable models for treating a variety of issues in cosmology. We would predict that unless and until another, equally tractable game is found, the popularity of inflationary cosmology will persist even in the face of conceptual and empirical anomalies....
>
> Whatever the fate of inflationary cosmology, philosophers interested in scientific explanation should be drawn to a case where, ostensibly, a major program of scientific research was launched not because the standard big bang model proved empirically inadequate,

> nor because it could not offer explanations of phenomena in its intended domain of application, but because the explanations were deemed to be unsatisfying. The demands of inflationary cosmologists were for explanations that use a common cause mechanism and are robust in the sense of being insensitive to initial conditions.
>
> (Earman and Mosterin 1999: 45–6)

The defender of the teleological argument sympathizes with inflationary theorists' dissatisfaction with the explanations afforded by standard Big Bang cosmology in so far as they might be thought to require us to regard the special initial conditions as explanatory stopping points. The difference between them is that whereas the inflationary theorist seeks to modify the standard model in hopes of defining a common cause mechanism that obviates the necessity of special initial conditions, the design theorist will attempt to provide a causal explanation of the special conditions themselves, not indeed a scientific explanation in terms of natural laws and further conditions, but a personal explanation in terms of an agent and his volitions.[2]

At the end of the day, it is important to realize that the inflationary *Ansatz* has failed to arrive at explanations that, even if accepted, are insensitive to highly special initial conditions. In other words, inflationary scenarios seem to require the same sort of fine-tuning that theorists had hoped these models had eliminated. This problem becomes most acute with respect to the value of the cosmological constant Λ, which may be analyzed as consisting of two components, bare Λ, which was the term introduced by Einstein into his gravitational field equations, and quantum Λ, which signifies the energy density of the true vacuum or of the false vacuum in inflationary scenarios. The total cosmological constant Λ_{tot} is usually taken to be zero. But this requires that the energy density of the true vacuum be tuned to zero *by hand*; there is no understanding of why this value should be so low. Worse, inflation requires that Λ_{tot} was once quite large, though zero today; this assumption is without any physical justification. Moreover, in order to proceed appropriately, inflation requires that bare Λ and quantum Λ cancel each other out with an enormously precise though inexplicable accuracy. A change in the strengths of either α_G or α_w by as little as one part in 10^{100} would destroy this cancellation on which our lives depend. If, in line with recent tests indicating an acceleration of the cosmic expansion, bare $\Lambda > 0$, then yet another fine-tuning problem unsolved by inflation arises. The density parameter Ω can be analyzed as consisting of two components, Ω^M, or the contribution of matter to the density parameter, and Ω^Λ, or the contribution of the cosmological constant to the parameter. If $\Lambda > 0$, then in order to arrive at a present value of Ω^Λ on the order of 1, the ratio Ω^Λ/Ω^M must have been exquisitely small in the very early universe, a ratio that is unconstrained by inflationary cosmology. There will also be other physical quantities unconstrained by inflationary scenarios. For example, the value of S seems to be wholly unrelated to Ω_0, H_0, or inflationary scenarios. Thus, fine-tuning is far from eliminated even if inflationary cosmology were embraced.

The inference to design

The discovery of cosmic fine-tuning has led many scientists to conclude that such a delicate balance cannot be dismissed as coincidence but cries out for explanation. In a sense more easy to discern than to articulate, this fine-tuning of the Universe seems to manifest the presence of a designing intelligence. John Leslie, the philosopher who has most occupied himself with these matters, can speak here only informally of the need for what he calls a "tidy explanation." A tidy explanation is one that not only explains a certain situation but also reveals in doing so that there is something to be explained. Leslie provides a whole retinue of charming illustrations of tidy explanations at work.[3] Suppose, for example, that you are playing cards and your opponent lays down his hand of 8, 6, 5, 4, 3. At first it appears to be mere trash, and you have no reason to attribute it to anything but chance. But then you realize that in the game you are playing that combination is the perfect winning hand, which happens to beat narrowly your own very strong hand, that there is a million dollars on the table, and that your opponent has been known to cheat. Now, as Leslie says, you immediately become suspicious! Or, again, suppose Bob is given a new car for his birthday. There are millions of license plate numbers, and it is therefore highly unlikely that Bob would get, say, CHT 4271. Yet that plate on his birthday car would occasion no special interest. But suppose Bob, who was born on 8 August 1949, finds BOB 8849 on the license plate of his birthday car. He would be obtuse if he shrugged this off with the comment, "Nothing remarkable about that!" Or, again, think of the silk merchant whose thumb just happened to be covering the moth hole in the drape of silk he sold you. What would you think of the explanation, "Well, his thumb had to be somewhere on the cloth, and any location is equally improbable…"?

Leslie believes that design similarly supplies a tidy explanation of the fine-tuning of the Universe for intelligent life. He concludes:

> The moral must now be plain. Our universe's elements do not carry labels announcing whether they are in need of explanation. A chief (or the only?) reason for thinking that something stands in such need, that is, for justifiable reluctance to dismiss it as how things just happen to be, is that one in fact glimpses some tidy way in which it might be explained.
>
> (Leslie 1989: 10)

The strength of Leslie's reasoning is that in everyday life we do intuitively see the need for and employ tidy explanations for various situations. We may not be able to articulate why such an explanation is called for, but we sense it clearly.

Still, it would be desirable to have a more rigorous formulation of the grounds for inferring design, since in that case one's argument would be all the stronger. A key insight into the problem came in 1965 when the Russian probability theorist Andrei Kolmogorov discovered that, although one cannot discern a random from a non-random series of coin tosses on the basis of the resources of probability theory alone (since any result of a sequence of tosses is equally improbable), nevertheless the difference could be determined by employing the resources of computational complexity theory. Kolmogorov mapped the series of coin tosses onto series of binary numbers and calculated the computational complexity of each series in terms of the length of the shortest computer program capable of generating that series of 0s and 1s. Kolmogorov argued that the shorter the generating program is, the lower the computational complexity of the series is and, hence, the less random the series is. By contrast, a series with a longer generating program will have higher computational complexity and, hence, greater randomness. Series having sufficiently low randomness cannot have been the product of sheer chance.

Now Kolmolgorov's insight has been incorporated into a fully-fledged theory of design inference by William Dembski.[4] Dembski furnishes the following ten-step Generic Chance Elimination Argument, which delineates the common pattern of reasoning that underlies chance-elimination arguments:

(1) A subject S learns that an event E has occurred.

(2) By examining the circumstances under which E occurred, S finds that a chance process characterized by the chance hypothesis **H** and the probability measure **P** could have been operating to produce E.

(3) S identifies a pattern D that delimits the event E.

(4) S calculates the probability of the event D* given the chance hypothesis **H**, that is **P**(D* | **H**) = p.

(5) In accord with how important it is for S to avoid a "false positive" (i.e. attributing E to something other than the chance hypothesis **H** in case **H** actually was responsible for E), S fixes a set of probabilistic resources Ω characterizing the relevant ways D* (and by implication E) might have occurred and been specified given the chance hypothesis **H**.

(6) Using the probabilistic resources **Ω**, S identifies the saturated event D^*_Ω and calculates (or approximates) the associated saturated probability $p_\Omega(= \mathbf{P}(D^*_\Omega \mid \mathbf{H}))$.

(7) S finds that the saturated probability p_Ω is sufficiently small.

(8) S identifies side information **I** and confirms that **I** satisfies the conditional independence condition, that is, that for any subinformation **J** generated by **I**, **J** is conditionally independent of E given **H**, that is, $\mathbf{P}(E \mid \mathbf{H} \,\&\, \mathbf{J}) = \mathbf{P}(E \mid \mathbf{H})$.

(9) With respect to a bounded complexity measure $\mathbf{\Phi} = (\phi, \lambda)$ that characterizes S's problem-solving capability, S confirms that D and **I** together satisfy the tractability condition, that is, that the problem of formulating the pattern D from the side information **I** is tractable, or equivalently, $\mathbf{\Phi}(D \mid \mathbf{I}) < \lambda$.

(10) S is warranted in inferring that E did not occur according to the chance hypothesis **H**.[5]

Dembski's analysis formalizes what Leslie had grasped in an intuitive way. What makes an explanation a tidy one is not simply the fact that the *explanandum* is some improbable event, but the fact that the event also conforms to some independently given pattern, resulting in what Dembski calls "specified complexity." It is this specified complexity that tips us off to the need for an explanation in terms of more than mere chance.

The teleological argument from the fine-tuning of the Universe can be formulated along the lines of Dembski's Chance Elimination Argument. Corresponding to step (1) is the discovery by the astrophysicist that various constants and physical quantities present at the inception of the Universe possess certain values. This will be the event E that the proponent of the teleological argument maintains is best explained in terms of design. The point of step (2) is to eliminate physical necessity as the preferred explanation of the various constants and quantities. By eliminating physical necessity the astrophysicist will narrow the alternatives to the chance hypothesis **H** or to design. In the case at hand, the astrophysicist will have to determine the plausibility of there being some theory that would reveal that the constants and quantities possess the values they do of physical necessity, that is to say, no other universe having similar constants and quantities but with different values could exist. If there is no such theory and the Universe is not the product of design, then **H** will be the hypothesis that the values possessed by the constants and quantities are the result of sheer accident. Some might arise due to random, symmetry-breaking quantum transitions from the early condition of the Universe described by various Grand Unified Theories; others may simply be brute givens obtaining at the origin of the Universe.

Corresponding to step (3) will be the astrophysicist's identification of the fine-tuning of the constants and quantities necessary for the existence of intelligent, carbon-based life. The fine-tuning constitutes a pattern D that not only delimits E, but matches E, that is to say, the event of the occurrence of the initial conditions of the Universe is identical to the patterned event D* of the occurrence of fine-tuning. In step (4) the astrophysicist calculates the probability p of the fine-tuning event's occurring by chance alone. This will involve the probability not only of each value's occurring by chance, but also of the co-occurrence of separate values, since the fine-tuning often involves ratios between separate values.

Step (5) involves fixing the probabilistic resources available for the occurrence of fine-tuning. Here the issue of the Many-Worlds Hypothesis arises, for the defender of chance will seek to augment his probabilistic resources **Ω** as greatly as possible in order to secure a reasonable probability or even a guarantee of the chance occurrence of cosmic fine-tuning. The defender of the design hypothesis will look skeptically upon unwarranted attempts to expand **Ω** beyond the one universe we know to exist. If we reject the Many-Worlds Hypothesis, then the so-called saturated event $D^*_\Omega = E_\Omega$, which comprises all the ways D* might occur relative to the probabilistic resources, will just be D*, since **Ω** is the Universe; therefore the associated saturated probability p_Ω will be identical to $p = \mathbf{P}(D^* \mid \mathbf{H}) = \mathbf{P}(E \mid \mathbf{H})$. Because of the uniqueness of the Universe and of the event E in step (1) (namely, the existence of fine-tuning), step (6) thus becomes redundant.

Step (7) requires that the probability p be sufficiently small. The probability of the initial conditions of the Universe occurring by chance falls well below the bound that Dembski sets in his Law of Small Probabilities. The question raised in step (8) is whether the pattern discerned in step (3) is detachable from the event of the various constants and quantities possessing certain values. Here the side information **I** is supplied to the astrophysicist by biologists and doctors, who know the conditions under which human existence is possible. Given this knowledge, physicists could have predicted the fine-tuning of the Universe prior to its discovery. Thus, both the conditional independence and tractability conditions in steps (8) and (9) are satisfied. As a result the astrophysicist will be warranted in inferring that the initial conditions of the Universe are not the result of chance.

Less formally, the teleological argument will look like this:

(1) One learns that the physical constants and quantities given in the Big Bang possess certain values.
(2) Examining the circumstances under which the Big Bang occurred, one finds that there is no theory that would render physically necessary

the values of all the constants and quantities, so they must be attributed to sheer accident.

(3) One discovers that the values of the constants and quantities are fantastically fine-tuned for the existence of intelligent, carbon-based life.

(4) The probability of each value and of all the values together occurring by chance is vanishingly small.

(5) There is only one universe; it is illicit in the absence of evidence to multiply one's probabilistic resources (i.e. postulate a World Ensemble of universes) simply to avert the design inference.

(6) Given that the Universe has occurred only once, the probability that the constants and quantities all possess the values they do remains vanishingly small.

(7) This probability is well within the bounds needed to eliminate chance.

(8) One has physical information concerning the necessary conditions for intelligent, carbon-based life (e.g. a certain temperature range, the existence of certain elements, certain gravitational and electromagnetic forces, etc.).

(9) This information about the finely tuned conditions requisite for a life-permitting universe is independent of the pattern discerned in step (3).

(10) One is 'warranted in inferring' that the physical constants and quantities given in the Big Bang are not the result of chance.

The design argument examined

Is the teleological argument a sound and persuasive argument? Regardless of whether one adopts Dembski's analysis of design inferences or chooses some alternate approach, the key to detecting design will in any case be eliminating the two competing hypotheses of physical necessity and chance. Step (2) of the argument formulated along Dembski's lines aims to eliminate the hypothesis of physical necessity and steps (3)–(9) are intended to eliminate the hypothesis of chance. Given the uncontroversial step (1), can these alternatives be excluded?

The issue raised by step (2) with respect to the fine-tuning of the Universe will be whether there is some unknown theory that would explain the way the Universe is. According to this alternative, the Universe has to be the way it is, and there was really no chance or little chance of the Universe's not being life-permitting. Now on the face of it, this alternative seems extraordinarily implausible. It requires us to believe that a life-prohibiting universe is virtually physically impossible. But surely it does seem possible. If the primordial matter and anti-matter had been differently proportioned, if the Universe had expanded just a little more slowly, if the entropy of the Universe were marginally greater – any of these adjustments and more would have prevented a life-permitting universe, yet all seem perfectly possible physically. The person who maintains that the Universe must be life-permitting is taking a radical line that requires strong proof. But there is none; this alternative is simply put forward as a bare possibility.

Moreover, there is good reason to reject this alternative. First, there are models of the Universe that are different from the model of the existing universe. As John Leslie explains,

> The claim that blind necessity is involved – that universes whose laws or constants are slightly different "aren't real physical possibilities"...is eroded by the various physical theories, particularly theories of random symmetry breaking, which *show* how a varied ensemble of universes might be generated.
>
> (1989: 202)

If, as Leslie intimates, quantum indeterminacy is ontic, rather than merely epistemic, then it *must* be possible for the Universe to be different than it is, since a number of physical variables depend upon quantum processes that are random in nature.

Second, even if the laws of nature were necessary, one would still have to supply initial conditions. As Paul Davies states,

> Even if the laws of physics were unique, it doesn't follow that the physical universe itself is unique...the laws of physics must be augmented by cosmic initial conditions....There is nothing in present ideas about "laws of initial conditions" remotely to suggest that their consistency with the laws of physics would imply uniqueness. Far from it....
>
> It seems, then, that the physical universe does not have to be the way it is: it could have been otherwise.
>
> (1992: 169)

The extraordinarily low entropy condition of the early universe would be a good example of an arbitrary quantity that seems to have just been put in at the creation as an initial condition. Sometimes it is said that we really do not know how much certain constants and quantities could have varied from their actual values. But this admitted uncertainty becomes less important when the number of variables to be fine-tuned is high. For example, the chances of all fifty known variables being finely tuned, even if each variable has a 50 percent chance of being its actual value, is less than three out of 10^{17}.

Finally, if there is a single, physically possible universe, then the existence of this incredibly complex world-machine might be itself powerful evidence that a designer exists. Some theorists call the hypothesis that the Universe must be life-permitting "the strong anthropic principle," and it is often taken as indicative of God's existence. As physicists Barrow and Tipler write, "The

Strong Anthropic Principle...has strong teleological overtones. This type of notion was extensively discussed in past centuries and was bound up with the question of evidence for a Deity" (1986: 28). Thus, the alternative of physical necessity is not very plausible to begin with and is perhaps indicative of design.

What, then, about the alternative of chance? Step (3) of the argument is, I think, uncontroversial. The pattern will be that range of values close to the present values that are life-permitting.

Step (4) raises philosophical issues concerning probability. For example, it is sometimes alleged that it is meaningless to speak of the probability of our finely tuned universe existing because there is, after all, only one universe. Therefore, one cannot significantly speak of how frequently universes turn out to be finely tuned like ours. I must confess that I have never understood why some people find this objection persuasive. John Barrow (1988) provides the following illustration that makes quite clear the sense in which our life-permitting universe is improbable. Take a sheet of paper and place upon it a red dot. That dot represents our universe. Now alter slightly one or more of the finely tuned constants and physical quantities that have been the focus of our attention. As a result we have a description of another universe, which we may represent as a new dot in the proximity of the first. If that new set of constants and quantities describes a life-permitting universe, make it a red dot; if it describes a universe that is life-prohibiting, make it a blue dot. Now repeat the procedure arbitrarily many times until the sheet is filled with dots. What one winds up with is a sea of blue with only a few pinpoints of red. That is the sense in which it is overwhelmingly improbable that the Universe should be life-permitting. There are simply vastly more life-prohibiting universes in our local area of possible universes than there are life-permitting universes.

It might be objected that we do not know if all these possible universes are equally probable. But this is merely to repeat the objection already dealt with in step (2), the claim that the actual range of possible values for a certain constant or quantity may be very narrow. As we saw, even if that were the case, when one has many variables requiring fine-tuning, the probability of a life-permitting universe existing is still very small. Moreover, in the absence of any physical reason to think that the values are constrained, we are justified in assuming a principle of indifference to the effect that the probability of our universe existing will be the same as the probability of any other universe existing that is represented on our sheet.[6]

It might be demanded why we should consider only universes represented on the sheet. Perhaps universes are possible that have wholly different physical variables and natural laws and are life-permitting. Perhaps these would contain forms of life vastly different from life as we know it. The teleologist need not deny the possibility, for such worlds are irrelevant to his argument. All one needs to show is that our universe is highly improbable within the local group of possible worlds. John Leslie gives the illustration of a fly resting on a large, blank area of the wall (Leslie 1989: 17). A shot is fired, and the bullet strikes the fly. Now even if the rest of the wall outside the blank area is covered with flies, such that a randomly fired bullet would probably hit one, nevertheless it remains highly improbable that a single, randomly fired bullet would strike the solitary fly within the large, blank area. In the same way, we need only concern ourselves with the universes represented on our sheet in order to determine the conditional probability of the universe's being finely tuned for intelligent, carbon-based life.[7]

Step (4) also raises issues pertinent to the so-called anthropic principle. As formulated by Barrow and Tipler, the anthropic principle states:

> The observed values of all physical and cosmological quantities are not equally probable, but they take on values restricted by the requirement that there exist sites where carbon-based life can evolve and by the requirement that the Universe be old enough for it to have already done so.
>
> (1986: 15)

Unfortunately this statement of the principle is very unclear. But the thrust of it seems to be that the observed fine-tuning of the Universe is not really improbable, due to the selection effect arising from our role as observers. Barrow and Tipler regard the anthropic principle as "just a restatement…of one of the most important and well-established principles of science: that it is essential to take into account the limitations of one's measuring apparatus when interpreting one's observations" (1986: 23). For example, if we were calculating the fraction of galaxies that lie within certain ranges of brightness, our observations would be biased towards the brighter ones, since we cannot see the dim ones so easily. Or again, a rat catcher may say that all rats are bigger than six inches because that is the size of his traps. Similarly, any observed properties of the Universe that may initially appear astonishingly improbable can only be seen in their true perspective after we have accounted for the fact that we would be unable to observe certain properties, were they to obtain, because we can only observe those compatible with our own existence:

> The basic features of the Universe, including such properties as its shape, size, age, and laws of change must be *observed* to be of a type that allows the evolution of the observers, for if intelligent life did not evolve in an otherwise possible universe, it is obvious that no one would be asking the reason for the observed shape, size, age, and so forth of the universe.
>
> (Barrow and Tipler 1986: 1–2)

Thus, our own existence acts as a selection effect in assessing the various features of the Universe. For example, a life form that evolved on an Earth-like planet "must necessarily see the Universe to be at least several billion years old and...several billion light years across," for this is the time necessary for the production of the elements essential to life and so forth (Barrow and Tipler 1986: 3).

Barrow and Tipler contend that the anthropic principle has "far-reaching implications" (1986: 2). The implication is that we ought not to be surprised at observing the Universe to be as it is and that therefore no explanation of its fine-tuning need be sought. Thus they say: "No one should be surprised to find the universe to be as large as it is" (Barrow and Tipler 1986: 18). Or again: "on Anthropic grounds, we should expect to observe a world possessing precisely three spatial dimensions" (Barrow and Tipler 1986: 247). Or again:

> We should emphasize once again that the enormous improbability of the evolution of intelligent life in general and *Homo sapiens* in particular does not mean we should be amazed we exist at all....Only if an intelligent species does evolve is it possible for its members to ask how probable it is for an intelligent species to evolve.
>
> (Barrow and Tipler 1986: 566)

If the probability at stake in the teleological argument is epistemic probability (which is a measure of the degree to which we may rationally expect some proposition to be true), then Barrow and Tipler's statements may be taken as the claim that the fine-tuning of the Universe is not, despite appearances, really improbable after all.

If this is their claim, then it is based on confusion. They have confused the true claim

(1) if observers who have evolved within a universe observe its fundamental constants and quantities, it is highly probable that they will observe them to be fine-tuned to their existence

with the false claim

(2) it is highly probable that a universe exists that is finely tuned for the existence of observers who have evolved within it.

An observer who has evolved within a universe should regard it as highly probable that he will find the basic conditions of that universe fine-tuned for his existence; but he should not infer that it is therefore highly probable that such a fine-tuned universe exists. It is true that

(3) we should not be surprised that we do not observe that the fundamental features of the Universe are not fine-tuned for our own existence.

For if the fundamental features of the Universe were not fine-tuned for our existence, we should not be here to notice it. Hence, it is not surprising that we do not observe such features. But it does not follow that

(4) we should not be surprised that we do observe that the fundamental features of the Universe are fine-tuned for our existence.

This can be clearly seen by means of another illustration borrowed from John Leslie (1989: 13–14). Suppose you are dragged before a firing squad of 100 trained marksmen, all of them with rifles aimed at your heart, to be executed. The command is given; you hear the deafening sound of the guns. *And you observe that you are still alive*, that all of the 100 marksmen missed! Now while it is true that

(5) you should not be surprised that you do not observe that you are dead,

nonetheless it is equally true that

(6) you should be surprised that you do observe that you are alive.

Since the firing squad's missing you altogether is extremely improbable, the surprise expressed in (6) is wholly appropriate, though you are not surprised that you do not observe that you are dead, since if you were dead you could not observe it. Similarly, while we should not be surprised that we do not observe that the fundamental features of the Universe are not fine-tuned for our existence, it is nevertheless true that

(7) we should be surprised that we do observe that the fundamental features of the Universe are fine-tuned for our existence,

in view of the enormous improbability that the Universe should possess such features.

The reason that the falsity of (7) does not follow from (3) is that subimplication fails for first-order predicate calculus. For (3) may be schematized as

3'. $\sim S{:}\ (x)\ ([Fx \bullet \sim Tx] \rightarrow \sim Ox)$

where "S:" is an operator expressing "we should be surprised that," "*F*" is "is a fundamental feature of the Universe," "*T*" is "is fine-tuned for our existence," and "*O*" is "is observed by us." And (7) may be schematized as

7'. S: $\exists(x)\,(Fx \cdot Tx \cdot Ox)$

It is clear that the object of surprise in (7') is not equivalent to the object of surprise in (3'); therefore the truth of (3') does not entail the negation of (7').

Therefore, the use of the anthropic principle to stave off our surprise at the fine-tuning of the Universe fails. It does not follow from the anthropic principle that our surprise at the fine-tuning of the Universe is unwarranted or that the existence of a life-permitting universe is not highly improbable.

Most anthropic theorizers now recognize that the anthropic principle can only legitimately be employed when it is conjoined to a Many-Worlds Hypothesis, according to which a World Ensemble of concrete worlds exist, actualizing a wide range of possibilities. This takes us to step (5) of our argument, for the Many-Worlds Hypothesis is essentially an effort on the part of partisans of the chance hypothesis to multiply the probabilistic resources in order to reduce the improbability of the occurrence of fine-tuning. The very fact that detractors of design have to resort to such a remarkable hypothesis underlines the point that cosmic fine-tuning is not explicable in terms of physical necessity alone or in terms of sheer chance in the absence of a World Ensemble. The Many-Worlds Hypothesis is a sort of backhanded compliment to the design hypothesis in its recognition that fine-tuning cries out for explanation. But is the Many-Worlds Hypothesis as plausible as the design hypothesis?

It seems not. In the first place, it needs to be recognized that the Many-Worlds Hypothesis is no more scientific, and no less metaphysical, than the hypothesis of a Cosmic Designer. As the scientist-theologian John Polkinghorne says, "People try to trick out a 'many universe' account in sort of pseudo-scientific terms, but that is pseudo-science. It is a metaphysical guess that there might be many universes with different laws and circumstances" (1995: 6). But as a metaphysical hypothesis, the Many-Worlds Hypothesis is arguably inferior to the design hypothesis because the design hypothesis is simpler. According to Ockham's razor, we should not multiply causes beyond what is necessary to explain the effect. But it is simpler to postulate one Cosmic Designer to explain our universe than to postulate the infinitely bloated and contrived ontology of the Many-Worlds Hypothesis. Only if the Many-Worlds theorist could show that there exists a single, comparably simple mechanism for generating a World Ensemble of randomly varied universes would he be able to elude this difficulty. But no one has been able to identify such a mechanism. Therefore, the design hypothesis is to be preferred.

Second, there is no known way for generating a World Ensemble. No one has been able to explain how or why such a collection of universes should exist. Moreover, those attempts that have been made require fine-tuning themselves. For example, although some cosmologists appeal to inflationary theories of the Universe to generate a World Ensemble, we have seen that inflation itself requires fine-tuning. As Robert Brandenburger of Brown University writes, "The field which drives inflation…is expected to generate an unacceptably large cosmological constant *which must be tuned to zero by hand*. This is a problem which plagues *all* inflationary universe models."[8]

Third, there is no evidence for the existence of a World Ensemble *apart from the concept of fine-tuning itself*. But fine-tuning is equally evidence for a Cosmic Designer. Indeed, the hypothesis of a Cosmic Designer is again the better explanation because we have independent evidence of the existence of such a Designer in the form of the other arguments for the existence of God.

Fourth, the Many-Worlds Hypothesis faces a severe challenge from biological evolutionary theory.[9] First, a bit of background. The nineteenth-century physicist Ludwig Boltzmann proposed a sort of Many-Worlds Hypothesis in order to explain why we do not find the Universe in a state of "heat death" or thermodynamic equilibrium (1964: 446–8). Boltzmann hy-pothesized that the Universe as a whole *does*, in fact, exist in an equilibrium state, but that over time fluctuations in the energy level occur here and there throughout the Universe, so that by chance alone there will be isolated regions where disequilibrium exists. Boltzmann referred to these isolated regions as "worlds." We should not be surprised to see our world in a highly improbable disequilibrium state, he maintained, since in the ensemble of all worlds there must exist by chance alone certain worlds in disequilibrium, and ours just happens to be one of these.

The problem with Boltzmann's daring Many-Worlds Hypothesis was that if our world were merely a fluctuation in a sea of diffuse energy, then it is overwhelmingly more probable that we should be observing a much tinier region of disequilibrium than we do. In order for us to exist, a smaller fluctuation, even one that produced our world instantaneously by an enormous accident, is inestimably more probable than a progressive decline in entropy to fashion the world we see. In fact, Boltzmann's hypothesis, if adopted, would force us to regard the past as illusory, everything having the mere appearance of age, and the stars and planets as illusory, mere "pictures" as it were, since that sort of world is vastly more probable given a state of overall equilibrium than a world with genuine, temporally and spatially distant events. Therefore, Boltzmann's Many-Worlds Hypothesis has been universally rejected by the scientific community, and the present disequilibrium is usually taken to be just a result of the initial low-entropy condition mysteriously obtaining at the beginning of the Universe.

Now a precisely parallel problem attends the Many-Worlds Hypothesis as an explanation of fine-tuning. According to the prevailing theory of biological evolution, intelligent life like ourselves, if it evolves at all, will do so as late in the lifetime of the Sun as possible. The less the time span available for the mechanisms of genetic mutation and natural selection to function, the lower the probability of intelligent life's evolving. Given the complexity of the human organism, it is overwhelmingly more probable that human beings will evolve late in the lifetime of the Sun rather than early. In fact Barrow and Tipler (1986: 561–5) list ten steps in the evolution of *Homo sapiens each of which* is so improbable that before it would occur the Sun would have ceased to be a main-sequence star and incinerated the Earth! Hence, if our universe is but one member of a World Ensemble, then it is overwhelmingly more probable that we should be observing a very old Sun rather than a relatively young one. If we are products of biological evolution, we should find ourselves in a world in which we evolve much later in the lifetime of our star. (This is the analogue to its being overwhelmingly more probable, according to the Boltzmann hypothesis, that we should exist in a smaller region of disequilibrium.) In fact, adopting the Many-Worlds Hypothesis to explain away fine-tuning also results in a strange sort of illusionism: it is far more probable that all our astronomical, geological, and biological estimates of age are wrong, that we really do exist very late in the lifetime of the Sun and that the Sun and the Earth's appearance of youth is a massive illusion. (This is the analogue of it's being far more probable, according to the Boltzmann hypothesis, that all the evidence of the old age of our universe is illusory.) Thus, the Many-Worlds Hypothesis is no more successful in explaining cosmic fine-tuning than it was in explaining cosmic disequilibrium.

The error made by the Many-Worlds Hypothesis is that it multiplies one's probabilistic resources without warrant. If we are allowed to do that, then it seems that *anything* can be explained away.[10] For example, a card player who gets four aces every time he deals could explain this away by saying, "there are an infinite number of universes with poker games going on in them, and therefore in some of them someone always by chance gets four aces every time he deals, and – lucky me! – I just happen to be in one of those universes." This sort of arbitrary multiplying of one's probabilistic resources would render rational conduct impossible. Thus, the Many-Worlds Hypothesis collapses and along with it the alternative of chance that it sought to rescue.

In step (6) of the argument, we see that in the absence of a World Ensemble, the so-called saturated probability of the occurrence of fine-tuning just remains the probability of the occurrence of fine-tuning. Since there is only one universe, there is only one replication of this event; it is unique. Thus, the very uniqueness of the Universe, to which partisans of chance appealed in an effort to undermine the significance of the notion of the probability of the Universe being fine-tuned for life, here returns to render a further calculation in light of wider probabilistic resources redundant.

Step (7) is incontrovertible if what we have said until now is generally correct. Dembski sets a universal probability bound $\delta = 1/2 \times 1/10^{150}$. If the specified probability of some event occurring is less than δ relative to the probabilistic resources, then we may be certain that that event never happens by chance. Unfortunately, Dembski's computation of δ is inapplicable to our case, since it is calculated on the basis of such factors as the number of elementary particles in the Universe, the rate at which physical states can change, and so on, considerations that are inapplicable when one is considering the initial state of the Universe. But if the chances of events occurring are so fantastically low that it would not occur even given the resources of 10^{80} elementary particles, 10^{45} changes per second, and 10^{25} seconds, then it would also not occur when these quantities are reduced to (virtually) zero. Minimally, we can say that the probability of the chance hypothesis on the fine-tuning of the Universe is so low that, unless the design hypothesis is almost impossible, then the design hypothesis is vastly more probable on the fine-tuning of the Universe and therefore rationally to be preferred.[11]

Step (8) I take to be uncontroversial. One can determine the range of life-permitting physical conditions independently of knowledge of the initial state of the Universe. The probability of the Universe's initial state being such as it was is not affected by the additional information that such a state is life-permitting. Moreover, the information we possess about the requisite conditions for the existence of intelligent, carbon-based life is sufficient to formulate a pattern that we may use to delimit the initial state of the Universe.

Similarly, step (9) seems unproblematic. Given what we know about carbon-based life, the difficulty of formulating the pattern used to delimit the initial state of the Universe is not great. The pattern of life-permitting conditions thus fulfills all the conditions for being a specification of the occurrence of the initial state of the Universe. The fine-tuning of the Universe is therefore a specified event of very small probability.

It therefore follows that the fine-tuning of the Universe is due neither to physical necessity nor to chance. It is therefore due to design. The implication of this hypothesis is that there exists a Cosmic Designer who fine-tuned the initial conditions of the Universe for intelligent life. Such a hypothesis supplies a personal explanation of the fine-tuning of the Universe. Is this explanation implausible?

Detractors of design sometimes object that the Designer Himself remains unexplained. It is said that an intelligent Mind also exhibits complex order, so that if the Universe needs an explanation, so does its Designer. If the Designer does not need an explanation, why think that the Universe does?

This popular objection is based on a misconception of the nature of explanation. It is widely recognized that in order for an explanation to be the best, one need not have an explanation of the explanation (indeed, such a requirement would generate an infinite regress, so that everything becomes inexplicable).[11] If the best explanation of a disease is a previously unknown virus, doctors need not be able to explain the virus in order to know that it caused the disease. If archaeologists determine that the best explanation of the existence certain artifacts is a lost tribe of ancient people, we need not be able to explain the tribe's origin in order to say justifiably that the tribe produced the artifacts. If astronauts should find traces of intelligent life on some other planet, we need not be able to explain such extraterrestrials in order to recognize that they are the best explanation. In the same way, the design hypothesis' being the best explanation of fine-tuning doesn't depend on our being able to explain the Designer.

Moreover, the complexity of a mind is not really analogous to the complexity of the Universe. A mind's *ideas* may be complex, but a mind itself is a remarkably simple thing, being an immaterial entity not composed of parts. Furthermore, a mind, in order to be a mind, must have certain properties like intelligence, consciousness, and volition. These are not contingent properties that it might lack, but are essential to its nature. So it's difficult to see any analogy between the contingently complex universe and a mind. Detractors of design have evidently confused a mind's thoughts (which may be complex) with the mind itself (which is pretty simple). Postulating an uncreated Mind behind the cosmos is thus not at all like postulating an undesigned cosmos.

Conclusion

It seems to me that the teleological argument for a Designer of the cosmos based on the fine-tuning of the initial state of the Universe is thus both a sound and persuasive argument. It can be formulated as follows:

(1) The fine-tuning of the initial state of the Universe is due to either physical necessity, chance, or design.
(2) It is not due to physical necessity or chance.
(3) Therefore, it is due to design.

Dembski's analysis of the design inference provides a fruitful approach to the defense of premise (2). But we should not think that the argument depends crucially on his approach; other approaches have been made as well.[12] In each case, similar evidence and arguments such as I have laid out will appear within different frameworks to support the teleological argument for a Designer of the Universe.

Notes

1 I depend on the impressive compilations by Barrow and Tipler (1986) and Leslie (1988, 1989). Detailed discussion and documentation may be found there.
2 For a good discussion of these two types of explanation, see Swinburne (1991: 32–48).
3 Look at "stories" in the Index of Concepts in Leslie (1989: 225).

4 Dembski (1998: 167–74). For discussion see Fitelson *et al.* (1999), Sober (1999), Dembski (forthcoming), Dembski (2002), Collins (2001), and Dembski (2001).
5 Less formally we may represent Dembski's analysis as follows:

(1) One learns that some event has occurred.
(2) Examining the circumstances under which the event occurred, one finds that the event (if not the result of intelligent design) could only have been produced by a certain chance process (or processes).
(3) One identifies a pattern that characterizes the event.
(4) One calculates the probability of the event given the chance hypothesis.
(5) One determines what probabilistic resources were available for producing the event via the chance hypothesis.
(6) On the basis of the probabilistic resources, one calculates the probability of the event occurring by chance once out of all the available opportunities to occur.
(7) One finds that the above probability is sufficiently small.
(8) One identifies a body of information that is independent of the event's occurrence.
(9) One determines that one can formulate the pattern referred to in step (3) on the basis of this body of independent information.
(10)One is warranted in inferring that the event did not occur by chance.

6 See Lipton (1991).
7 See Collins (1999).
8 For further discussion of this issue see Timothy McGrew, Lydia McGrew, and Eric Vestrup (in this volume) as well as Collins (1999: 68–70).
9 Robert Brandenburger, personal communication.
10 I am indebted to Robin Collins for this point. Although Collins disagrees that this undesirable consequence follows, it does seem to me that the Many-Worlds theorist commits what Ian Hacking has called the Inverse Gambler's Fallacy. Just as the fallacy-prone gambler thinks that because the chances of rolling double-sixes are one in thirty-six and he has rolled the dice thirty-five times without double-sixes he therefore has a high probability of rolling double sixes on the next throw, so the Many-Worlds theorist thinks that the chances of our world's being fine-tuned are greater if there have already been a huge number of rolls (i.e. many worlds). But the odds of the gambler rolling double-sixes or our world being finely tuned are not affected by the other's rolls that have taken place. We still want to know why *this* universe is fine-tuned. To say that the probability that *this universe is fine-tuned* is high because there are many other universes is to commit the Gambler's Fallacy, just as the gambler who thinks that because thirty-five rolls of the dice have been made without double-sixes therefore the chances are high that *this roll will yield double-sixes*. To think that because the Universe's fine-tuning appears astonishingly improbable there must exist other universes is to commit the Inverse Gambler's Fallacy, just as the gambler who thinks that because double-sixes have just been thrown there must be many other gamblers also throwing dice.
11 See Collins (1999: 51–3).
12 See Lipton (1991).
13 Collins (1999) prefers a Bayesian approach; see also Ratzsch (2001).

References

Barrow, J.D. (1988) *The World within the World*, Oxford: Clarendon Press.

Barrow, J.D. and Tipler, F.J. (1986) *The Anthropic Cosmological Principle*, Oxford: Clarendon Press.

Boltzmann, L. (1964) *Lectures on Gas Theory*, trans. S.G. Brush, Berkeley: University of California Press.

Collins, R. (2001) "An evaluation of William A. Dembski's *The Design Inference*: A review essay," *Christian Scholar's Review* 30(3): 329–41.

—— (1999) "A scientific argument for the existence of God: The fine-tuning design argument," in M.J. Murray (ed.) *Reason for the Hope Within*, Grand Rapids, Michigan: Wm. B. Eerdmans, pp. 67–72.

Davies, P. (1992) *The Mind of God*, New York: Simon & Schuster.

Dembski, W. (forthcoming) "Intelligent design and its theoretical underpinnings: A response to Elliott Sober."

—— (2002) *No Free Lunch: Why Specified Complexity Cannot be Purchased without Intelligence*, Lanham, Maryland: Rowman & Littlefield.

—— (2001) "Detecting design by eliminating chance: A response to Robin Collins," *Christian Scholar's Review* 30: 343–57.

—— (1998) *The Design Inference: Eliminating Chance through Small Probabilities*, Cambridge: Cambridge University Press.

Earman, J. and Mosterin, J. (1999) "A critical look at inflationary cosmology," *Philosophy of Science* 66: 1–49.

Fitelson, B., Stephens, C., and Sober, E. (1999) "How not to detect design – critical notice: William A. Dembski, *The Design Inference*," *Philosophy of Science* 66: 472–88.

Guth, A., (1997) *The Inflationary Universe*, Reading Massachusetts: Addison-Wesley.

Hawking, S., (1988) *A Brief History of Time*, New York: Bantan Books.

Kolmogorov, A. (1965) "Three approaches to the quantitative definition of information," *Problemy Peredachi Informatsii* (in translation) 1(1): 3–11.

Leslie, J. (1989) *Universes*, London: Routledge.

—— (1988) "The prerequisites of life in our universe," in G.V. Coyne, M. Heller, and J. Zycinski (eds) *Newton and the New Direction in Science*, Vatican City State: Vatican Observatory Press, pp. 229–58.

Lipton, P. (1991) *Inference to the Best Explanation*, London: Routledge.

Penrose, R. (1981) "Time-asymmetry and quantum gravity," in C.J. Isham, R. Penrose, and D.W. Sciama (eds) *Quantum Gravity 2*, Oxford: Clarendon Press pp. 245–72.

Polkinghorne, J. (1995) *Serious Talk: Science and Religion in Dialogue*, Harrisburg, Pennsylvania: Trinity Press International.

Ratzsch, D. (2001) *Nature, Design, and Science: The Status of Design in Natural Science*, Albany, New York: State University of New York Press.

Sober, E. (1999) "Testability," *Proceedings and Addresses of the American Philosophical Association* 73(2): 47–76.

Swinburne, R. (1991) *The Existence of God*, Oxford: Clarendon Press.

Tennant, F.R. (1935) *Philosophical Theology*, vols 1 and 2, Cambridge: Cambridge University Press.

9

EVIDENCE FOR FINE-TUNING[1]

Robin Collins

Introduction

Perhaps the most widely discussed argument from design in the last thirty years has been that based on the fine-tuning of the cosmos for life. The literature presenting the evidence for fine-tuning is fairly extensive, with books by theoretical physicist Paul Davies (1982), physicists John Barrow and Frank Tipler (1986), astrophysicist Martin Rees (2000), and philosopher John Leslie (1989) being some of the most prominent. Yet despite this abundance of literature, several leading scientists are still skeptical of the purported evidence of fine-tuning. Nobel Prize-winning physicist Steven Weinberg, for instance, says that he is "not impressed with these supposed instances of fine-tuning" (1999: 46). Other physicists, such as MIT's astrophysicist Alan Guth, have presented similar reservations.[2] As explicated in the Appendix, there is some basis for this skepticism. The arguments for some of the most widely cited cases of purported fine-tuning are highly problematic.

To counter-act this form of skepticism, I will present six of what seem to me to be among the strongest cases of fine-tuning. The way in which I judge the strength of a particular case of purported fine-tuning is primarily based on how secure the physical calculations or types of reasoning are behind the case of fine-tuning in question. The existence of secure cases of fine-tuning, along with philosopher John Leslie's comment that "clues heaped upon clues can constitute weighty evidence despite doubts about each element in the pile" (1988: 300), should go a long way towards answering these critics. Before looking at the evidence, however, we first need a rough definition of fine-tuning, and some important criteria for when a parameter of physics can be considered fine-tuned in the sense relevant to the design argument (or the argument for many universes).

Definition of and criteria for fine-tuning

As a first approximation, we can think of the claim that a parameter of physics is "fine-tuned" as the claim that the range of values, r, of the parameter that is life-permitting is very small compared with some non-arbitrarily chosen theoretically "possible" range of values R. The degree of fine-tuning could then be defined as the ratio of the width of the life-permitting region to the comparison region. (Hereafter, when we speak of life, the life-permitting region, etc., the type of life we have in mind is life of comparable intelligence to human beings since it is the existence of intelligent observers that is relevant in fine-tuning arguments.) In each of the cases below, we will indicate how the physical situation itself suggests a plausible lower bound for the overall range R, and hence a plausible lower bound for the degree of fine-tuning. A full treatment of the fine-tuning, however, would need to justify the choice of the lower bound of the overall range R in considerably more depth, but that is beyond the scope this chapter.[3]

The inference to design (or to many universes) does not require that a parameter be fine-tuned in the full sense as defined above. First, all that needs to be shown is that conditions would be much less *optimal* for the evolution of intelligent life if a parameter were to have fallen outside of some narrow range r (as compared to the theoretically "possible" range R). The reason for this has to do with the nature of the inference to design (or to many universes). The inference to design or many universes typically involves two steps. First, the claim is made that it is very coincidental, or surprising, for some parameter of physics to fall within the life-permitting range (instead of somewhere else in the theoretically "possible" range R) under the non-design, non-many-universe hypothesis, but *not* surprising or coincidental under the design or many-universe hypothesis. Then a general rule of confirmation is implicitly or explicitly invoked, according to which if a body of evidence E is highly surprising or coincidental under hypothesis H_2, but not under hypothesis H_1, then that body of evidence E confirms H_1 over H_2: that is, E gives us significant reason to prefer H_1 over H_2.[4] The surprisingness or coincidental character of the values of the parameters of physics, however, would still remain if their actual values were merely *optimal* for the evolution of intelligent life, and thus the soundness of the above inference would be unaffected. Furthermore, this optimality criterion largely avoids the objections based on the possibilities of non-carbon-based life-forms. A change in a parameter that decreased the likelihood of carbon-based intelligent life-forms would clearly be less optimal for intelligent life unless it resulted in a compensating increase in the likelihood of other kinds of intelligent life, such as those based on silicon or liquids other than water. But this is highly unlikely, given the well-known difficulties involved in the existence of any kind of alternative to carbon-based life.

Second, all we actually need to show is that a parameter falls near the edge of the life-permitting region, not that the life-permitting region is small compared to some non-arbitrarily defined region R. For example, as I will discuss below, we only have well-developed reasons to believe that a *relatively small* decrease in the weak-force strength would severely inhibit the possibility of life. Thus at present we only have a solid argument for what I will call *one-sided* fine-tuning of the weak force, instead of what could be called *two-sided* fine-tuning, in which either a decrease or an increase in the strength of the weak force would be severely life-inhibiting. (The way we defined fine-tuning above was essentially as two-sided fine-tuning.)

The basic reason we only need evidence for one-sided fine-tuning is that it still seems highly coincidental for a parameter to fall very *near* the edge of the life-permitting region (instead of somewhere else in the comparison region R) under the non-design, non-many-universes hypothesis.[5] However, it does not seem highly coincidental under the joint hypothesis of design *and* two-sided fine-tuning (or many universes and two-sided fine-tuning). The reason for this is that the existence of two-sided fine-tuning implies that all life-permitting values are near the edge (since the life-permitting region is so small), and the design hypothesis renders it not coincidental that the parameter is in the life-permitting region. So, taken together, these two hypotheses remove the coincidence of a parameter falling near the edge of the life-permitting region. Thus, by the rule of inference mentioned above, the existence of one-sided fine-tuning confirms the joint hypothesis of design and two-sided fine-tuning over the non-design, non-many-universes hypothesis.

It is worth noting that these explanations of one-sided fine-tuning (design and two-sided fine-tuning, and many universes and two-sided fine-tuning) are at least in part testable, since they lead us to expect the existence of two-sided fine-tuning in those cases in which there is a significant degree of one-sided fine-tuning. This distinction between one- and two-sided fine-tuning will be very important, since in some of the most important cases of fine-tuning, we only have well-developed arguments for one-sided fine-tuning.

Six solid cases of fine-tuning

Now we are ready to consider six solid cases of fine-tuning. The cases that I discuss are those in which we can make a quantitative estimate of the degree of fine-tuning. There are many other significant, more qualitative, cases such as those presented by Michael Denton (1998) that we will not discuss here. We will begin with the cosmological constant.

The cosmological constant

The smallness of the cosmological constant is widely regarded as the single greatest problem confronting current physics and cosmology. The cosmological constant, Λ, is a term in Einstein's equation that, when positive, acts as a repulsive force, causing space to expand and, when negative, acts as an attractive force, causing space to contract. Apart from some sort of extraordinarily precise fine-tuning or new physical principle, today's theories of fundamental physics and cosmology lead one to expect that the vacuum – that is, the state of space–time free of ordinary matter fields – has an extraordinarily large energy density. This energy density in turn acts as an effective cosmological constant, thus leading one to expect an extraordinarily large effective cosmological constant, one so large that it would, if positive, cause space to expand at such an enormous rate that almost every object in the Universe would fly apart, and would, if negative, cause the Universe to collapse almost instantaneously back in on itself. This would clearly make the evolution of intelligent life impossible.

What makes it so difficult to avoid postulating some sort of highly precise fine-tuning of the cosmological constant is that almost every type of field in current physics – the electromagnetic field, the Higgs fields associated with the weak force, the *inflaton* field hypothesized by inflationary cosmology, the *dilaton* field hypothesized by superstring theory, and the fields associated with elementary particles such as electrons – contributes to the vacuum energy. Although no one knows how to calculate the energy density of the vacuum, when physicists make estimates of the contribution to the vacuum energy from these fields, they get values of the energy density anywhere from 10^{53} to 10^{120} higher than its maximum life-permitting value, Λ_{max}.[6] (Here, Λ_{max} is expressed in terms of the energy density of empty space.)

Although each field contributes in a different way to the total vacuum energy, for the purposes of illustration we will look at just two examples. As our first example, consider the inflaton field of inflationary cosmology. Inflationary Universe models postulate that the inflaton field had an enormously high energy density in the first 10^{-35} to 10^{-37} seconds of our universe (Guth 1997: 185), causing space to expand by a factor of around 10^{60}. By about 10^{-35} seconds or so, however, the value of the inflaton field fell to a relatively small value corresponding to a *local minimum* of energy of the inflaton field. Since the initial energy density was anywhere from $10^{53}\Lambda_{max}$ to $10^{123}\Lambda_{max}$, depending on the inflationary model under consideration, theoretically the local minimum of the inflaton field could be anything from zero to $10^{53}\Lambda_{max}$, or even $10^{123}\Lambda_{max}$ (see Sahni and Starobinsky 1999: section 7.0; Rees 2000: 154).[7] The fact that it is less than Λ_{max}, therefore, suggests a high degree of fine-tuning, to at least one part in 10^{53}.[8]

A similar sort of fine-tuning occurs in the case of the symmetry breaking of the weak force in the widely accepted Weinberg–Salem–Glashow electroweak theory. According to this theory, the electromagnetic force and the weak force acted as one force prior to symmetry breaking of a postulated Higgs field in the very early universe when temperatures were still extremely high. Before symmetry breaking, the vacuum energy of the Higgs field had its maximum value V_0. This value was approximately $10^{53}\Lambda_{max}$. After symmetry breaking, the Higgs field falls into some local minimum of energy density, which theoretically could be anywhere from zero to $10^{53}\Lambda_{max}$, being solely determined by V_0 and other free parameters of the electroweak theory.[9] Once again, the fact that this energy density is less than Λ_{max}, instead of somewhere else in the range zero to $10^{53}\Lambda_{max}$, suggests an extraordinarily high degree of fine-tuning.

To account for the near-zero value of the cosmological constant one could hypothesize some unknown physical principle or mechanism which requires that the cosmological constant be zero. One problem with this hypothesis is that recent cosmological evidence from distant supernovae strongly indicates that the effective cosmological constant is not exactly zero (Sahni and Starobinsky 1999; Krauss 1999). Thus the principle or mechanism could not simply be one which specifies that the cosmological constant must be zero, but would have to be one that specified that it be less than some small upper bound. This hypothesis, however, seems to simply relocate the cosmological constant problem to that of explaining why this upper bound is less than Λ_{max} instead of being much, much larger.[10] Second, current inflationary cosmologies require that the effective cosmological constant be relatively large at very early epochs in the Universe, since it is a large cosmological constant that drives inflation. Thus any mechanism that forces it to be zero or near-zero now must allow for it to be large in early epochs. Accordingly, if there is a physical principle that accounts for the smallness of the cosmological constant, it must be (1) attuned to the contributions of every particle to the vacuum energy, (2) only operative in the later stages of the evolution of the cosmos (assuming inflationary cosmology is correct), and (3) something that drives the cosmological constant extraordinarily close to zero, but not exactly zero, which would itself seem to require fine-tuning. Given these constraints on such a principle, it seems that, if such a principle exists, it would have to be "well-designed" (or "fine-tuned") to yield a life-permitting cosmos. Thus, such a mechanism would most likely simply reintroduce the issue of design at a different level.[11]

These difficulties confronting finding a physical principle or mechanism for forcing the cosmological constant to be near-zero have led many cosmologists, most notably Steven Weinberg, to search reluctantly for an anthropic many-universes explanation for its apparent fine-tuning (Weinberg 1987, 1996).

The strong and electromagnetic forces

The strong force is the force that keeps the nucleons – that is, the protons and neutrons – together in an atom. The effect on the stability of the atom of decreasing the strong force is straightforward, since the stability of elements depends on the strong force being strong enough to overcome the electromagnetic repulsion between the protons in a nucleus. A 50 percent decrease in the strength of the strong force, for instance, would undercut the stability of all elements essential for carbon-based life, with a slightly larger decrease eliminating all elements except hydrogen (Barrow and Tipler 1986: 326–7).

Another effect of decreasing the strength of the strong force is that it would throw off the balance between the rates of production of carbon and oxygen in stars, as discussed immediately below. This would have severe life-inhibiting consequences. Although various life-inhibiting effects are claimed for increasing the strength of the strong force, the arguments are not nearly as strong or well developed, except for the one below involving the existence of carbon and oxygen. Furthermore, the argument most commonly cited – namely, that it would cause the binding of the diproton, which would in turn result in an all-helium universe – appears faulty (see the section entitled "The strong force" in the Appendix). At present, therefore, we have a solid argument that a de-crease in the strength of the strong force would be life-forbidding, along with a significant and well-developed argument for two-sided fine-tuning based on the joint production of carbon and oxygen in stars, as explicated in the next section.

Now the forces in nature can be thought of as spanning a range of G_0 to $10^{40}G_0$, at least in one widely used dimensionless measure of the strengths of these forces (Barrow and Tipler 1986: 293–5). (Here, G_0 denotes the strength of gravity, with $10^{40}G_0$ being the strength of the strong force.) If we let the theoretically possible range R of force strengths in nature be the total range of force strengths, then it follows that the degree of one-sided fine-tuning of the strong force is insignificant, being about one part in two by the formula given in note 5. Of course, one might think that it is likely that the theoretically possible range is much larger than given above, hence making the one-sided fine-tuning much more significant.

Finally, around a fourteen-fold increase in the electromagnetic force would have the same effect on the stability of elements as a 50 percent decrease in the strong force (Barrow and Tipler 1986: 327). Now in the dimensionless units mentioned above, the strength of the electromagnetic force is considered to have a value of approximately $10^{37}G_0$, and hence the upper bound of the life-permitting region is approximately $14 \times 10^{37}G_0$. Consequently, as shown in note 5, this yields a one-sided fine-tuning of approximately one part in a hundred or less.

Carbon production in stars

The first significantly discussed, and probably most famous, case of fine-tuning involves the production of carbon and oxygen in stars. Since both carbon and oxygen play crucial roles in life-processes, the conditions for complex, multicellular life would be much less optimal without the presence of these two elements in sufficient quantities. (For a fairly complete presentation of the reasons for this, see Michael Denton 1998: Chs 5 and 6.) Yet a reasonable abundance of both carbon and oxygen appears to require a fairly precise adjustment of the strong nuclear force, as we will now see.

Carbon and oxygen are produced by the processes of nuclear synthesis in stars via a delicately arranged process. At first, a star burns hydrogen to form helium. Eventually, when enough hydrogen is burnt, the star contracts, thereby increasing the core temperature of the star until helium ignition takes place, which results in helium being converted to carbon and oxygen. This process occurs by helium nuclei colliding first to form beryllium 8 (^{8}Be), which is a metastable nuclei with a half-life of 10^{-17} seconds. During ^{8}Be's short lifespan, it can capture another helium nucleus to form carbon 12. Some of the carbon 12 that is formed is then burnt to oxygen 16 by collisions with other helium nuclei.

Helium burning in stars thus involves two simultaneous reactions:

(1) the carbon-producing reaction chain, $^{4}\text{He} + {}^{4}\text{He} \rightarrow {}^{8}\text{Be}$, $^{8}\text{Be} + {}^{4}\text{He} \rightarrow {}^{12}\text{C}$,

and

(2) the oxygen-producing reaction, $^{12}\text{C} + {}^{4}\text{He} \rightarrow {}^{16}\text{O}$.

Now in order for appreciable amounts of *both* carbon and oxygen to be formed, the rates of these two processes must be well adjusted. If, for example, one were drastically to increase the rate of carbon production – say by a thousand-fold – without increasing the rate of oxygen production, most of the helium would be burnt to carbon before significant quantities of it had a chance to combine with carbon to form oxygen. On the other hand, if one decreased the rate of carbon synthesis by a thousand-fold, very little carbon would be produced, since most of the carbon would be burnt to oxygen before it could accumulate in significant quantities.

Astrophysicist Sir Fred Hoyle was the first to notice that this process involved several coincidences that allowed for this balance between the rate of synthesis of carbon and that of oxygen: namely, the precise position of the 0^{+} nuclear "resonance" states in carbon, the opportune positioning of a resonance state in oxygen, and the fact that ^{8}Be has an anomalously long lifetime of 10^{-17} seconds as compared to the $^{4}\text{He} + {}^{4}\text{He}$ collision time of 10^{-21} seconds (Barrow and Tipler 1986: 252).

Among other factors, the position of these resonance states, along with the lifetime of ^{8}Be, is dependent on the strengths of the strong nuclear force and the electromagnetic force. A quantitative treatment of the effect of changes in either the strong force or the electromagnetic force on the amount of carbon and oxygen produced in stars has been performed by three astrophysicists – H. Oberhummer, A. Csótó, and H. Schlattl (Oberhummer *et al.* 2000a). Using the latest stellar evolution codes, they calculated the effects on the production of carbon and oxygen in stars of a small decrease, and a small increase, in the strength of either the strong force or the electromagnetic force. Their codes took into account the effect of changes in the strength of the strong force and the electromagnetic force on the relevant resonance levels of both carbon and oxygen, along with the effect of a change in temperature of helium ignition. They also examined a wide variety of different types of stars in which carbon and oxygen are produced. Based on this analysis, the authors conclude that

> [A] change of more than 0.5% in the strength of the strong interaction or more than 4% in the strength of the Coulomb [electromagnetic] force would destroy either nearly all C or all O in every star. This implies that irrespective of stellar evolution the contribution of each star to the abundance of C or O in the ISM [interstellar medium] would be negligible. Therefore, for the above cases the creation of carbon-based life in our universe would be strongly disfavored.
>
> (Oberhummer *et al.* 2000a: 90)

The exact amount by which the production of either carbon or oxygen would be reduced by changes in these forces is thirty- to a thousand-fold, depending on the stellar evolution code used and the type of star examined (Oberhummer *et al.* 2000a: 88).

One hitch in the above calculation is that no detailed calculations have been performed on the effect of further increases or decreases in the strong and electromagnetic forces that go far beyond the 0.5 and 4 percent changes, respectively, presented by Oberhummer *et al.* For instance, if the strong nuclear force were decreased sufficiently, new carbon resonances might come into play, thereby possibly allowing for new pathways to become available for carbon or oxygen formation. In fact, an additional 10 percent decrease or increase would likely bring such a new resonance of carbon into play. A 10 percent increase could also open up another path-way to carbon production during Big Bang nucleosynthesis via ^{5}He or ^{5}Li, both of which would become bound. Apart from detailed calculations, it is difficult to say what the abundance ratio would be if such resonances or alternative pathways came into play (Oberhummer *et al.* 2000b). We can say, however, that decreases or increases from 0.5 to 10 percent would magnify the disparities in the oxygen/carbon ratios by magnifying the relevant disparities in the rate of carbon synthesis and oxygen synthesis. Thus we have a small island of life-permitting values with a width of 1 percent, with a distance of 10 percent between it and the next nearest possible, though not likely, life-permitting island. This would leave a two-sided fine-tuning of one part in ten for the strong force (or similarly for the electromagnetic force), which is significant without being enormous: for example, if one had six independent cases of one-in-ten two-sided fine-tuning, one would have a total two-sided fine-tuning of one part in a million.

Another hitch is the amount of carbon, or oxygen, actually needed for intelligent life to evolve. Even very small amounts of carbon or oxygen in the interstellar medium could, as a remote possibility, become concentrated in sufficient quantities to allow for complex life to evolve, though the existence of intelligent life would almost certainly be much less likely under this scenario. So it seems one can conclude with significant confidence that such changes in the strong nuclear force would make intelligent life much less likely, and thus that in our universe the strong interaction is optimized (or close to being optimized) for carbon-based life, giving an abundance ratio of carbon to oxygen of the same order (C:O about 1:2).

Overall, therefore, I conclude that the argument for the fine-tuning of the strong force and the electromagnetic force for carbon and oxygen production, though not straightforward, seems to be on fairly solid ground because of the detailed calculations that have been performed. Nonetheless, more work needs to be done on the two "hitches" cited above to make it completely solid.

The proton / neutron mass difference

The neutron is slightly heavier than the proton by about 1.293 MeV. If the mass of the neutron were increased by another 1.4 MeV – that is, by one part in 700 of its actual mass of about 938 MeV – then one of the key steps by which stars burn their hydrogen to helium could not occur. The main process by which hydrogen is burnt to helium in stars is proton–proton collision, in which two protons form a coupled system, the diproton, while flashing past each other. During that time, the two-proton system can undergo a decay via the weak force to form a deuteron, which is a nucleus containing one proton and one neutron. The conversion takes place by the emission of a positron and an electron neutrino:

p + p → deuteron + positron + electron neutrino + 0.42 MeV of energy.[12]

About 1.0 MeV more energy is then released by positron/electron annihilation, making a total energy release of 1.42 MeV. This process can occur because the deuteron is less massive than two protons, even though the neutron itself is more massive. The reason is that the binding energy of the strong force between the proton and neutron in the deuteron is approximately 2.2 MeV, thus overcompensating by about 1 MeV for the greater mass of the neutron. If the neutron's mass were increased by around 1.42 MeV, however, then neither this reaction nor any other reaction leading to deuterium could proceed, because those reactions would become endothermic instead of exothermic (that is, they would absorb energy instead of producing it). Since it is only via the production of deuterium that hydrogen can be burnt to helium, it follows that (apart from a remote possibility considered in note 13), if the mass of the neutron were increased beyond 1.4 MeV, stars could not exist.[13]

On the other hand, a small decrease in the neutron mass of around 0.5 to 0.7 MeV would result in nearly equal numbers of protons and neutrons in the early stages of the Big Bang, since neutrons would move from being energetically disfavored to being energetically favored (Hogan 1999: equation 19; Barrow and Tipler 1986: 400). The protons and neutrons would then combine to form deuterium and tritium, which would in turn fuse via the strong force to form ^{4}He, resulting in an almost all-helium universe. This would have severe life-inhibiting consequences, since helium stars have a lifetime of at most 300 million years and are much less stable than hydrogen-burning stars, thus providing much less time and stability for the evolution of beings of comparable intelligence to ourselves. A decrease in the neutron mass beyond 0.8 MeV, however, would result in neutrons becoming energetically favored, along with free protons being converted to neutrons, and hence an initially all-neutron universe (Hogan 1999: equation 20; Barrow and Tipler 1986: 400). Contrary to what Barrow and Tipler argue, however, it is unclear to what extent, if any, this would have life-inhibiting effects (see the section entitled "The proton/neutron mass difference" in the Appendix).

So the above argument establishes a one-sided fine-tuning of the neutron/proton mass difference. Since the maximum life-permitting mass difference is 1.4 MeV, and the mass of the neutron is in the order of 1,000 MeV, by the formula presented in note 5 the degree of one-sided fine-tuning relative to the neutron mass is at least one part in 700, or less, given that the lower bound of the total theoretically possible range of variation in the neutron mass, R, is in the order of the neutron mass itself – that is, 1,000 MeV.

Another plausible lower bound of the theoretically possible range R is given by the range of quark masses. According to the Standard Model of particle physics, the proton is composed of two *up* quarks and one *down* quark (uud), whereas the neutron is composed of one up quark and two down quarks (udd). Thus we could define the neutron and proton in terms of their quark constituents. The reason the neutron is heavier than the proton is that the down quark has a mass of 10MeV, which is 4 MeV more than the mass of the up quark. This overcompensates by about 1.3 MeV for the 2.7 MeV contribution of the electric charge of the proton to its mass. (Most of the mass of the proton and neutron, however, is due to gluon exchange between the quarks (Hogan 1999: section III-A).) The quark masses range from 6 MeV for the up quark to 180,000 MeV for the top quark (Peacock 1999: 216). Thus a 1.42 MeV increase in the neutron mass – which would correspond to a 1.42 MeV increase in the down quark mass – is only a mere one part in 126,000 of the total range of quark masses, resulting in a lower bound for one-sided fine-tuning of about one part in 126,000 of the range of quark masses. Furthermore, since the down quark mass must be greater than zero, its total life-permitting range is 0 to 11.4 MeV, providing a total two-sided fine-tuning of about one part in 18,000 of the range of quark masses.

The weak force

One of the major arguments for the fine-tuning of the weak force begins by considering the nuclear dynamics of the early stages of the Big Bang. Because of the very high temperature and mass/energy density during the first seconds, neutrons and protons readily converted via the weak force into each other through interactions with electrons, positrons, neutrinos, and anti-neutrinos. The rate of this interconversion was dependent on, among other things, the temperature, the mass/energy density, the mass difference between the proton and neutron, and the strength of the weak force.

Because the neutron is slightly heavier than the proton, at thermal equilibrium the number of neutrons will always be less than the number of protons: that is, the ratio of neutrons to protons will always be less than one. This ratio will depend on the equilibrium temperature, via what is known as the Maxwell–Boltzmann distribution. The result is that the higher the temperature (that is, the more energy available to convert protons into neutrons), the closer the ratio will be to one, since the difference in rest mass between the neutron and proton becomes less and less significant as the energy available for interconversion becomes greater.

As the Universe expands, however, the density of photons, electrons, positrons, and neutrinos needed to bring about this interconversion between protons and neutrons rapidly diminishes. This means that at some point in the Big Bang expansion, the rate of interconversion becomes effectively zero, and hence the interconversion is effectively shut off. If one were to imagine suddenly shutting off the interaction at some point, one could see that the ratio of neutrons to protons would be frozen at or near the equilibrium value for the temperature at which the interaction was shut off. The temperature at which such a shut-off effectively occurs is known as the freeze-out temperature, T_f. It determines the ratio of neutrons to protons. The higher the T_f, the closer the ratio will be to one.

Since the interconversion between protons and neutrons proceeds via the weak force, it is highly dependent on the strength of the weak force. The stronger the weak force, the greater the rate of interconversion at any given temperature and density. Thus, an increase in the weak force will allow this interaction to be non-negligible at lower temperatures, and hence cause the freeze-out temperature to decrease. Conversely, a decrease in the weak force will cause the freeze-out temperature to increase. Using the fact that the freeze-out temperature T_f is proportional to $g_w^{(-2/3)}$, where g_w is the weak-force coupling constant (Davies 1982: 63), it follows that a thirty-fold decrease in the weak force would cause the freeze-out temperature to increase by a factor of ten. This would in turn cause the neutron/proton ratio to become 0.9 (Davies 1982: 64). Thus almost all of the protons would quickly combine with neutrons to form deuterium and tritium, which, as in the case of the hydrogen bomb, would almost immediately fuse to form ^{4}He during the very early stages of the Big Bang. Consequently, stars would be composed almost entirely of helium. As is well known, helium stars have a maximum lifetime of around only 300 million years and are much less stable than hydrogen-burning stars such as the Sun. This would make conditions much, much less optimal for the evolution of intelligent life.

Thus we have a good case for a one-sided fine-tuning of the weak force. Although there are some reasons to think that a significant increase in the weak force might have intelligent-life-inhibiting effects, they are currently not nearly as convincing. Finally, since, in the dimensionless units mentioned in the section above, the weak force has a strength of about $10^{31}G_0$, relative to the total range of forces (see the section above), this one-sided fine-tuning of the weak force is quite impressive, being around one part in 10^9 of the total range of forces.[14]

Gravity

The main way in which significantly increasing the strength of gravity would have an intelligent-life-inhibiting effect has to do with the strength of a planet's gravitational pull. If we increased the strength of gravity on Earth a billion-fold, for instance, the force of gravity would be so great that any land-based organism anywhere near the size of human beings would be crushed. (The strength of materials depends on the electromagnetic force via the fine-structure constant, which would not be affected by a change in gravity.) As Martin Rees notes, "In an imaginary strong gravity world, even insects would need thick legs to support them, and no animals could get much larger" (Rees 2000: 30). Of course, organisms that exist in water would experience a severely diminished gravitational force if the density of the organism were very close to that of water. It is unlikely, however, that technologically advanced organisms such as ourselves could evolve in a water-based environment given that the overall density of the organism would need to be very close to that of water.

Furthermore, even if aquatic organisms did evolve, such a drastic increase in gravity would still present a problem: any difference in density between various parts of the organism, or between the organism and the surrounding water, would be amplified a billion-fold from what it would be in our world. This would create enormous gravitational differentials. For example, if the liquid inside the organism were one part in a thousand less salty than the surrounding ocean, it would experience a gravitational pull of around a million times the equivalent force on Earth of a land-based organism of the same mass. This would certainly preclude the possibility of bones or cartilage. Inserting air pockets into the cartilage or bone to compensate would cause the organism to be crushed under an enormous pressure of about 1,500,000 kilograms per square centimeter (or about 5,000,000 pounds per square inch) a mere one centimeter below the surface.

The above argument assumes that the planet on which intelligent life formed would be Earth-sized. Could intelligent life comparable to ourselves develop on a much smaller planet? The answer appears to be "no." A planet with a gravitational pull of a thousand times that of Earth – which would itself make the existence of organisms of our brain size very improbable – would have a diameter of about forty feet (or twelve meters). This is certainly not large enough to sustain the sort of large-scale ecosystem necessary for carbon-based organisms of comparable intelligence to human beings to evolve.[15]

Finally, as shown in the section entitled 'Gravity' in the Appendix, stars with lifetimes of more than a billion years (as compared to our Sun's lifetime of 10 billion years) could not exist if gravity were increased by more than a factor of 3,000. This would have significant intelligent-life-inhibiting consequences.

Of course, an increase in the strength of gravity by a factor of 3,000 is significant, but compared to the total range of strengths of the forces in nature (which span a range of 0 to $10^{40}G_0$ as we saw above), this still amounts to a *one-sided* fine-tuning of approximately one part in 10^{36}. On the other hand, if the strength of gravity were zero (or negative, making gravity repulsive), no stars or other solid bodies could exist. Accordingly, the intelligent-life-permitting values of the gravitational force are restricted to at least the range 0 to 3×10^3G_0, which is one part in 10^{36} of the total range of forces. This means that there is a *two-sided* fine-tuning of gravity of at least one part in 10^{36}.

Conclusion

We have examined six apparently solid cases of the fine-tuning of the constants of physics, though we did not examine many other cases of this type of fine-tuning. Furthermore, we did not examine more qualitative cases of fine-tuning, such as those extensively discussed by Michael Denton (1998) – for example, the many special properties of elements such as carbon and oxygen that allow for carbon-based life. Nor did we look at the way in which the Universe has just the right laws for life – for example, if any one of the four forces (gravity, electromagnetism, the strong force, and the weak force) did not exist, life would not be possible.[16] These other cases of apparent fine-tuning further bolster the case for the Universe being delicately arranged for the existence of complex life-forms such as ourselves.

One might wonder what effect the development of some Grand Unified Theory that explains the above cases of fine-tuning would have on the case for design or many universes. Even if such a theory were developed, it would still be a huge coincidence that the Grand Unified Theory implied just those values of these parameters of physics that are life-permitting, instead of some other values. As astrophysicists Bernard Carr and Martin Rees note, "even if all apparently anthropic coincidences could be explained [in terms of some Grand Unified Theory], it would still be remarkable that the relationships dictated by physical theory happened also to be those propitious for life" (1979: 612). It is very unlikely, therefore, that these cases of fine-tuning and others like them will lose their significance with the further development of science.[17]

Appendix: seriously problematic claims in the literature

In this Appendix we will discuss several of the most seriously problematic claims of fine-tuning that are prominent, and often repeated, in the literature. What this shows is that one must demand careful calculations and examination of assumptions before relying on any purported claim of fine-tuning.

The strong force

The most commonly cited problematic assertion in the literature concerns the effect of increasing the strength of the strong force. This assertion goes back to various (perhaps misinterpreted) statements of Freeman Dyson in his *Scientific American* article "Energy in the Universe" (1971: 56) and is repeated in various forms by the most prominent writers on the subject (Barrow and Tipler 1986: 321–2; Davies 1982: 70–1; Rees 2000: 48–9; Leslie 1989: 34). The argument begins with the claim that, since the diproton is unbound by a mere 93 KeV, an increase of the strong force by a few percent would be sufficient to cause it to be bound (Barrow and Tipler 1986: 321). Then it is claimed that, because of this, all the hydrogen would have been burnt to helium in the Big Bang, and hence no long-lived stable stars would exist. According to Barrow and Tipler, for example, if the diproton were bound, "all the hydrogen in the Universe would be burnt to He^2 during the early stages of the Big Bang and no hydrogen compounds or long-lived stable stars would exist today" (1986: 322).

The first problem with this line of reasoning is that ^2He (that is, the diproton) would be unstable, and would decay relatively quickly to deuterium (heavy hydrogen) via the weak force. So, the binding of the diproton would not have resulted in an almost all-^2He universe. On the other hand, one might ask whether the resulting deuterium and any remaining ^1H and ^2He could fuse via the strong force ultimately to form ^4He, thus resulting in an all-helium universe, as Davies seems to suggest.The problem with this latter suggestion is that none of these authors present, or reference, any calculations of the half-life of the diproton. Preliminary calculations by nuclear physicist Richard Jones at the University of Connecticut yield a *lower bound* for the half-life of around 13,000 seconds, with the actual half-life estimated to be within one or two orders of magnitude of this figure (private communication). As Barrow and Tipler note, however, there is only a short window of time of approximately 500 seconds when the temperature and density of the Big Bang are high enough for significant deuterium to be converted to ^{4}He (1986: 398). Since only a small proportion of diprotons would have been able decay in 500 seconds, little deuterium would have been formed to convert to ^{4}He.

Of course, most of these diprotons would eventually decay to form deuterium, resulting in predominantly deuterium stars. Stars that burnt deuterium instead of hydrogen would be considerably different from ours. Preliminary calculations performed by astrophysicist Helmut Schlattl using the latest stellar evolution codes show that deuterium stars with the same mass as the Sun would have lifetimes of around 300 million years, instead of the Sun's 10 billion years (private communication). This would seriously hamper the evolution of intelligent life. On the other hand, a deuterium star with a 10 billion-year lifetime would have a mass of 4 percent that of the Sun, and a luminosity of 7 percent that of the Sun, with a similar surface temperature (private communication). This would require that any planet containing carbon-based life be about four times closer to its Sun than we are to ours. It is unclear whether such a star would be as conducive to life.

Gravity

In discussing a strong-gravity world in which gravity is a million times stronger, Martin Rees claims that

> The number of atoms needed to make a star (a gravitationally bound fusion reactor) would be a billion times less in this imagined universe.... Heat would leak more quickly from these "mini-stars": in this hypothetical strong-gravity world, stellar lifetimes would be a million times shorter. Instead of living for ten billion years, a typical star would live for about 10,000 years. A mini-Sun would burn faster, and would have exhausted its energy before even the first steps in organic evolution had got under way.
>
> (Rees 2000: 30–1)

Although Rees's claim that stars would have a shorter lifetime appears to be correct, his claim that the lifetimes would be a million times shorter is highly questionable. We begin our analysis by looking at a commonly used simple model of a star, in which the star is assumed to have uniform density. In such a star, the condition of hydrostatic equilibrium dictates that

$$T_c \propto GM/R \propto GDR^2$$

(Hanson and Kawaler 1994: 20, equation 1.53)

where $\propto$ represents proportional to, G is the gravitational constant, T_c is the central temperature of the star, M is its mass, R is its radius, and D is its density. (I am assuming the star is composed of a single material, hydrogen.)

The lifetime of our model star is

$$L_i \propto M/[(T_s)^4 R^2];$$

that is, the lifetime is proportional to the mass – which determines the total amount of nuclear fuel – divided by the amount of energy radiated from the surface, which is proportional to the fourth power of the surface temperature T_s times the surface area R^2.

Dividing L_i by T_c, we obtain

$$L_i/T_c \propto 1/[G(T_s)^4 R],$$

or

$$L_i \propto T_c/[GR(T_s)^4].$$

Thus, according to this simple model, if we increased the gravitational constant by a million-fold, the conditions of hydrostatic equilibrium could be met by a star with the same lifetime, surface temperature, and core temperature, but with one-millionth the radius. The density would increase by a million-fold, however, as required by the first relation linking T_c with G, D, and R.

A star actively burning nuclear fuel cannot remain stable, however, if its internal density is more than about 100 times that of the density at the core of the Sun: the gas would be in an electron degenerate state in which quantum effects become predominant (see Clayton 1983: 88–9). Since the pressure of an electron degenerate gas does not increase with temperature, it is not stable when fusion is taking place; it lacks the usual mechanism – cooling by expansion resulting from an increase in temperature – for preventing a runaway fusion reaction. Thus, according to our simple star model, if G is increased by more than a factor of 100, we cannot compensate by further increasing D while holding T_c, T_s, and L_i constant; instead, we must hold D constant and hence vary either T_c, T_s, or L_i, to obtain a stable star. We will now consider what happens when we allow L_i to vary while holding T_c, T_s, and D constant.

As originally pointed out to me by physicist Richard Jones, if we hold T_s, T_c, and D constant, the above equations dictate that an x-fold increase in gravity must result in a $\sqrt{x}$ decrease in radius, and hence a $\sqrt{x}$ decrease in L_i.

It follows, therefore, that for increases in G beyond a factor of 100, L_i must decrease by a factor of at least $\sqrt{y}$ for stable stars to exist with surface temperatures and core temperatures similar to our Sun, where y is the factor by which G has increased beyond a factor of 100. Thus, according to the above simple star model, a stable star could exist in Rees's strong gravity world with the same surface and core temperature as our Sun, but it could have a lifetime of at most one-hundredth that of our Sun. (The y factor in this case would be 10,000.) Our simple star model, however, only provides a guide of one or two orders of magnitude to what happens when G is increased. More sophisticated calculations performed by astrophysicist Helmut Schlattl (private communication) using the latest stellar evolution codes and taking into account electron degeneracy show that stars with lifetimes greater than a billion years could not exist if the strength of gravity were increased by more than a factor of 3,000. This would seriously reduce the probability of the evolution of life-forms of comparable intelligence to that of humans. Thus, a 3,000-fold increase in the strength of gravity could be taken as the upper bound of the optimal range for the existence of intelligent life.

The proton/neutron mass difference

Barrow and Tipler (and others) have argued that a small decrease in the neutron mass relative to the proton would also eliminate the possibility of life. Specifically, they argue that if the difference between the neutron mass and the proton mass were less than the mass of the electron, then protons would spontaneously convert to neutrons by the weak force via electron capture. Thus, they claim, the Universe would simply consist of neutrons, which in turn

> [W]ould lead to a World in which stars and planets could not exist. These structures, if formed, would decay into neutrons by pe^- [that is, proton–electron] annihilation. Without electrostatic forces to support them, solid bodies would collapse rapidly into neutron stars…or black holes…if that were to happen no atoms would ever have formed and we would not be here to know it.
>
> (Barrow and Tipler 1986: 400)

As appealing as this line of reasoning initially sounds, apart from detailed calculations that Barrow and Tipler neither present nor reference, it is highly questionable. They are correct in asserting that protons would initially convert to neutrons. They neglect to consider, however, that the reaction

$$n + n \rightarrow d + e^- + \text{anti-neutrino}$$

could take place. The reason is that the deuteron has a mass 2.2 MeV less than the sum of two neutrons, as can be seen by the fact that the dineutron just barely fails to be bound, yet the deuteron is bound by 2.2 MeV (Barrow and Tipler 1986: 321). Hence the reaction allows for a conversion of 0.511 MeV for an electron, some amount for a neutrino, and some for the kinetic energy of the electron. Similar processes of conversion would happen in larger conglomerations of neutrons: neutrons would be converted to protons to fill lower energy levels that are already filled with neutrons as much as the Pauli exclusion principle will allow. Since these sorts of conversions appear to be allowed, the only effects we can immediately deduce that a moderate decrease of the neutron mass would have are that stars would burn very differently and that stable nuclei, including hydrogen, would shift towards having a higher proportion of neutrons than we presently find. I know of no current well-developed argument, however, that these effects would inhibit the existence of intelligent life. This is an area that needs further exploration.

Notes

1 This chapter is a condensed version of Chapter 2 of a book I am working on tentatively entitled *The Well-tempered Universe: God, Fine-tuning, and the Laws of Nature*.

2 I base this on Guth's presentation, and a personal conversation with Guth, at the 'Nature of Nature' conference at Baylor University, April 2000.

3 It should be noted here that, just because some physical quantity is fine-tuned, it does not follow that every function of that quantity will be fine-tuned. So, for instance, if the strength of the gravitational force – given by the gravitational constant G – is fine-tuned, there will always exist an infinite number of functions F(G) of this force strength that are not fine-tuned. For the purposes of this chapter, we simply note that when we talk about fine-tuning, we are always referring to the fine-tuning of the parameters that are actually considered in physics (such as G), not arbitrary functions of those parameters.

4 When spelled out in terms of what is known as *epistemic* probability, this principle is often called *the likelihood principle*. According to the likelihood principle, evidence E confirms hypothesis H_1 over hypothesis H_2 if $P(E/H_1) > P(E/H_2)$, where $P(E/H_1)$ and $P(E/H_2)$ represent the conditional epistemic probability of E on H_1 and H_2, respectively. There is a growing literature on epistemic probability. Good authors to begin with are Swinburne (1973: Ch. 1) and Plantinga (1993: Chs 8 and 9).

5 Of course, the nearness is relative to the width, W_R, of the theoretically "possible" range R. Quantitatively, we could define the degree of nearness, or of *one-sided* fine-tuning, as n/R, where n represents the distance the parameter falls from the known (or closest) edge of the life-permitting range. For example, as calculated below, the upper bound of the life-permitting range of strength of the electromagnetic force is approximately $14 \times 10^{37}G_0$, whereas the electromagnetic force's actual strength is $10^{37}G_0$. Thus, $n = 14 \times 10^{37}G_0 - 10^{37}G_0 = 13 \times 10^{37}G^0$. Given that the theoretically possible range is the range of force strengths (0 to $10^{40}G_0$), then $W_R = 10^{40}G_0$. Hence the degree of one-sided fine-tuning is $(13 \times 10^{37}G_0)/10^{40}G_0 \approx 10^{-2}$, or one part in a hundred. (Actually, since we are considering W_R to be the lower bound for the width of the theoretically

"possible" range, the degree of one-sided fine-tuning should be considered to be less than or equal to one part in a hundred.)

6 There are many good discussions of the cosmological constant problem. See, for example, Sahni and Starobinsky (1999: sections 5–7) and Cohn (1998: section II).

7 If one allows negative energies, then theoretically this lower bound for the width of the range R would be even greater, going from below zero to $10^{53}\Lambda_{max}$, or even $10^{123}\Lambda_{max}$.

8 To be absolutely precise, all that the existence of life requires is that the total cosmological constant, Λ_{tot}, be within the life-permitting range. But $\Lambda_{tot} = \Lambda_{vac} + \Lambda_{bare}$, where Λ_{vac} represents the contribution to the cosmological constant from the vacuum energy of all the fields combined, and Λ_{bare} represents the "intrinsic" value of the cosmological constant apart from any contribution from the vacuum energy. Thus the contribution of any given field, such as the inflaton field, to the vacuum energy could be much greater than Λ_{max}, if such a contribution were almost cancelled out by the other contributions to the cosmological constant. But to get such a precise cancellation would itself require some sort of extraordinary fine-tuning or new principle of physics.

9 Expressed in terms of the equations of electroweak theory, the value of the Higgs field in this local minimum is $V = V_0 - u^4/4\lambda$, where the values of the experimentally determined free parameters of the electroweak theory yield a value for $u^4/4\lambda$ of approximately $10^{53}\Lambda_{max}$ (Sahni and Starobinsky 1999: section 6). In order to get $|V| < \Lambda_{max}$, V_0 and $u^4/4\lambda$, which theoretically are independent parameters, must almost have the same values, to within one part in 10^{53}; that is, $|V_0 - u^4/4\lambda| < \Lambda_{max}$, which implies that $|V_0 - u^4/4\lambda|/V_0 < 10^{-53}$. (Here $|\ |$ represents absolute value.) Letting 0 to V_0 represent the lower bound of the theoretically possible range R for the values of V_0 and $u^4/4\lambda$, we get a fine-tuning of at least one part in 10^{53}.

10 A principle requiring it to be zero does not run into the same problem because zero is a natural, non-arbitrary number and thus the sort of value we would expect a principle to require.

11 One currently popular proposed mechanism to partly circumvent the cosmological constant problem is what has been called "quintessence," which is a postulated field that tracks the matter fields of the Universe in such a way that it allows for a large cosmological constant in the very early universe, as required by inflation, but results in a very small positive cosmological constant during early stages of our current epoch (see Ostriker and Steinhart 2000). As astrophysicist Lawrence Krauss notes, however, all theoretical proposals for quintessence seem *ad hoc* (1999: 59). Furthermore, not only must the quintessence potential have a special ("well-designed") mathematical form, but the parameters of the potential seem to require extreme fine-tuning (Kolda and Lyth 1999; Sahni and Starobinksy 1999: sections 8.3 and 11). But even if this fine-tuning of the parameters can eventually be circumvented, the history of such attempts strongly suggests that the properties and mathematical form of the quintessence would have to be further constrained (or "fine-tuned") to achieve this, as illustrated by a recent proposal by K. Dimopoulos and J.W.F. Valle (2001). This seems merely to reintroduce the issue of fine-tuning and design at a different level, though perhaps in a mitigated way.

12 Two other processes by which hydrogen can be burnt to helium are the helium-catalyzed nuclear reactions, in which helium serves as a catalyst for hydrogen burning, and the carbon–nitrogen–oxygen (CNO) cycle, in which carbon, nitrogen, and oxygen serve as catalysts. These processes, however, are dependent on the existence of the $p + p \rightarrow$ deuterium reaction to produce the initial abundance of helium, carbon, nitrogen, and oxygen in the Big Bang or in stars themselves.

13 As a very remote possibility, a star that might be able to support the evolution of life on a nearby planet as well as our Sun could result from hydrogen burning occurring by alternative reactions as a protostar collapses and further heats up, with some of the initial stages of the reaction being endothermic, but the overall reaction being exothermic. If carbon-based life is to exist, however, some isotope of carbon must be able to exist. The binding energy *per neutron* (i.e. the total binding energy divided by the total number of neutrons) of the various isotopes of carbon has a maximum value of 15 MeV (Harvey 1969: 428–9). This means that if the mass of the neutron were increased by more than 15 MeV, or approximately 1/70 of the total neutron mass of 938 MeV, then carbon could not exist. Thus 15 MeV is an absolute upper bound for a (carbon-based) life-permitting increase in the neutron mass even under this remote scenario.

14 The weak force is approximately a billion times weaker than the strong force. The first crucial step in the conversion of hydrogen to helium in stars ($p + p \rightarrow$ deuterium) is mediated by the weak force. This has led some authors (e.g. Leslie 1989: 34; Rozental 1988: 76–7) to claim that, if the strength of the weak force were greatly increased, the rate of fusion in stars would become so great that stars would blow up like hydrogen bombs. This claim is incorrect. Stars are complex equilibrium systems whose fusion rates are highly dependent on temperature, since, in order for fusion to have any chance of occurring, the nuclei must have enough kinetic energy to overcome the electrostatic repulsion between them. Thus a large increase in the strength of the weak force would simply cause a compensating decrease in the internal temperature of the star. It would change the character of the star, however, most likely making it much like the deuterium-burning stars (where hydrogen burning is mediated by the strong force) discussed in the section entitled 'The strong force' in the Appendix.

15 To see this, note that the mass, M_p, of such a planet equals $4Dr^3/3$, where D is the average density of the planet and r is its radius. Consequently, assuming D remains constant, the mass of a planet is proportional to r^3: $M_p \propto r^3$, where $\propto$ represents proportionality. Thus, since by Newton's law of gravity, $F = GM_O M_p/r^2$, where M_O is the mass of the organism under consideration, F is the force of gravity on that organism, and G is the gravitational constant, it follows that the force on any organism is proportional to $GDM_O r^3/r^2 = GDM_O r$. Hence, a billion-fold increase in G would require a compensating billion-fold decrease in r in order for the force on an organism of mass M_O to remain the same, given that D remained approximately the same.

16 If gravity did not exist, masses would not clump together to form stars or planets, and hence no carbon-based life would exist. If the strong force didn't exist, protons and neutrons could not bind together and hence no atoms with an Satomic number greater than hydrogen would exist. If the electromagnetic force didn't exist, there would be no chemistry. Similarly, other laws and principles are necessary for complex life. If the Pauli exclusion principle, which dictates that no two fermions can occupy the same quantum state, did not exist, all electrons would occupy the lowest atomic orbit, eliminating complex chemistry. If there were no quantization principle, which dictates that particles can only occupy certain discrete allowed quantum states, there would be no atomic orbits and hence no chemistry, since all electrons would be sucked into the nucleus.

17 This work was supported by a year-long fellowship from the Pew Foundation, several grants from the Discovery Institute, and a grant from Messiah College. I would especially like to thank nuclear astrophysicist Heinz Oberhummer at the Institute of Nuclear Physics of the Vienna University of Technology, astrophysicist Helmut Schlattl at the Max Planck Institut für Astrophysik, and nuclear physicist Richard Jones at the University of Connecticut for helpful discussions

and comments, and generously performing various calculations for me. Dr Oberhummer, Dr Schlattl, and Dr Jones also graciously read through a near-final draft of the chapter, though any remaining errors are my own. This work would not have been possible without their help. Physicists Wytse van Dijk and Yuki Nogami at McMaster University also helped in some of the initial calculations. Finally, I would like to thank Neil Manson for putting this anthology together and for helping in the editing process.

References

Barrow, J. and Tipler, F. (1986) *The Anthropic Cosmological Principle*, Oxford: Oxford University Press.

Carr, B.J. and Rees, M.J. (1979) "The anthropic cosmological principle and the structure of the physical world," *Nature* 278 (12 April 1979): 605–12.

Clayton, D. (1983) *Principles of Stellar Evolution and Nucleosynthesis*, Chicago: University of Chicago Press.

Cohn, J.D. (1998) "Living with lambda," available online at http://xxx.lanl.gov/abs/astro-ph/9807128v2 (26 October 1998).

Davies, P. (1982) *The Accidental Universe*, Cambridge: Cambridge University Press.

Denton, M. (1998) *Nature's Destiny: How the Laws of Biology Reveal Purpose in the Universe*, New York: The Free Press.

Dimopoulos, K. and Valle, J.W.F. (2001) "Modelling quintessential inflation," available online at http://www.lanl.gov/abs/astro-ph/0111417v1 (21 November 2001).

Dyson, F. (1971) "Energy in the universe," *Scientific American* September 1971: 51–9.

Guth, A. (1997) *The Inflationary Universe: The Quest for a New Theory of Cosmic Origins*, New York: Helix Books.

Hanson, C.J. and Kawaler, S.D. (1994) *Stellar Interiors: Physical Principles, Structure, and Evolution*, New York: Springer-Verlag.

Harvey, B. (1969) *Introduction to Nuclear Physics and Chemistry*, 2nd edn, Englewood Cliffs, New Jersey: Prentice-Hall.

Hogan, C. (1999) "Why the universe is just so," available online at http://xxx.lanl.gov/abs/astro-ph/9909295 (16 September 1999).

Kolda, C. and Lyth, D. (1999) "Quintessential difficulties," available online at http://www.lanl.gov/abs/hep-ph/9811375 (25 May 1999).

Krauss, L. (1999) "Cosmological antigravity," *Scientific American* January 1999: 53–9.

Leslie, J. (1989) *Universes*, New York: Routledge.

—— (1988) "How to draw conclusions from a fine-tuned cosmos," in Robert Russell, William R. Stoeger, S.J., and George V. Coyne, S.J. (eds) *Physics, Philosophy and Theology: A Common Quest for Understanding*, Vatican City State: Vatican Observatory Press, pp. 297–312.

Oberhummer, H., Csótó, A., and Schlattl, H. (2000a) "Fine-tuning of carbon based life in the universe by triple-alpha process in red giants," *Science* 289(5,476) (7 July 2000): 88–90.

—— (2000b) "Bridging the mass gaps at A = 5 and A = 8 in nucleosynthesis," available online at http://www.lanl.gov/abs/nucl-th/0009046 (18 September 2000).

Ostriker, J. and Steinhardt, P. (2000) "The quintessential universe," *Scientific American* 284(1) (January 2000): 47–53.
Peacock, J. (1999) *Cosmological Physics*, Cambridge: Cambridge University Press.
Plantinga, A. (1993) *Warrant and Proper Function*, Oxford: Oxford University Press.
Rees, M. (2000) *Just Six Numbers: The Deep Forces that Shape the Universe*, New York: Basic Books.
Rozental, I.L. (1988) *Big Bang Big Bounce: How Particles and Fields Drive Cosmic Evolution*, New York: Springer-Verlag.
Sahni, V. and Starobinsky, A. (1999) "The case for a positive cosmological lambda-term," available online at http://www.lanl.gov/abs/astro-ph/9904398 (28 April 1999).
Swinburne, R. (1973) *An Introduction to Confirmation Theory*, London: Methuen & Co. Ltd.
Weinberg, S. (1999) "A designer universe?" *The New York Review of Books* 46(14) (21 October 1999): 46–8.
—— (1996) "Theories of the cosmological constant," available online at http://www.lanl.gov/abs/astro-ph/9610044 (7 October 1996).
—— (1987) *Physical Review Letters* 59: 2,607.

10

PROBABILITIES AND THE FINE-TUNING ARGUMENT

A skeptical view

Timothy McGrew, Lydia McGrew, and Eric Vestrup

Contemporary cosmological design arguments consist broadly of two main types: the Life Support Argument (LSA), which urges that the relatively local features of the cosmos (our Galaxy, the Sun, the Earth, the Moon) are unusually hospitable to life, and the Fine-Tuning Argument (FTA), which takes its cue from the fact that a multitude of physical constants must apparently take precise values or stand in exacting ratios to each other in order to make our universe as a whole hospitable to organic life. The apparent "fine-tuning" of the constants suggests to the advocates of the FTA that the Universe was itself designed with carbon-based life in mind.

Much of the critical discussion of the FTA in the current literature centers on two factors. First, some critics question the material adequacy of the argument: whether the relevant physical constants are indeed related in such exacting ratios or can be relaxed while remaining, in some sense, life-friendly (Shapiro and Feinberg 1982; Barrow and Tipler 1986). Second, some suggest that the argument derives specious plausibility from a selection effect. Life-friendly universes, so this criticism runs, may be rare, but the existence of conscious life in life-friendly universes is not such a puzzle. Like the lucky winner of a lottery who finds himself bemused and wonders if there has been some mistake, the denizens of a life-friendly universe who marvel at their own good fortune are forgetting that anyone else would find the situation equally astonishing (Carter 1974).

These are interesting arguments, but they leave the formal aspects of the FTA unquestioned. By contrast, we have serious doubts about the use of probabilities in the FTA – in particular, the calculation of the odds against a life-friendly universe "on chance." This phrase, as a moment's reflection reveals, cannot have an obvious, everyday meaning: the image of a barrel full of universes from which one is selected at random does not inspire confidence. In the literature the talk of probabilities is generally supported with analogies and stories designed to pump the intuition in favor of the FTA. Many of these stories describe plausible evidence for inferring design or agent intervention, but we contend that they are crucially disanalogous to the problem of the origin of the Universe "on chance." The FTA, as it stands in the literature, has serious formal flaws that may well be insuperable.

The structure of the argument

The literature on the FTA overflows with examples of "fine-tuned" constants (Leslie 1989, 1998; Ross 1998), and it would be superfluous to survey them here. Typically, however, they take the form of placing upper and lower bounds on the constants in question. Advocates of the FTA assume, plausibly enough, that the list of constants, and hence of their ratios, is finite, though there is some argument as to whether the various parameters are independent of each other. In each case, the field of possible values for the parameters appears to be an interval of real numbers unbounded at least in the upward direction. There is no *logical* restriction on the strength of the strong nuclear force, the speed of light, or the other parameters in the upward direction. We can represent their possible values as the values of a real variable in the half-open interval $[0, \infty)$, and the set of logically possible joint values for K independent constants can be represented as a K-dimensional real-valued space $R_+{}^K$. If we consider each mathematically distinct set of possible values for the parameters to denote a distinct possible type of universe, then we can think of this space as representing all of the possible types of universes that have no types of constraints other than those found in ours.

Among the universes thus represented, which ones will be friendly to life? There is some vagueness surrounding this question, since it is not clear just what life-friendliness amounts to (Manson 2000). At least one universe qualifies: the point that represents the various constants actually exhibited in our universe is clearly life-friendly. But in some of the alternative universes considered – say, those in which the strength of the gravitational force is slightly greater – the upshot is not to rule out life absolutely but rather to lessen the odds of its arising and surviving.

The point can, however, be waived: let us grant that there is a plausible convention we may adopt as to the line of demarcation between life-friendly and life-unfriendly universes, and that this convention will give us a range, perhaps even a narrow one, within which each variable will have to fall in order for the universe to be life-friendly. It seems quite plausible (and seems to be tacitly granted in discussions of the FTA) that, for each constant, the life-friendly region will have an interior of positive Lebesgue measure – that the "inside" of the region will have positive "volume." These regions need not satisfy any other conditions for our discussion. However, for graphical simplicity, we can without loss of generality treat the points lying within these intervals in all K dimensions as what mathematicians affectionately term a "ball" – a set that includes the points representing our universe and those universes sufficiently similar to ours to meet the criterion we have adopted.[1]

Might the life-friendly region end up being just the singular point corresponding to our own universe? This seems extremely unlikely. Since the variables in question range over the reals, we may reasonably assume that there are infinitely many possible universes that are arbitrarily similar to ours though mathematically distinct – universes in which the constants differ from those in ours by amounts so small that the physical implications remain negligible even in the large-scale effects. In such universes, admittedly, some highly elegant symmetry principles and conservation laws might be violated (though the violations could be too small to be physically detected), but these principles appear to be contingent rather than logically necessary. It is certainly incumbent on anyone who would contest this possibility to explain why *only* this universe, and not one arbitrarily similar to it, could sustain life.

With this model in view, we can state the FTA in a more rigorous form. Assume for the sake of the argument that no particular range of values for the parameters is more likely than any other – an application of the Principle of Indifference designed to reflect our lack of information regarding universes. The measure, in R_+^K, of the multi-dimensional "ball" is vanishingly small: like a grain of sand lost in an endless Euclidean universe, it takes up no finite proportion of the possible space.[2] If the measure of the ball represents the probability in question, then it would seem that the odds against a life-friendly universe on chance are, to put it mildly, overwhelming. Advocates of the FTA maintain that the odds of a life-friendly universe under the creative auspices of an intelligent designer are a good deal better. But in that case, so runs the argument, the evidence for the fine-tuning of the Universe must count as evidence favoring design over chance.

In his delightful book *Universes*, John Leslie tries to illustrate the form of the argument by telling stories in which our suspicions are aroused because chance provides a poor explanation for the coincidences at hand. In the Fishing Story, to which he returns many times throughout the book, you catch a fish that is exactly 23.2576 inches long from a murky lake. In itself this is not very surprising: no doubt every fish needs to have some length. But then you note that your fishing apparatus was capable of catching only fish of this length, plus or minus one part in a million. You might hypothesize many explanations for this astonishing coincidence – that the lake is stuffed with so many different sizes of fish that you were bound to find one that fit sooner or later; that a benevolent deity desired to provide you with a fish supper; and so forth. But surely, says Leslie, one explanation that should be ruled out at once is the idea that the lake contained only one fish, which *just happened* to be the ideal size for you to catch. As with fish, so with universes: there may be many universes, of which ours is a singularly life-friendly specimen; there may be a benevolent creator who desired to create a universe that could sustain life. But one explanation Leslie thinks we should certainly reject is the notion that our universe is the only one and just happened, by chance, to be so constituted as to sustain life (Leslie 1989: 9ff.).

The normalizability problem

The Fishing Story solicits our assent by describing a situation where our intuitive judgment of the relevant probabilities leans heavily against chance. But before we can port those intuitions over to the FTA, we need to examine the mathematical model in more detail. When we do so, it turns out that the analogy between the FTA and the Fishing Story breaks down – and it breaks down precisely where probabilities are invoked.

The critical point is that the Euclidean measure function described above is not normalizable. If we assume every value of every variable to be as likely as every other – more precisely, if we assume that, for each variable, every small interval of radius e on R has the same measure as every other – there is no way to "add up" the regions of R_+^K so as to make them sum to one. If they have any sum, it is infinite.

This is more than a bit of mathematical esoterica. Probabilities make sense only if the sum of the logically possible disjoint alternatives adds up to one – if there is, to put the point more colloquially, some sense that attaches to the idea that the various possibilities can be put together to make up 100 percent of the probability space. But if we carve an infinite space up into equal finite-sized regions, we have infinitely many of them; and if we try to assign them each some fixed positive probability, however small, the sum of these is infinite.

Unfortunately, there is not a great deal of discussion on this point in the philosophical literature since the normalization problem has not been raised with sufficient clarity. Leslie does consider some worries about ratioing infinite quantities, but he deflects them with an analogy. The bull's-eye of a target, he points out, contains infinitely many mathematical points (as does, he might have added, the rest of the target). Since this fact makes it no easier to hit the bull's-eye, Leslie dismisses concern about his use of probabilities as mathematical pettifogging (Leslie 1989: 11).

But in view of the foregoing discussion we can see that this analogy is flawed. The target itself has finite area; we can therefore integrate over the area and compare regions without running into normalizability problems. Leslie clearly does not recognize the way in which the bull's-eye and the real-valued parameter space are disanalogous, for he goes on in the next paragraph to speak of life-permitting possibilities constituting "only a thousandth of the range of possibilities under consideration" – an expression that only makes sense because normalizability is not a problem with the target.

What moves us to adopt such a picture in the first place? The culprit is the Principle of Indifference. Working from bare logical possibilities, it seems unreasonable to suggest that any one range of values for the constants is more probable *a priori* than any other similar range – we have no right to assume that one sort of universe is more probable *a priori* than any other sort. But this very feature rules out this measure function as a basis for assigning probabilities.

The problem is not simply that there are infinitely many points both within the ball and outside of it: there are mathematical techniques for coming to grips with that problem. The difficulty lies in the fact that there is no way to establish ratios of regions in a non-normalizable space. As a result, there is no meaningful way in such a space to represent the claim that one sort of universe is more probable than another. Put in non-mathematical language, treating all types of universes evenhandedly does not provide a *probabilistic* representation of our ignorance regarding the ways that possible universes vary among themselves – whatever that means.

Rescues and replies

One way to try to get around the problem is simply to give up a probabilistic interpretation of the narrow intervals and rely instead on an intuitively plausible argument from extrapolation. A small area in a larger finite region, like the bull's-eye at the center of a target, is relatively less likely to be hit (at random) as its size diminishes in relation to the rest of the target. If we think of an *infinitely* large target with a finite bull's-eye, we seem to have the limiting case of low probabilities. Even if we cannot represent this as a ratio of areas in a strict probabilistic sense, are we not entitled to take the "ratio" of a finite to an infinite measure as a basis for the FTA?

Unfortunately, this rescue, if permitted, would achieve far too much. For using such reasoning we can also underwrite what we shall call the "Coarse-Tuning Argument" (CTA). Suppose that the open set of life-friendly universes contained a ball in which the various parameters, rather than being constrained to within tiny intervals around those that characterize our own universe, could take any values within a few billion orders of magnitude of our values. It is hard to imagine anyone's being surprised at the existence of a life-friendly universe under such circumstances. Yet the "ball" in this case is isomorphic to the ball in the FTA: both of them have measure zero in R_+^K. In consequence, any inference we can draw from fine-tuning is not only paralleled by a CTA, it also has precisely the same probabilistic force. So if we are determined to invoke the Principle of Indifference regarding possible universes, we are confronted with an unhappy conditional: if the FTA is a good argument, so is the CTA. And conversely, if the CTA is not a good argument, neither is the FTA.

A natural response to this is to blame the Principle of Indifference. If we do not insist on treating all equal intervals for all parameters as equally probable, then we can perfectly well speak of the probability that a particular parameter falls within a given interval by invoking density functions that integrate (or can be scaled so as to integrate) to unity. Loosely speaking, this is similar to using a converging series in which an infinite number of terms (say, 1/2, 1/4, 1/8,..., $1/2^n$, $1/2^{n+1}$,...) may have a finite sum (in this case, 1), because there is no finite positive lower bound on the size of the terms. By the standard mathematical device of taking the probability in question to be proportional to the area beneath that region of the density function, we can bring the FTA back within the pale of respectable probability theory.

And so we can. But which density function shall we now choose; which intervals shall we favor? There are myriads, continua, of such functions. What is worse, there are infinitely many density functions that pack as great a proportion of the probability mass as you like into the ball; and from the standpoint of any of these functions the existence of a life-friendly universe on chance is practically *inevitable*. True, there are also infinitely many density functions that render the measure of the ball arbitrarily small. But now we have traded our initial problem for another: how to choose, without arbitrariness, which density function or family of density functions is the "right" one. And every such function will be biased in favor of some regions over others – a bias that is difficult to justify in a representation of the bare possibilities.

How does this analysis square with the intuitions prompted by the Fishing Story? Fish, as many a lazy summer's day out on the lake has taught us, vary widely in size: their mature lengths are approximately normally distributed, and relatively few are within one part in a million of 23.2576 inches – or one part in a million of any other specific length. From such homely data we construct reasonable expectations regarding the lengths of fish, and it is those expectations that give the example its plausibility. The data mitigate against any expectations that render a length of 23.2576 inches, plus or minus a few microns, overwhelmingly likely for the next fish we shall catch. But such data are precisely what we lack regarding universes.

This illustrates a second way in which Leslie's bull's-eye analogy is misleading. Thanks to our ample experience of projectile motion, we have a non-arbitrary means of adjudicating disputes regarding non-equivalent density functions over the target. Experience indicates that a flat prior is reasonable for randomly flung darts – any area of a given size is as likely to be hit as any other if the thrower is sufficiently unskilled.

An important attempt to salvage the FTA can be found in Leslie's Fly on the Wall Story (Leslie 1989: 17–18; cf. 158ff.). Suppose that a fly occupies an otherwise vacant stretch of a wall. Perhaps far away there are regions densely covered with flies, but in the local area around this fly the wall is clear. A shot rings out and the lonely fly is struck. Surely, argues Leslie, this requires explanation regardless of the presence or absence of many flies far away. For *in the area of interest* there was a very small chance of hitting a fly at random. Carried over to the FTA, the Fly on the Wall Story is a plea for working with a normalizable space.

The Fly on the Wall Story can be pressed into service for two distinct purposes. On the one hand, the wall may represent a space of possible *types of parameters*.[3] Obviously, if we stray far from the types of parameters we find in our own universe we may not be in any position to speak of the life-friendliness of universes. (Would a small repulsive force between photons be life-unfriendly? What if it were combined with tweaks of half a dozen other constants and a few utterly new forces?) From this standpoint, our willingness to restrict the discussion to R_+^K rather than introducing a possibly infinite number of new parameters represents an attempt to take the Fly on the Wall Story seriously.

On the other hand, the wall in Leslie's story may represent a range of values for the parameters we actually have. In this case, however, there is a serious difficulty in determining how wide a range we ought to survey, how much of the "wall" we ought to take into account. On what basis should we restrict our focus to the area that is amenable to current theoretical discussion? There is, of course, a good pragmatic argument for discussing only those possible universes whereof we are, in some sense, qualified to speak. But there is a serious gap between this sensible pragmatic advice and the epistemic force that the FTA is supposed to have. What we need is an *epistemic* rationale for working with the local region rather than the whole of R_+^K. Without such a restriction, we are back to the normalizability problem. And no one has yet succeeded in articulating a convincing reason for limiting the field.

Some proponents of the FTA will object that all of this is needlessly fussy, that we need no airtight philosophical criterion in order to see that any function that does not make the anthropic coincidences surprising is an unreasonable function.[4] It is certainly true that a number of people not otherwise known for their theological interests have been greatly impressed with the apparent narrowness of the intervals for life-friendly universes. But in our opinion it is profoundly unsatisfying to stipulate that we can just "tell" which functions are reasonable and which are not. Reasonable people have conflicting intuitions here. The point of the argument was supposed to be that objective results in modern cosmology virtually compel disbelief in a chance origin of the Universe. If, at a critical point, the argument turns on a subjectively variable sense of which assessments of probabilities are reasonable, a sense that cannot be adjudicated in terms of any more fundamental criteria, then the FTA is effectively forceless. To retreat to the point where the argument rests on unargued intuitions is to deprive it of anything more than devotional significance.

A wholly different approach involves appealing to inflationary cosmology to give us an empirical probability distribution over universes. Waiving questions about the empirical affidavits of inflationary schemes, however, this will give advocates of the FTA no comfort, for two reasons. First, the probabilities generated all depend on some very strong extrapolations of indifference assumptions, extrapolations exorbitant enough to warrant some skepticism. Second and more importantly, in inflationary scenarios the "universes" in question are all physically real; ours is just one universe among an enormous number of others that are separating from each other faster than they are expanding. But this undermines the need to appeal to design in the first place: the appearance of design is just a feature of our "universe," which is a small but statistically predictable outlier among the vast horde of coexisting "universes."

It is natural to wonder whether a critique of this kind can be ported over to the LSA, thereby demolishing in one fell swoop both types of cosmological design argument.[5] In our opinion these particular problems do not necessarily affect the LSA. If the prior probability of a hospitable planet on chance is taken to be, for example, the odds that a randomly selected star will have a planet of the appropriate mass, axial tilt, and so forth – in short, conditions suitable for the flourishing of life – then there is no reason in principle that these odds cannot be calculated from astronomical data. We do not have to arrive at such odds by direct appeal to the Principle of Indifference.

There are, of course, additional questions that need to be addressed if the LSA is to be convincing, most urgently questions regarding the likelihood of conditions hospitable to carbon-based life in particular given the existence of a designer. But the difference between the two arguments illustrates nicely the distinction between calculating the odds of encountering by chance a certain type of object *within* a universe and calculating the odds on chance of a universe itself.

If universes were as experimentally tractable as amino acids or as profusely displayed to our gaze as binary star systems, then we might in principle be able to collect sufficient statistical information to make informed claims about the relative frequency of life-friendly parameters. But they are not. In consequence, we are in no position to speak of what we might have expected instead of the universe we have. Arguments for design have to be framed within our universe and in terms of its laws if they are to have probative force.[6]

Notes

1 The mathematics will not be affected if the points constituting the boundary of the ball are taken to be external to the region; the only crucial assumption here is that these regions have an interior with positive measure.

It is conceivable that the life-friendly zone might take a topologically more interesting form if the loss of life-friendliness contingent on increasing the value of one parameter could be restored by increasing or decreasing others. We might then have a life-friendly "foam" in R_+^K, a continuous region of universes favorable to life but sensitive to the alteration of individual parameters. This possibility tends, however, to lessen the force of the FTA by increasing the range of possible life-friendly universes. In what follows we will ignore it.

2 The argument could, of course, have some force even if the probability of a life-friendly universe on the chance hypothesis were merely somewhat low rather than vanishingly small, provided that it is lower than the probability on the design hypothesis.

3 Leslie seems to endorse this interpretation (1989: 17).

4 We owe this objection to Rob Koons in extended correspondence.

5 John Leslie points out in private conversation that another version of what might be called an FTA, in which the ratio of two constants encodes a detailed message, may also escape the normalizability problem. This argument raises fresh issues and lies beyond our scope here, but it is worth pointing out that no one has seriously advanced the claim that the ratios of the constants actually encode such a message.

6 Thanks to John Leslie for constructive criticism that improved the paper, Robin Collins for discussions of inflationary cosmology, Rob Koons for much vigorous correspondence over the FTA, and the participants at the two places where earlier versions have been presented: the "Design and its Critics" conference at Concordia University in June 2000 and the Notre Dame summer symposium on Cosmic Fine-tuning in July 2000, particularly Neil Manson, Mike Thrush, John Mullen, Bradley Beach, and Peter and Alisa Bokulich.

References

Barrow, J. and Tipler, F. (1986) *The Anthropic Cosmological Principle*, Oxford: Clarendon Press.

Carter, B. (1974) "Large number coincidences and the anthropic principle in cosmology," in Leslie, J. (ed.) (1998), pp. 131–9

Leslie, J. (ed.) (1998) *Modern Cosmology and Philosophy*, New York: Prometheus Books.

—— (1989) *Universes*, New York: Routledge.

Manson, N. (2000) "There is no adequate definition of 'fine-tuned for life'," *Inquiry* 43: 341–51.

Ross, H. (1998) "Big Bang model refined by fire," in W. Dembski (ed.) (1998) *Mere Creation*, Downer's Grove, Illinois: InterVarsity Press, pp. 363–84.

Shapiro, R. and Feinberg, G. (1982) "Possible forms of life in environments very different from earth," in Leslie, J. (ed.) (1998), pp. 254–61 .

Part III

MULTIPLE UNIVERSES

11

OTHER UNIVERSES

A scientific perspective[1]

Martin Rees

Many "universes"?

We do not know whether there are other universes. Perhaps we never shall. But I want to argue that "Do other universes exist?" can be posed in a form that makes it a genuine scientific question. Moreover, I shall outline why it is an interesting question; and why, indeed, I already suspect that the answer may be "yes."

First, a pre-emptive and trivial comment: if you define "the Universe" as "everything there is," then by definition there cannot be others. I shall, however, follow the convention among physicists and astronomers, and define "the Universe" as "the domain of space–time that encompasses everything that astronomers can observe." Other "universes," if they existed, could differ from ours in size, content, dimensionality, or even in the physical laws governing them. It would be neater, if other universes existed, to redefine the whole enlarged ensemble as "the Universe," and then introduce some new term – for instance, "the metagalaxy" – for the domain to which cosmologists and astronomers have access. But so long as these concepts remain so conjectural, it is best to leave the term "Universe" undisturbed, with its traditional connotations, even though this then demands a new term, "the multiverse," for a (still hypothetical) ensemble of universes.

Current theories – based on well-defined (albeit often untested) assumptions – have expanded our conceptual horizons, bringing the multiverse within the scope of cosmological discourse. Our entire universe, stretching 10 billion light years in all directions, could (according to one widely studied model) have inflated from an infinitesimal speck; moreover, this "inflationary" growth could have led to a universe so large that its extent requires a million-digit number to express it. But even this vast expanse may not be everything there is: patches where inflation doesn't end may grow fast enough to provide the seeds for other Big Bangs. If so, our Big Bang wasn't the only one, but could be part of an eternally reproducing cosmos.

Other lines of thought (distinct from the generic concept of inflation) also point towards a possible multiplicity of universes. Some theorists conjecture, for instance, that, whenever a black hole forms, processes deep inside it trigger another universe that creates a "new" space–time disjoint from our own. Alternatively, if there were extra spatial dimensions that weren't tightly rolled up, we may be living in one of many separate universes embedded in a higher dimensional space. The entire history of our universe could be just an episode, one facet, of the infinite multiverse. I shall try to argue that this is a genuinely testable and scientific hypothesis. But before doing so, let me briefly sketch why the multiverse concept seems attractive.

A special recipe?

Obviously we can never fully delineate all the contingencies that led from a Big Bang to our own birth here 13 billion years later. But the outcome depended crucially on a recipe encoded in the Big Bang, and this recipe seems to have been rather special. A degree of fine-tuning – in the expansion speed, the material content of the Universe, and the strengths of the basic forces – seems to have been a prerequisite for the emergence of the hospitable cosmic habitat in which we live. Let us consider some prerequisites for a universe containing organic life of the kind we find on Earth.

It must be very large compared to individual particles and very long-lived compared with basic atomic processes. Indeed, this is surely a requirement for any hypothetical universe that a science fiction writer could plausibly find interesting. If atoms are the basic building blocks, then clearly nothing elaborate could be constructed unless there were huge numbers of them. Nothing much could happen in a universe that was too short-lived; an expanse of time, as well as space, is needed for evolutionary processes.

Even a universe as large and long-lived as ours could be very boring. It could contain just black holes, or inert dark matter, and no atoms at all. It could even be completely uniform and featureless. Moreover, the laws must allow the variety of atoms required for complex chemistry.

Three interpretations of the apparent "tuning"

If our existence depends on a seemingly special cosmic recipe, how should we react to the apparent fine-tuning? There appears to be a choice between three options: we can dismiss it as happenstance; we can acclaim it as the workings of providence; or (my preference) we can conjecture that our universe is a specially favored domain in a still vaster multiverse. Let's consider them in turn.

Happenstance (or coincidence)

Maybe a fundamental set of equations, which some day will be written on T-shirts, fixes all key properties of our universe uniquely. It would then be an unassailable fact that these equations permitted the immensely complex evolution that led to our emergence.

But I think there would still be something to wonder about. It is not guaranteed that simple equations permit complex consequences. To take an analogy from mathematics, consider the beautiful pattern known as the Mandelbrot set. This pattern is encoded by a short algorithm, but has infinitely deep structure; tiny parts of it reveal novel intricacies however much they are magnified. In contrast, you can readily write down other algorithms, superficially similar, that yield very dull patterns. Why should the fundamental equations encode something with such potential complexity, rather than the boring or sterile universe that many recipes would lead to?

One hardheaded response is that we couldn't exist if the laws had boring consequences. We manifestly are here, so there is nothing to be surprised about. I'm afraid this leaves me unsatisfied. I'm impressed by a well-known analogy given by the philosopher John Leslie. Suppose you are facing a firing squad. Fifty marksmen take aim, but they all miss. If they hadn't all missed, you wouldn't have survived to ponder the matter. But you wouldn't leave it at that. You'd still be baffled and you'd seek some further reason for your luck. Likewise, I think we would need to know why the unique recipe for the physical world should permit consequences as interesting as those we see around us (and which, as a byproduct, allow us to exist).

Providence (or design)

Two centuries ago William Paley introduced the famous analogy of the watch and the watchmaker – adducing the eye, the opposable thumb, and so on as evidence of a benign Creator. This line of thought fell from favor, even among most theologians, in post-Darwinian times. But the seemingly biophilic features of basic physics and chemistry can't be as readily dismissed as the old claims for design in living things. Biological systems evolve in symbiosis with their environment, but the basic laws governing stars and atoms are given, and nothing biological can react back on them to modify them. A modern counterpart of Paley, John Polkinghorne, interprets our fine-tuned habitat as "the creation of a Creator who wills that it should be so."

A special universe drawn from an ensemble (or multiverse)

If one doesn't believe in providential design, but still thinks the fine-tuning needs some explanation, there is another perspective – a speculative one, however. There may be many "universes" of which ours is just one. In the others, some laws and physical constants would be different. But our universe wouldn't be just a random one. It would belong to the unusual subset that offered a habitat conducive to the emergence of complexity and consciousness. The analogy of the watchmaker could be off the mark.

Instead, the cosmos maybe has something in common with an "off the shelf" clothes shop: if the shop has a large stock, we're not surprised to find one suit that fits. Likewise, if our universe is selected from a multiverse, its seemingly designed or fine-tuned features wouldn't be surprising.

Are questions about other universes part of science?

Science is an experimental or observational enterprise, and it is natural to be troubled by assertions that invoke something inherently unobservable. Some might regard the other universes as being in the province of metaphysics rather than physics. But I think they already lie within the proper purview of science. It is not absurd or meaningless to ask "Do unobservable universes exist?" even though no quick answer is likely to be forthcoming. The question plainly can't be settled by direct observation, but evidence can be sought that could lead to an answer.

There is actually a blurred transition between the readily observable and the absolutely unobservable, with a very broad gray area in between. To illustrate this, one can envisage a succession of horizons, each taking us further than the last from our direct experience:

The limit of present-day telescopes

There is a limit to how far out into space our present-day instruments can probe. Obviously, there is nothing fundamental about this limit; it is constrained by current technology. Many more galaxies will undoubtedly be revealed in the coming decades by bigger telescopes now being planned. We would obviously not demote such galaxies from the realm of proper scientific discourse simply because they haven't been seen yet. When ancient navigators speculated about what existed beyond the boundaries of the then-known world, or when we speculate now about what lies below the oceans of Jupiter's moons Europa and Ganymede, we are speculating about something real – we are asking a scientific question. Likewise, conjectures about remote parts of our universe are genuinely scientific, even though we must await better instruments to check them.

The limit in principle at the present era

Even if there were absolutely no technical limits to the power of telescopes, our observations are still bounded by a horizon, set by the distance that any signal, moving at the speed of light, could have traveled since the Big Bang. This horizon demarcates the spherical shell around us at which the redshift would be infinite. There is nothing special about the galaxies on this shell, any more than there is anything special about the circle that defines your horizon when you're in the middle of an ocean. On the ocean, you can see

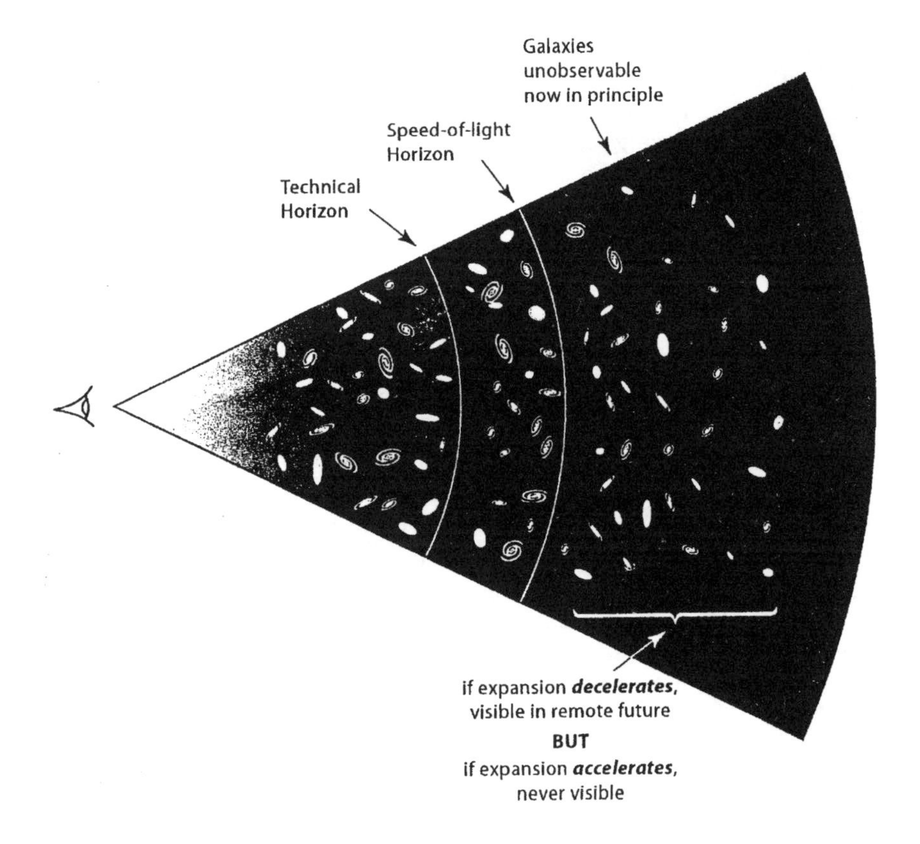

Figure 11.1 Horizons of observability

farther by climbing up your ship's mast. But our cosmic horizon can't be extended unless the Universe changes, so as to allow light to reach us from galaxies that are now beyond it.

If our universe were decelerating, then the horizon of our remote descendants would encompass extra galaxies that are beyond our horizon today. It is, to be sure, a practical impediment if we have to await a cosmic change taking billions of years, rather than just a few decades (maybe) of technical advance, before a prediction about a particular distant galaxy can be put to the test. But does that introduce a difference of principle? Surely the longer waiting-time is a merely quantitative difference, not one that changes the epistemological status of these far-away galaxies.

Never-observable galaxies from "our" Big Bang

But what about galaxies that we can never see, however long we wait? It is now believed that we inhabit an accelerating universe. As in a decelerating universe, there would be galaxies so far away that no signals from them have yet reached us. Yet if the cosmic expansion is accelerating, we are now receding from these remote galaxies at an ever-increasing rate, so if their light hasn't yet reached us, it never will. Such galaxies aren't merely unobservable in principle now – they will be beyond our horizon forever. But if a galaxy is now unobservable, it hardly seems to matter whether it remains unobservable forever, or whether it would come into view if we waited a trillion years (and I have argued above that the latter category should certainly count as real).

Galaxies in disjoint universes

The never-observable galaxies would have emerged from the same Big Bang as we did. But suppose that, instead of causally-disjoint regions emerging from a single Big Bang (via an episode of inflation) we imagine separate Big Bangs. Are space–times completely disjoint from ours any less real than regions that never come within our horizon in what we'd traditionally call our own universe? Surely not – so these other universes too should count as real parts of our cosmos.

This step-by-step argument (those who don't like it might dub it a slippery slope argument!) suggests that whether other universes exist or not is a scientific question. So how might we answer it?

Scenarios for a multiverse

At first sight, nothing seems more conceptually extravagant – more grossly in violation of Occam's razor – than invoking multiple universes. But this concept follows from several different theories (albeit all speculative). Linde, Vilenkin, and others have performed computer simulations depicting an "eternal" inflationary phase where many universes sprout from separate Big Bangs into disjoint regions of space–time. Guth and Smolin have, from different viewpoints, suggested that a new universe could sprout inside a black hole, expanding into a new domain of space and time inaccessible to us. And Randall and Sundrum suggest that other universes could exist, separated from us in an extra spatial dimension; these disjoint universes may interact gravitationally, or they may have no effect whatsoever on each other. In the hackneyed analogy where the surface of a balloon represents a two-dimensional universe embedded in our three-dimensional space, these other universes would be represented by the surfaces of other balloons. Any bugs confined to one, and with no conception of a third dimension, would be unaware of their counterparts crawling around on another balloon. Other universes would be separate domains of space and time. We couldn't even meaningfully say whether they existed before, after, or alongside our own, because such concepts make sense only in so far as we can impose a single measure of time, ticking away in all the universes.

Guth and Harrison have even conjectured that universes could be made in the laboratory by imploding a lump of material to make a small black hole. Is our entire universe perhaps the outcome of some experiment in another universe? If so, the theological arguments from design could be resuscitated in a novel guise. Smolin speculates that a daughter universe may be governed by laws that bear the imprint of those prevailing in its parent universe. If that new universe were like ours, then stars, galaxies, and black holes would form in it; those black holes would in turn spawn another generation of universes, and so on, perhaps *ad infinitum*.

Parallel universes are also invoked as a solution to some of the paradoxes of quantum mechanics – most notably in the "many worlds" theory first advocated by Everett and Wheeler in the 1950s. This concept was prefigured by Stapledon, as one of the more sophisticated creations of his *Star Maker*:

> Whenever a creature was faced with several possible courses of action, it took them all, thereby creating many…distinct histories of the cosmos. Since in every evolutionary sequence of this cosmos there were many creatures and each was constantly faced with many possible courses, and the combinations of all their courses were innumerable, an infinity of distinct universes exfoliated from every moment of every temporal sequence.

None of these scenarios has been simply dreamed up out of the air: each has a serious, albeit speculative, theoretical motivation. However, one of them, at most, can be correct. Quite possibly none are; there are alternative theories that would lead just to one universe.

Firming up any of these ideas will require a theory that consistently describes the extreme physics of ultra-high densities, how structures on extra dimensions are configured, and so on. But consistency is not enough; there must be grounds for confidence that such a theory isn't a mere mathematical construct, but applies to external reality. We would develop such confidence if the theory accounted for things we can observe that are otherwise unexplained.

At the moment, we have an excellent framework, called the Standard Model, which accounts for almost all subatomic phenomena that have been observed. But the formulae of the Standard Model involve numbers that can't be derived from the theory but have to be inserted from experiment. Perhaps in the future physicists will develop a theory that yields insight into (for instance) why there are three kinds of neutrinos, and into the nature of the nuclear and electric forces. Such a theory would thereby acquire credibility. If the same theory, applied to the very beginning of our universe, were to predict many Big Bangs, then we would have as much reason to believe in separate universes as we now have for believing inferences from particle physics about quarks inside atoms, or from relativity theory about the unobservable interiors of black holes.

Universal laws, or mere bylaws?

"Are the laws of physics unique?" is a less poetic version of Einstein's famous question, "Did God have any choice in the creation of the universe?" The answer determines how much variety the other universes – if they exist – might display. If there were something uniquely self-consistent about the actual recipe for our universe, then the aftermath of any Big Bang would be a rerun of our own universe. But a far more interesting possibility (which is certainly tenable in our present state of ignorance of the underlying laws) is that the underlying laws governing the entire multiverse may allow variety among the universes. Some of what we call "laws of nature" may in this grander perspective be local bylaws, consistent with some overarching theory governing the ensemble, but not uniquely fixed by that theory.

As an analogy, consider the form of snowflakes. Their ubiquitous six-fold symmetry is a direct consequence of the properties and shape of water molecules. But snowflakes display an immense variety of patterns because each is molded by its microenvironment; how each flake grows is sensitive to the fortuitous temperature and humidity changes during its growth. If physicists achieved a fundamental theory, it would tell us which aspects of nature were direct consequences of the bedrock theory (just as the symmetrical template of snowflakes is due to the basic structure of a water molecule) and which are (like the distinctive pattern of a particular snowflake) the outcome of accidents. The accidental features could be imprinted during the cooling that follows a Big Bang – rather as a piece of red-hot iron becomes magnetized when it cools down, but with an alignment that may depend on chance factors.

The cosmological numbers in our universe, and perhaps some of the so-called "constants" of laboratory physics as well, could be "environmental accidents" rather than uniquely fixed throughout the multiverse by some final theory. Some seemingly fine-tuned features of our universe could then only be explained by "anthropic" arguments, which are analogous to what all observers or experimenters do when they allow for selection effects in their measurements. If there are many universes, most of which are not habitable, we should not be surprised to find ourselves in one of the habitable ones!

Testing multiverse theories here and now

We may one day have a convincing theory that tells us whether a multiverse exists and whether some of the so-called "laws of nature" are just parochial bylaws of our cosmic patch. But while we are waiting for that theory – and it could be a long wait – the "off the shelf" clothes shop analogy can already be checked. It could even be refuted: this would happen if our universe turned out to be even more specially tuned than our presence requires. Let me give two quite separate examples of this style of reasoning.

First, Boltzmann argued that our entire universe is an immensely rare "fluctuation" within an infinite and eternal time-symmetric domain. There are now many arguments against this hypothesis, but even when it was proposed one could already have noted that fluctuations in large volumes are far more improbable than in smaller volumes. So it would be overwhelmingly more likely (if Boltzmann were right) that we would be in the smallest fluctuation compatible with our existence. Whatever our initial assessment of Boltzmann's theory, its probability would plummet as we came to realize the extravagant scale of the cosmos.

Second, even if we knew nothing about how stars and planets formed, we would not be surprised to find that our Earth's orbit wasn't highly eccentric: if it had been, water would boil when the Earth was at perihelion, and freeze at aphelion – a harsh environment unconducive to our emergence. However, a modest orbital eccentricity is plainly not incompatible with life. If it had turned out that the Earth moved in a near-perfect circle, then we could rule out a theory that postulated anthropic selection from orbits whose eccentricities had a "Bayesian prior" that was uniform in the range [0,1].

We could apply this style of reasoning to the important numbers of physics (for instance, the cosmological constant lambda) to test whether our universe is typical of the subset that could harbor complex life. The methodology requires us to decide what values are compatible with our emergence. It also requires a specific theory that gives the relative Bayesian priors for any particular value. For instance, in the case of lambda, are all values equally probable? Are low values favored by the physics? Or are there a finite number of discrete possible values? With this information, one can then ask if our actual universe is typical of the subset in which we could have emerged. If it is a grossly atypical member even of this subset (not merely of the entire multiverse) then we would need to abandon our hypothesis.

As another example of how multiverse theories can be tested, consider Smolin's conjecture that new universes are spawned within black holes, and that the physical laws in the daughter universe retain a memory of the laws in the parent universe: in other words, there is a kind of heredity. Smolin's concept is not yet bolstered by any detailed theory of how any physical information (or even an arrow of time) could be transmitted from one universe to another. It has, however, the virtue of making a prediction about our universe that can be checked. If Smolin were right, universes that produce many black holes would have a reproductive advantage that would be passed on to the next generation. Our universe, if an outcome of this process, should therefore be near-optimum in its propensity to make black holes, in the sense that any slight tweaking of the laws and constants would render black hole formation less likely. (I personally think Smolin's prediction is unlikely to be borne out, but he deserves our thanks for presenting an example that illustrates how a multiverse theory can in principle be vulnerable to disproof.)

These examples show that some claims about other universes may be refutable, as any good hypothesis in science should be. We cannot confidently assert that there were many Big Bangs – we just don't know enough about the ultra-early phases of our own universe. Nor do we know whether the underlying laws are "permissive." Settling this is a challenge to twenty-first century physicists. But if the laws are permissive, then anthropic explanations would become legitimate – indeed they'd be the only sort of explanation we will ever have for some important features of our universe.

What we have traditionally called "the Universe" may be the outcome of one Big Bang among many, just as our Solar System is merely one of many planetary systems in our Galaxy. Just as the pattern of ice crystals on a freezing pond is an accident of history, rather than being a fundamental property of water, so some of the seeming constants of nature may be arbitrary details rather than features uniquely defined by the underlying theory. The quest for exact formulae for what we normally call the constants of nature may consequently be as vain and misguided as was Kepler's quest for the exact numerology of planetary orbits. And other universes will become part of scientific discourse, just as "other worlds" have been for centuries. Nonetheless (and here I gladly concede to the philosophers), any understanding of why anything exists – why there is a universe (or multiverse) rather than nothing – remains in the realm of metaphysics.

Notes

1 This chapter is based on a presentation made at the symposium "Our Universe – and Others?" held at Darwin College, Cambridge University, on 26 April 2000. The other main presentation at this seminar, by D.H Mellor, appears as the next chapter in this book. The themes presented in this chapter are discussed more fully in *Our Cosmic Habitat* (Princeton New Jersey: Princeton University Press, 2001).

12

TOO MANY UNIVERSES[1]

D.H. Mellor

Universes and the multiverse

In his talk, and the book it is based on (Rees 1997), Martin Rees argues that whether there are universes other than our own is a scientific question to which he suspects the answer is "yes." To this I could agree, but only up to a point, the point being that there could be scientific evidence for the theories he mentions that postulate other universes. But Martin also has what I shall argue is a spurious non-scientific reason for suspecting that other universes exist, namely that he thinks they would make the fact that our universe supports life less surprising than it would otherwise be. Whether, without that anthropic support, we have enough reason to take these theories seriously I do not know.

Before rebutting Martin's anthropic arguments, I need to settle some terminology. I agree with him that, to avoid trivializing our debate, "our universe" (as I shall call it) must not mean "everything there is." But nor can it quite mean, as he suggests, "the domain of space–time that encompasses everything that astronomers can observe." For, first, our universe may have parts that we cannot observe, perhaps because (as he himself suggests) the acceleration of our universe's expansion will stop light from them ever reaching us. More seriously, Martin's definition entails that no other universe could contain astronomers, since, if it did, that would automatically make it part of our universe! (And we cannot exclude these otherworldly astronomers because they are not in our universe without an independent definition of "our universe" that will then make Martin's redundant.) So what I propose to call "our universe," and what I think Martin really means, is everything, past, present, and future, in the single space–time whose earliest point is our Big Bang.

Next, since I shall still need a term for everything that exists in some space–time or other, and it would beg the question to call that "the multiverse," I shall call it "the Universe." The question then is this: Does the Universe contain more than our universe? Specifically, does it contain other space–times with different contents, laws, and/or initial conditions, most of which would not, unlike ours, permit life as we know it? That is the multiverse hypothesis, which Martin thinks the existence of life in our universe gives us some reason to accept.

His case for this hypothesis rests on the "fine-tuning – in the expansion speed, the material content of the universe, and the strengths of the basic forces – [which] seems to have been a prerequisite for the emergence of the hospitable cosmic habitat in which we live." To the fact of this "seemingly special cosmic recipe" Martin offers us three responses: "we can dismiss it as happenstance; we can acclaim it as the workings of providence; or…we can conjecture that our universe is a specially favored domain in a still vaster multiverse." Setting providence aside, he prefers the last hypothesis because he thinks it does, as happenstance does not, explain the fine-tuning of our universe. Is he right?

First, a point about Martin's other universes, which may be understood in two apparently different ways. The first way takes them to be as actual as our universe, thus making it just a part (if not a spatial part, or even a temporal part) of all there actually is. The second way makes them merely *possible* universes, which might have been actual but in fact are not. This may seem wrong, since the multiverse hypothesis says that these other universes exist, which for most of us is the same as saying they are actual. But there is a view, forcefully argued by David Lewis (1986), according to which all possible universes exist, just as ours does, and all that our calling ours "actual" means is that it is the one we happen to be in.

Both readings of Martin's universes seem to me conceivable. Which is right Martin does not say; but then he may not need to say. For both readings make the Universe – i.e. everything that exists in some space – time or other – include far more universes than ours, and differ only in how inclusive they allow a single universe to be. For on the all-actual reading, one universe can grow out of another (e.g., as Martin suggests, from a black hole), so that two universes can have parts related to each other in time if not in space (or not in our everyday three-dimensional space). Whereas for Lewis, who takes a single universe to include the whole of its time as well as its space, the contents of two space–times linked in this way would constitute one universe with a more complex space–time structure.

Existence, location, and ultimate explanations

For our purposes, however, it does not matter how many universes a multiverse contains, given that, by definition, it contains more than one. For that is what enables multiverse theories to replace questions of *existence* with questions of *location*, which is what seems to let them do what Martin wants, and what he thinks one-universe theories cannot do, namely render the fine-tuning of our universe unsurprising.

The idea is this. If our universe is all there is, we cannot explain why it has the features that permit life. For example, early conditions in our universe may give physical explanations of later ones, and the very earliest, initial conditions may even explain the later laws and values of physical constants that life requires. But initial conditions themselves can obviously not be explained in this way, since – by definition – nothing is earlier than they are. And if there is no physical explanation of our universe's initial conditions, there is no ultimate physical explanation of what they explain, and in particular no such explanation of the emergence of life. Hence the problem, which is of course no news to philosophers or theologians.

But for Martin a multiverse poses no such problem, since it contains all possible features in some universe or other. It is, as he puts it, rather like

> an "off the shelf" clothes shop: if the shop has a large stock, we're not surprised to find one suit that fits. Likewise, if our universe is selected from a multiverse, its seemingly designed or fine-tuned features wouldn't be surprising.

In other words, the question now is not why we *exist*, but why we exist *where* we exist, namely in a universe with such-and-such features.

To change the question in this way is like turning the question of why there are fish (say) into the question of why they live where they do, namely in water: to which the obvious answer is that water, unlike dry land, has what fish need. The situation is similar with the multiverse. It lets us turn the hard question of why there is life at all into the relatively easy one of why there is life in our universe: to which the obvious answer is that our universe has what life needs.

But one-universe theories can answer *that* question just as well as multiverse theories can, because their answer is the same. Take our fish again. The explanation of why fish live in water (because water has what fish need) is the same whether there is dry land or not, i.e. whether this explanation of fish is locational or existential. Likewise, the features of our universe that explain why it permits life are the same whether there are other universes or not, i.e. whether this explanation of the possibility of life is locational or existential. The only question then is this: can multiverse (water + land) theories meet the need, which one-universe (all-water) theories cannot meet, for an explanation of the fine-tuning of our universe (water) that allows life (fish) to exist in it? Martin says they can; I say they cannot.

Explanations and probabilities

The illusion that multiverse theories can explain the fine-tuning of our universe rests on a confounding of two kinds of possibility, *epistemic* and *physical*, of which only the latter enables the explanations that physical theories give. To see this, consider in general when and why we take events to need explaining. Generally we want explanations only of events that we think did not *have* to happen, since there seemed to be alternative possibilities. This is why the best explanations are those that eliminate all such alternatives, as when we discover deterministic causes that make it impossible for their effects not to happen.

When the possibility of an event not happening cannot be eliminated in this way, it may still be reduced. That is how indeterministic causes explain events, by reducing the possibility of their not happening by making the events more probable than they would otherwise have been. That, for example, is how smoking explains the cancers smokers get: not by making it impossible for them not to get cancer, but by reducing that possibility by raising their chances of getting cancer. And even when an event happens that is as improbable as it can be, as when a radioactive atom decays from its most stable state, it may still be explained, to some extent, by saying what the chance of that event was.

Some of what I have just said about how explanations depend on chances is controversial. What should not be controversial is that whenever chances do explain events, they can only do so because they are real physical probabilities, measuring real possibilities of the events of which they are chances. Merely epistemic probabilities, because they are not real features of our universe, but only measures of our knowledge or ignorance of what is going on in it, can explain nothing of it (although they can and do explain our reactions to it).

Thus suppose, for example, I am surprised to see a tossed coin land on edge. Suppose also that the lighting and my eyesight are good, I am sober, and there are no conjurers around. Then relative to such facts, which are what make my vision reliable, the epistemic probability that the coin did what I saw it do is very high. In other words, relative to the evidence of my senses, there is almost no epistemic possibility of the coin *not* having landed on edge. Yet this fact in no way *explains* the coin landing on edge, precisely because it tells us nothing about the real physical probability of that event. That is why the event still surprises me, despite its high epistemic probability, since I still think there was a much lower chance – a much smaller real possibility – of it happening than of it not happening.

Equally, of course, many events that I see clearly, and which therefore have a very high epistemic probability, I know independently to have a very high chance – as when I see a tossed coin *land*. That is an event that, unlike it landing on edge, I find unsurprising, and think needs no explaining, precisely because its physical probability is high. It is only events that I think have low chances, and therefore high chances of not happening, that I find surprising and think need explaining: their epistemic probability, high or low, is irrelevant.

A prerequisite of chances

What, then, gives an event a chance, a physical probability, that may if it is high enough give the event a physical explanation? The normal answer is that an earlier event (a coin being tossed, someone smoking, an atom being in a relatively stable or unstable state), together with laws of nature, gives the event in question its chance, and hence whatever explanation that value of chance can provide. This is why earlier events can explain later ones but not vice versa. It is also why an event like our Big Bang seems to have no precursors seems thereby rendered incapable of physical explanation, since there is, by hypothesis, nothing earlier that could give it any physical probability, high or low.

But suppose our Big Bang did have a precursor – say, a black hole in a parent universe. This might indeed give the initial conditions of our universe chances that could, if high enough, explain them and thereby tell us why our universe permits life. But then these conditions would not really be *initial* conditions: the real initial conditions would be those of our parent universe, or those of its parent universe, or…And now we face a dilemma. For on the one hand, there may be a first universe. Then, as its initial conditions can have no physical probability, and hence no physical explanation, they cannot give us any ultimate explanation of whatever later conditions and universes they explain. On the other hand, there may be no first universe: all the universes in the multiverse may have ancestors. Then while the initial conditions of each universe *may* have a physical probability, there are no absolutely initial conditions that could give us a physical explanation of all the others and hence of the emergence of life. So in neither case do we get the ultimate explanation that one-universe theories are criticized for not supplying.

To this Martin might retort that multiverse theories are not trying to give *physical* explanations of the life-friendly features of our universe. But then they must explain these features in some other way. But no other credible way exists, as I now propose to show by looking more closely at the stock argument for multiverses.

An improbable argument

The basic premise of the argument for multiverses is this: it is surprising that a single universe should have the very improbable features, including the initial conditions, which enable it to contain life. But what does 'improbable' mean here? It cannot mean physically improbable, since the initial conditions of a single universe have no physical probability, high or low. So 'improbable' here can only mean epistemically improbable. Yet relative to the empirical evidence that tells us what the relevant features of our universe are, they are not at all epistemically improbable: on the contrary, they are – by definition –

epistemically very probable. Only if we ignore this evidence, and take the epistemic probability of these features relative only to logic, and perhaps a few basic assumptions of physics, can they be made to appear improbable. And that, I am willing to grant, for the sake of argument, is what they are.

Yet even granting this, what does this difference between two epistemic probabilities show? Compare my surprise at seeing a coin land on edge. Relative to my seeing it, this event has a very high epistemic probability. Relative to the coin's geometry, however, I may think its epistemic probability is very low – perhaps because I think that far fewer of a tossed coin's possible trajectories make it land on edge than make it land heads or tails, and, following Laplace (1820), that all these trajectories are, *a priori*, equally probable.

But however I derive it, this low *a priori* epistemic probability is not what makes me surprised to see a coin land on edge. What makes that event surprise me is my belief that it had a low *physical* probability, because of how I think the coin was tossed. What *would* remove my surprise, therefore, by explaining the coin landing on edge better than my assumption that it was fairly tossed does, would be my discovering that it was *placed* on edge, i.e. that, unknown to me, there was a mechanism that gave this event a high physical probability.

But this is not what Martin's multiverse provides. All it provides is a large set of possible initial conditions, and other relevant features of universes, over which something like a flat Laplacean probability distribution yields a very low probability of the subset of features that let a universe support life. But as this low *a priori* probability is merely epistemic, no one should be surprised that, relative to the *a posteriori* evidence provided by physics, the same features have a very high epistemic probability. For to say that such evidence increases an epistemic probability is just to say that it tells us something we did not know before: in this case, what the relevant features of our universe are. But then, as our coin analogy shows, their high *a posteriori* epistemic probability in no way implies a high physical probability, any more than their low *a priori* probability implies a low physical probability. So, by the same token, if these features of a single universe seem incapable of explanation, that is not because they have a low *a priori* epistemic probability, but because they include features, like initial conditions, which have no physical probability, high or low, at all.

Facing the firing squad

The fact is that multiverse theories could only explain the fine-tuning of our universe by giving it a physical probability high enough to provide a physical explanation of it; yet that, as we have seen, they neither do nor claim to do. To see that this is what they would have to do, take the example of John Leslie's that Martin cites:

> Suppose you are facing a firing squad. Fifty marksmen take aim, but they all miss. If they had not all missed, you would not have survived to ponder the matter. But you would not leave it at that. You'd still be baffled, and you'd seek some further reason for your luck.

Well, maybe you would, but only because you thought the ability of the firing squad, the accuracy of their weapons, and their intention to kill you made their firing together a mechanism that gave your death a very high physical probability.

So now suppose there is no such mechanism. Imagine, as Russell (1927) did, that our universe (including all our memories and other present traces of the past) started five minutes ago, with these fifty bullets coming past you, but with no prior mechanism to give their trajectories any physical probability, high or low. Suppose in other words that these trajectories really were among the *initial* conditions of our universe. If you thought that, should you really be baffled and seek some further reason for your luck? I say not, and I say also that, if you were still baffled, it should not reduce your bafflement to be told that the initial conditions of many other universes include similar swarms of bullets, the vast majority of which end up hitting people! If that information affected you at all – which I do not think it should – it should make you more baffled, not less, that your swarm missed you.

I think therefore that the anthropic intuitions that have led Martin and others to favor multiverse theories are simply mistaken. They are like the intuitions behind the Gambler's Fallacy that (for example) the longer an apparently normal coin goes on always landing heads, the more likely it is to land tails next time. That intuition, common though it may be among unsuccessful gamblers, we know is just wrong. For if a coin repeatedly landing heads tells you anything, what it tells you is that the coin is biased towards heads and so more, not less, likely to land heads next time than you previously thought.

In short, what the intuition behind the Gambler's Fallacy needs is not an explanation of why it is right, since it isn't. What anyone with that intuition needs is not a theory to justify it but some kind of therapy to remove it. The same goes for anthropic intuitions about the alleged improbability of the features of our universe that enable it to support life. Martin should not be trying to explain and justify these intuitions by postulating other universes. Rather, he should be taking to heart Thomas Carlyle's alleged response to one Margaret Fuller's reported remark that she accepted the universe: "Gad," said Carlyle, "she had better." And so had Martin.

Notes

1 This chapter is based on a response to Martin Rees's talk at the symposium "Our Universe – and Others?" held at Darwin College, Cambridge University, on 26 April 2000.

References

Laplace, P.S. de (1951 [1820]) *A Philosophical Essay on Probabilities*, trans. F.W. Truscott and F.L. Emory, New York: Dover.

Lewis, D.K. (1986) *On the Plurality of Worlds*, Oxford: Blackwell.

Rees, M. (1997) *Before the Beginning: Our Universe and Others*, London: Simon & Schuster.

Russell, B. (1927) *An Outline of Philosophy*, London: Allen & Unwin.

13

FINE-TUNING AND MULTIPLE UNIVERSES

Roger White

Introduction

John Leslie (1989) argues vigorously that the fact that our universe meets the extremely improbable yet necessary conditions for the evolution of life supports the thesis that there exist very many universes. This view has found favor with a number of philosophers such as Derek Parfit (1998), J.J.C. Smart (1989), and Peter van Inwagen (1993).[1] My purpose is to argue that it is a mistake. First let me set out the issue in more detail.

The Universe is said to be extraordinarily "fine-tuned" for life. The inhabitability of our universe depends on the precise adjustment of what seem to be arbitrary, contingent features. Had the boundary conditions in the initial seconds of the Big Bang and the values of various fundamental constants differed ever so slightly we would not have had anything like a stable universe in which life could evolve. In the space of possible outcomes of a Big Bang, only the tiniest region consists of universes capable of sustaining life. Most either last only a few seconds, or contain no stable elements, or consist of nothing but black holes. This is a fairly standard story told by cosmologists – there is some controversy, concerning for instance the appropriate measure on the space of possible outcomes – but I will assume it is the right picture for the purpose of this discussion.[2] The situation can also be described using the following analogy. Nuclear bombs have been connected to a high-security combination lock, such that dozens of dials had to be adjusted with extreme precision to avoid detonating the bombs. Had any one dial differed ever so slightly from its actual position, the world would have been destroyed. In the absence of an explanation of why the dials were adjusted as they were (suppose they had been spun at random) we should find it astonishing that we were here to consider the matter.

In response to this seemingly remarkable state of affairs, philosophers and physicists have suggested various hypotheses involving multiple universes. By "universe" I do not mean "possible world." Rather, according to multiple-universe theories, the actual world consists of very many large, more or less isolated subregions (universes) either coexisting, or forming a long temporal sequence. The crucial feature of the various multiple-universe theories is that those physical parameters on which *inhabitability* depends are understood to be assigned randomly for each universe.[3]

How are multiple universes relevant to the puzzle? The basic idea is straightforward. For any improbable outcome of a trial (e.g. dealing a royal flush, hitting a hole in one, throwing a bull's-eye), if you repeat the trial enough times you can expect to get an outcome of that type eventually. If we suppose that our universe is just one of very many universes, randomly varying in their initial conditions and fundamental constants, it is to be expected that at least one of them is life-permitting. Add to this the fact that we could only find ourselves in a life-permitting universe and we seem to have satisfyingly accounted for what at first seemed amazing, removing the temptation to suppose that there was a Fine-Tuner who adjusted the physical constants for a purpose. It is widely thought, therefore, that the fact that our universe is fine-tuned for life *provides evidence* for the multiple-universe theory. In fact almost everyone who has written on the topic accepts that the fine-tuning facts count in favor of multiple universes, even if they are not persuaded that there are other universes.[4] But they are mistaken, or so I will argue. Perhaps there is independent evidence for the existence of many universes. But the fact that our universe is fine-tuned gives us no further reason to suppose that there are universes other than ours. I will examine the two main lines of reasoning found in the literature from fine-tuning to multiple universes to see where they go wrong.

Probabilistic confirmation

The first strategy takes a probabilistic approach to confirmation, according to which confirmation is the raising of probability. That is, evidence E confirms hypothesis H, given background knowledge K, if and only if

$$P(H|E \& K) > P(H|K).$$

A probabilistic understanding of confirmation supports the use of the common-sense principle that, as Leslie puts it, "observations improve your reasons for accepting some hypothesis when its truth would have made those observations more likely" (1989: 121). A theorem of the probability calculus that underlies this principle is

$$P1: P(H|E \& K) > P(H|K) \leftrightarrow P(E|H \& K) > P(E|{\sim}H \& K).$$

A related theorem that will prove useful is

$$P2: P(H|E \& K) = P(H|K) \leftrightarrow P(E|H \& K) = P(E|{\sim}H \& K).$$

In applications of probability to confirmation, controversy often arises concerning the assignment of prior probabilities. How are we to determine the probability of M, that there are many universes, prior to the fine-tuning evidence E? One possible reason for concern is that if $P(M|K)$ is extremely low, then $P(M|E \& K)$ might not be much higher, even if $P(E|M \& K)$ is much higher than $P(E|{\sim}M \& K)$. This need not concern us, however, for the question at hand is whether E provides *any* support for M at all. We may grant, for the sake of argument, that the multiple-universe hypothesis has a non-negligible prior probability, or is even quite probable. Principles P1 and P2 give us a handy test for whether the fine-tuning evidence E provides *any* evidence for M: it does so if and only if E is more likely given M than given its denial.

Now the appealing idea here is that a *single* life-permitting universe is exceedingly improbable, but if we suppose there are or have been very many universes, it is to be expected that eventually a life-permitting one will show up, just as if you throw a pair of dice long enough you can expect to get a double-six at some time (you cannot, of course, expect it on any particular throw). It is tempting, then, to suppose that the fine-tuning evidence confirms the multiple-universe theory by P1, since the latter raises the probability of the former.

But here we need to be clear about what our evidence is. For simplicity, let us suppose that we can partition the space of possible outcomes of a Big Bang into a finite set of equally probable configurations of initial conditions and fundamental constants: $\{T_1, T_2, \ldots, T_n\}$ (think of the universes as n-sided dice, for a very large n).[5] Let the variable 'x' range over the actual universes. Let *a* be our universe and let T_1 be the configuration that is necessary to permit life to evolve.[6] Each universe instantiates a single T_i, i.e. $(\forall x)(\exists i)T_i x$. Let m be the number of universes that actually exist, and let

$E = T_1 a$ = *a* is life-permitting;
$E' = (\exists x)T_1 x$ = some universe is life-permitting; and
M = m is large (the multiple-universe hypothesis).

It is important to distinguish E from the weaker E'. For while E' is more probable given M than it is given ~M, M has no effect on the probability of E. First let us consider E'. In general,

$$P((\exists x)T_i x|m = k) = 1 - (1 - 1/n)^k \text{ for any i.}^7$$

So

$$P((\exists x)T_i x|M) > P((\exists x)T_i x|{\sim}M) \text{ for any i.}$$

So

$$P((\exists x)T_1x|M) > P((\exists x)T_1x|{\sim}M).$$

That is

$$P(E'|M) > P(E'|{\sim}M).$$

E, on the other hand, is just the claim that *a* instantiates T_1, and the probability of this is just 1/n, regardless of how many other universes there are, since *a*'s initial conditions and constants are selected randomly from a set of n equally probable alternatives, a selection that is independent of the existence of other universes. The events that give rise to universes are not causally related in such a way that the outcome of one renders the outcome of another more or less probable. They are like independent rolls of a die. That is,

$$P(E|M) = P(T_1a|M) = 1/n = P(T_1a|{\sim}M) = P(E|{\sim}M)$$

Given M, it is likely that some universe instantiates T_1, and it is true that *a* instantiates some T_i, but it is highly improbable that the T_i instantiated by *a* is T_1, regardless of the truth of M. So by P2, P(M|E) = P(M), i.e. the fact that our universe is life-permitting does not confirm the multiple-universe hypothesis one iota. Perhaps the claim that it does results from a confusion between E and E'.

Ian Hacking (1987) has made a similar criticism with respect to J.A. Wheeler's oscillating-universe theory, according to which our universe is the latest of a long temporal sequence of universes. Hacking labels the mistake involved the Inverse Gambler's Fallacy, suggesting that it is related to the notorious Gambler's Fallacy. In the Gambler's Fallacy, after throwing a pair of dice repeatedly without getting a double-six, the gambler concludes that he has a much better chance of getting it on the next roll, since he is unlikely to roll several times without a double-six. In the Inverse Gambler's Fallacy, the gambler is asked "Has this pair of dice been rolled before?" He asks to see the dice rolled before he makes a judgment. They land double-six. He concludes that they probably have been rolled several times, since they are so unlikely to land double-six in one roll, but are quite likely to after several.

There is no doubt that Hacking has identified a fallacy here. He suggests that this is what is at work in the inference from the fine-tuning of our universe to Wheeler's hypothesis that ours is just the most recent in a long sequence of universes. We note that, against all odds, the Big Bang has produced a life-permitting universe – extremely unlikely in one shot, but highly likely after several. So we conclude that there have probably been many Big Bangs in the past. The mistake is in supposing that the existence of many other universes makes it more likely that *this* one – the only one that we have observed – will be life-permitting. The Inverse Gambler's Fallacy combines the Gambler's Fallacy with P1, so the usual antidotes to the gambler's reasoning should be instructive here also. Wheeler universes, like dice, "have no memories"; the individual oscillations are stochastically independent. Previous Big Bangs in the sequence have no effect on the outcome of any other Big Bang, so they cannot render it more likely to produce a life-permitting universe. Although Hacking does not mention them, similar points apply to models of coexisting universes. These universes are usually taken to be causally isolated, or, if there is any causal relation between them, it is not of a type that could increase the probability of this universe being life-permitting.

Our universe versus some universe

Let us now turn to a common response to the arguments above. I have been insisting that *a* is no more likely to be life-permitting no matter how many other universes there are, but of course the more universes there are, the more likely it is that *some* universe supports life. That is, M raises the probability of E' but not E. But now, the response goes, we know that E' is true since it follows from E. So E' confirms M even if E does not. In other words, our knowledge that *some* universe is life-permitting seems to give us reason to accept the multiple-universe hypothesis, even if our knowledge that *a* is life-permitting does not.[8]

We can quickly see that there is something going wrong here. A known proposition, the probability of which is not raised by the hypothesis, is being set aside in favor of a *weaker* proposition, the probability of which is raised by the hypothesis. The weaker proposition is then taken as evidence for the hypothesis. Suppose I'm wondering why I feel sick today, and someone suggests that perhaps Adam got drunk last night. I object that I have no reason to believe this hypothesis since Adam's drunkenness would not raise the probability of *me* feeling sick. But, the reply goes, it does raise the probability that *someone* in the room feels sick, and we know that this is true, since we know that you feel sick, so the fact that someone in the room feels sick is evidence that Adam got drunk. Clearly something is wrong with this reasoning. Perhaps if all I knew (by word of mouth, say) was that someone or other was sick, this would provide some evidence that Adam got drunk. But not when I know specifically that *I* feel sick. This suggests that in the confirming of hypotheses, we cannot, as a general rule, set aside a specific piece of evidence in favor of a weaker piece.

What has gone wrong here seems to be a failure to consider the *total evidence* available to us. If the extent of our knowledge was just E', then this would count as evidence for M, since $P(M|E') > P(M)$. But we also know E, and must not leave that out of our calculation of the probability of M. What matters is the probability of M given E' and E. But now since E entails E', (E' & E) is equivalent to E. So $P(M|E' \& E) = P(M|E)$. But, as we have seen above, $P(M|E)$ is just equal to $P(M)$. Hence $P(M|E' \& E) = P(M)$. So while the multiple-universe hypothesis may be confirmed by E' alone, it is not confirmed by E' in conjunction with the more specific fact E, which we also know. It does not matter in which order we calculate the relevance of E and E'; our confidence in M on the basis of our total evidence should remain the same as it is without considering E or E'.

Consider how this fits with our intuitions about the gambler's reasoning. Suppose on being asked how many times the pair of dice has been rolled, the gambler asks if a double-six has been rolled. Upon learning that one has, he is more confident than he was that the dice have been rolled a number of times. Here his reasoning is sound, for the more times the dice have been rolled, the greater the chance that a double-six has been rolled. However, when the gambler witnesses a single roll and is then more confident that the dice have been rolled before, he is clearly making a mistake. The difference is that, in the first case, the information he has gained is just that *some* roll or other landed double-six; in the second case, he witnesses a specific roll. Compare this with the case where astronomers discover that there have been some Big Bangs in addition to the one from which we came. They ask us to guess whether there have been just a few or very many of these additional Big Bangs. We might ask whether any had produced a universe containing life, and, on learning that one did, be more inclined to suppose that there have been many. This reasoning would be correct. But this is not our situation. Like the gambler in the second case we have simply witnessed a single Big Bang producing this universe. And no number of other Big Bangs can affect the probability of the outcome we observed.

Carter's hypothesis

Puzzlingly, Hacking believes that there is a version of the multiple-universe hypothesis which avoids the errors that we have been considering. He interprets Brandon Carter as proposing a set of coexisting universes instantiating *all* possible configurations of initial conditions and fundamental constants. Hacking argues that there is no fallacy of probability involved here since the inference is deductive: "Why do we exist? Because we are a possible universe, and all possible universes exist.…Everything in this reasoning is deductive. It has nothing to do with the inverse gambler's fallacy" (1987: 337).

I believe Hacking is making a similar mistake as that identified above. Carter's hypothesis can be represented as M*: $(\forall i)(\exists x)T_i x$. Now M* certainly entails E': $(\exists x)T_1 x$. But it does not entail, nor does it raise the probability of, E: $T_1 a$. From the hypothesis that each of the possible configurations of initial conditions and constants is instantiated in some actual universe, it follows that *some* universe meets the conditions required for life.

It by no means follows that *a* does. The situation here is parallel to the standard multiple-universe hypothesis M. Where M *raised the probability of* E', but not E, M* *entails* E', but does not entail E.

In saying that "our universe follows deductively from [M*]" Hacking (1987: 339) may mean to say that the existence of a universe of the same *type* as ours – one instantiating the same set of conditions and constants – follows deductively from M*, and this would certainly be correct. He may wish to maintain that it is the existence of a universe of our type that constitutes evidence for Carter's hypothesis. But if this move worked, we could likewise argue that this same fact confirms Wheeler's hypothesis, for the existence of a long sequence of universes does raise the probability that a universe of our type will exist at some time. Since Hacking, correctly in my view, finds fault with the argument for Wheeler's hypothesis, he should likewise find fault with the argument for Carter's.

The observational selection effect

Hacking's Inverse Gambler's Fallacy argument has received a series of replies and I will turn now to consider these. First, Leslie complains that "Hacking's story *involves no observational selection effect*" (1988: 270). An observational selection effect is a feature of a process that restricts the type of outcomes of an event that are observable. In the case of the Big Bang, had the Universe not instantiated T_1, neither we nor anyone else would be around to notice, since the necessary conditions for life would not have been met. So even though Big Bangs can so easily result in dud universes, no one ever has the misfortune of seeing one. In an attempt to show how such an effect can be crucial to the inference to multiple universes, a number of intriguing analogies have been suggested. I will focus on two analogies suggested by P.J. McGrath, as I believe they capture the essence of each of the stories suggested in the literature (my critique of these carries over to the other stories). In each case I will argue that the inference involved in the story is correct, but that the story is not analogous to our situation with respect to the Universe.

The first case involves an analogy with Wheeler's oscillating universe theory:

> *Case A*: Jane takes a nap at the beginning of a dice-rolling session, on the understanding that she will be awakened as soon as a double-six is rolled and not before. Upon waking she infers that the dice have been rolled a number of times.[9]

The reasoning here certainly seems legitimate, but it will pay us to be clear on why this is so. Note that it seems that even *before* she takes a nap, she should predict that she will awake after a number of rolls. This is roughly because it is unlikely that a double-six occurs in just a few rolls, and hence the first double-six is likely to occur *later* than a few rolls. Now, if it is reasonable to predict this *before* she takes the nap, it is just as reasonable to believe this afterward. But there is an implicit assumption involved here, namely that *there will be many rolls*, or at least as many as it takes to get a double-six.

It is not clear whether McGrath intended that this assumption be made in the story, but it is in fact necessary for his conclusion that she "is entitled to conclude, when roused, that it is probable that the dice have been rolled at least twenty-five times" (McGrath 1988: 266). How do we calculate the figure *twenty-five*? This calculation crucially depends on a prior probability distribution over hypotheses concerning the maximum number of times the dice rollers will roll. Suppose Jane knows that they are planning to roll just once, unless they happen to win the lottery that day, in which case they will roll many times. In this case Jane is certainly not entitled to the conclusion that McGrath suggests.

To consider the matter more carefully, we can let W = Jane is awakened, and partition this into two hypotheses: W_L = Jane is awakened in twenty-five rolls or more, and W_E = Jane is awakened in less than twenty-five rolls. The prior probability of there being *no* double-six in the first twenty-four rolls, $P(\sim W_E) = (35/36)^{24} \approx 0.5$. When Jane is roused and hence knows W is true, how confident should she be that twenty-five or more rolls have occurred?

$$
\begin{aligned}
P(W_L|W) &= P(W_L \,\&\, W)/P(W) && \\
&= P(W_L)/P(W) && \text{(since } W_L \text{ entails W)} \\
&= [P(W) - P(W_E)]/P(W) && \\
&= P(\sim W_E) && \text{if and } \textit{only if } P(W) = 1
\end{aligned}
$$

If P(W) is significantly less than one, then $P(W_L|W) < 0.5$. So Jane is entitled to conclude, when roused, that it is probable that the dice have been rolled at least twenty-five times, only on the assumption that the prior probability of her waking was close to one, i.e. that it was almost guaranteed that the dice would be rolled many times, or at least enough times for a double-six to appear.[10]

Now it should be clear that this assumption is not welcome in the case of our universe. It will be useful here to make use of some propositions that Hacking distinguishes for a different purpose:

> W_1: our universe is one of a large temporal sequence of universes; and
> W_2: our universe has been preceded by very many universes.[11]

W_2 is quite probable given W_1. For, on the basis of W_1, we know that there exists, speaking timelessly, a temporally ordered sequence of universes in space–time. But we do not know which position in the sequence our universe holds. Whenever we have a large sequence of objects, the probability that a particular object will be very early in the sequence will be very low. So if the sequence of universes entailed by W_1 is large enough, it renders W_2 highly probable (note that this reasoning has nothing to do with fine-tuning).[12] But of course we do not know that W_1 is the case. We only know that our universe is fine-tuned for life. The truth of W_1 is part of what we are trying to figure out. So McGrath's story is not relevant to the question at hand.

Now let us consider McGrath's second analogy, which is drawn with a model of coexisting universes:

> *Case B*: Jane knows that an unspecified number of players will simultaneously roll a pair of dice just once, and that she will be awakened if, and only if, a double-six is rolled. Upon waking she infers that there were several players rolling dice.[13]

Once again Jane's reasoning seems to be cogent. However, McGrath is mistaken in supposing that this case is essentially the same as Case A, and that, as before, Jane is entitled to infer that there were probably at least *twenty-five* players rolling dice. The judgment concerning the twenty-five rolls had to do with the position within a sequence at which the first double-six occurred. There is no such sequence in Case B, and in fact the reasoning should proceed along very different lines. The probability of Jane waking is raised by the multiple-rolls hypothesis, since she is to be awakened if and only if *some* player rolls a double-six. And the more players there are, the greater the chance that at least one of them will roll a double-six. There is no Inverse Gambler's Fallacy here. Jane's evidence is not about the outcome of a particular roll, but simply the fact that she is awake. And the probability of this fact *is* raised by the multiple-rolls hypothesis, given the policy of the dice rollers to wake her upon *any* double-six.

To see what is fishy about this case, however, let us compare it with the following:

> *Case B**: Jane knows that she is one of an unspecified number of sleepers each of whom has a unique partner who will roll a pair of dice. Each sleeper will be awakened if and only if *her* partner rolls a double-six. Upon waking, Jane infers that there are several sleepers and dice rollers.

Jane's reasoning here is unsound. She may of course have independent grounds for the multiple-rolls hypothesis, but her being awake adds nothing. The crucial difference here concerns the nature of the observational selection effect involved. In each case, if there is no double-six rolled then Jane will not awake. But in Case B, the converse holds also: if some double-sixes are rolled, then Jane will awake, whereas in Case B*, Jane's waking depends on a single roll. It is this *converse* observational selection effect at work in Case B that provides a link between the evidence (her being awake) and the multiple-rolls hypothesis. Since this is lacking in Case B*, the multiple-rolls hypothesis does not raise the probability of Jane's being awake. So Jane has no grounds on which to infer that there were many dice rollers.

The crucial question, therefore, is whether the case of our observership in the Universe involves a similar *converse* selection effect. It strikes me that it obviously does not. As Leslie admits, it is not as though we were disembodied spirits waiting for a Big Bang to produce some universe that could accommodate us. We are products of the Big Bang that produced this universe. It is certainly not sufficient, for us to exist in some universe β, that β is fine-tuned, or even that β is qualitatively exactly as α actually is. After all, if we postulate enough universes, the chances are that there exist several life-permitting universes, perhaps even universes with precisely the same initial conditions and fundamental constants as our universe, and containing human beings indistinguishable from us. But *we* do not inhabit these universes, other folks do. If we accept Kripke's (1980) thesis of the necessity of origins, we should hold that no other Big Bang could have possibly produced us. But even if this thesis is denied, even if it is metaphysically possible for us to have evolved in a different universe, or be products of a different Big Bang, we have no reason to suppose that we would exist if a different universe had been fine-tuned. In order for the multiple-universe hypothesis to render our existence more probable, there must be some mechanism analogous to that in Case B linking the multiplicity of universes with our existence. But there is no such mechanism. So the existence of lots of universes does not seem to make it any more likely that *we* should be around to see one. So the converse selection effect does not hold, and hence McGrath's analogy fails to vindicate the reasoning from the fact that we are alive to see a fine-tuned universe to the hypothesis that our universe is one of many.

Improbable and surprising events

Let us turn to the second and perhaps more tempting line of reasoning in support of the multiple-universe hypothesis. At some points, Leslie insists that although multiple universes do not render the fine-tuning of our universe, or even our existence, less *improbable*, they do render it less *surprising*, and it is the latter that is significant. The distinction between surprising and unsurprising improbable events is easily illustrated with examples. It is unsurprising that Jane won a lottery out of a billion participants, but it is surprising that Jim won three lotteries in a row each with a thousand participants (even though the probability in each case is one in a billion). It is unsurprising that a monkey types "nie348n sio 9q;c," but when she types "I want a banana!" we are astonished.

Now, it is a familiar theme in the philosophy of science that scientific knowledge often advances by making that which is puzzling understandable. We should not be content with events like a monkey typing English sentences; we must seek some account that makes these events understandable. It seems then that any theory that could remove the surprising nature of the fine-tuning data would thereby be confirmed. As Leslie suggests, "a fairly reliable sign of correctness is ability to reduce amazement" (1989: 141). And the multiple-universe theory does seem to do just that. For given enough universes it is unsurprising that there is a life-permitting one, and it is unsurprising that we happen to be in a life-permitting one since we could not be in any other kind. Doesn't the fact that this story satisfyingly accounts for what is otherwise puzzling make it plausible?

The idea here can be brought out in another way. That the Universe, by pure chance, should have such a fine adjustment of physical parameters to allow for the evolution of life would be *extraordinary*, and it is contrary to reason to believe in the extraordinary (like believing that a monkey wrote *Hamlet*, or that Rembrandt's works are entirely the result of randomly spilt paint). One way to avoid believing that an extraordinary coincidence has occurred is to accept that the Universe is the product of intelligent design; another way is to suppose that ours is one of very many universes. One or the other of these, it is argued, must be preferred to the "extraordinary fluke" hypothesis. So if the design hypothesis is not to your liking, the multiple-universe hypothesis is a plausible alternative.

This intuition is not entirely misguided. In many cases where a hypothesis renders an event less surprising, the hypothesis is thereby confirmed. For one way to make an event less surprising is to make it less *improbable*. And according to P1, raising the probability of an event is one way that a hypothesis can be confirmed. But according to the probabilistic account of confirmation this is the *only* way that a hypothesis is confirmed by the occurrence of an improbable event. I hope to remove the temptation to suppose that *any* hypothesis that reduces the surprisingness of an event is thereby confirmed, by considering a counter-example (ironically one of Leslie's) and by giving a satisfying probabilistic account of how a hypothesis can render an event less surprising without being confirmed.

The distinction between surprising and unsurprising improbable events is an important one that deserves much attention, yet it has received very little in the literature. There is not the space here to consider the matter in depth. I will sketch an account of surprisingness, drawing on suggestions by Paul Horwich (1982), which is adequate for the purposes of our discussion. The crucial feature of surprising events seems to be that they challenge our assumptions about the circumstances in which they occur. If at first we assume that the monkey is typing randomly, then her typing "nie348n sio 9q" does nothing to challenge this assumption. But when she types "I want a banana" we suspect that this was more than an accident. The difference is that in the second case there is some alternative but not wildly improbable hypothesis concerning the conditions in which the event took place, according to which the event is much more probable. On the assumption that the monkey is typing randomly, it is just as improbable that she types "nie348n sio 9q" as it is that she types "I want a banana." But that the second sequence is typed is more probable on the hypothesis that it was not merely a coincidence, that an intelligent agent had something to do with it, either by training the monkey or by rigging the typewriter, or something similar. There is no such hypothesis (except an extremely improbable *ad hoc* one) that raises the probability that the monkey would type the first sequence. Of course, by P1, the human intervention hypothesis is confirmed in the case of "I want a banana." So what makes the event surprising is that it forces us to reconsider our initial assumptions about how the string of letters was produced (of course, someone who already believes that the typewriter was rigged should not be surprised).

Why is it surprising that the Universe is fine-tuned for life? Perhaps because on the assumption that the Big Bang was just an accident it is extremely improbable that it would be life-permitting, but it is far more likely on the assumption that there exists an intelligent designer, for a designer might prefer to bring about a universe that is inhabitable by other intelligent creatures rather than a homogeneous cosmic soup. The event is surprising in that it forces us to question whether the Big Bang really was an accident (someone who already believes in a designer should not be surprised that the Universe is life-sustaining).[14]

Leslie's shooting analogy

To see the way that different hypotheses can affect the surprisingness of an event, consider one of Leslie's analogies.[15] You are alone in the forest when a gun is fired from far away and you are hit. If at first you assume that there is no one out to get you, this would be surprising. But now suppose you were not in fact alone but instead part of a large crowd. Now it seems there is less reason for surprise at being shot. After all, someone in the crowd was bound to be shot, and it might as well have been you.

Leslie suggests this as an analogy for our situation with respect to the Universe. Ironically, it seems that Leslie's story supports my case against his. For it seems that while knowing that you are part of a crowd makes your being shot less surprising, being shot gives you no reason at all to suppose that you are part of a crowd. Suppose it is pitch dark and you have no idea if you are alone or part of a crowd. The bullet hits you. Do you really have any reason at all now to suppose that there are others around you?

Let us examine the case more carefully. While it is intuitively clear that the existence of many people surrounding you should reduce the surprisingness of your being shot, there is no adequate account of why this is so. I will present an original analysis of this surprisingness reduction, which both helps us see why reduction of surprisingness need not involve confirmation, and serves as a model for a deeper understanding of the relation between fine-tuning data and multiple universes. Let

E = you are shot; and
D = the gunman was malicious and not shooting accidentally (the design hypothesis); and
M = you are part of a large crowd (the multiple-people hypothesis).

We begin with the assumption that you are alone and the gun was fired randomly. $P(E|\sim D \,\&\, \sim M)$ is very low, i.e. there is a slim chance that a randomly fired bullet would hit you, for there is a wide range in which the bullet could move, with equal intervals of roughly equal probability and with those in which the bullet hits you constituting only a small proportion. But $P(E|D \,\&\, \sim M)$ is greater, since, if there is no other interesting target about you, then a malicious shooter is more likely to aim at you. So

$$P(E|D \,\&\, \sim M) > P(E|\sim D \,\&\, \sim M),$$

and hence, by P1,

$$P(D|E \,\&\, \sim M) > P(D|\sim M);$$

i.e. the fact that you have been shot confirms the malicious-gunman hypothesis, on the assumption that you are alone. This is what makes your being shot surprising: it challenges you to reconsider whether the shooting really was accidental (if you already knew that the gunman was a psychopath, you should not be surprised at getting hit).

Now consider the case where you know that you are part of a crowd. $P(E|\sim D \,\&\, M)$ is still very low, for the same reason that $P(E|\sim D \,\&\, \sim M)$ is. But unlike $P(E|D \,\&\, \sim M)$, $P(E|D \,\&\, M)$ is not much higher than $P(E|\sim D \,\&\, M)$, if higher at all. The reason is that while a malicious shooter may be expected to shoot a person, there is little reason to suppose that he would intend to shoot *you* in particular (unless perhaps you are the President). The probability that he will shoot *someone* is high, given that there is a crowd there, but the probability that it will be *you* remains very low, regardless of whether the shooting is deliberate. So

$$P(E|D \,\&\, M) \approx P(E|\sim D \,\&\, M)$$

and hence

$$P(D|E \,\&\, M) \approx P(D|M);$$

i.e. the fact that you have been shot does not confirm the malicious-gunman hypothesis on the assumption that you are part of a crowd.

What happens here is that the multiple-people hypothesis M *screens off* the probabilistic support that D lends to E, and hence also screens off the support that E lends to D. That is, relative to M, E and D are probabilistically independent. So if you first assumed that you were alone, your being shot may count as evidence that the gunman was firing deliberately. But if you later discover that you are part of a large crowd (perhaps it was pitch dark before), there is no longer any reason to question your original assumption that the shooting was accidental. So the multiple-people hypothesis renders your having been shot less surprising.

However, the multiple-people hypothesis does not *raise the probability* that you would be shot. No matter how many people are about you, a randomly fired bullet has the same chance of hitting you. So $P(E|M \& {\sim}D) = P(E|{\sim}M \& {\sim}D)$. But now it follows by P2 that $P(M|E \& {\sim}D) = P(M|{\sim}D)$. So the multiple-people hypothesis is not confirmed by the fact you have been shot, on the assumption that the bullet was fired randomly.[16, 17]

Someone may still be tempted to suppose that being shot gives them some reason to suppose that there are many people about. For getting shot all alone in an open field from far away would be extraordinary, and we should not believe in the extraordinary. One way to avoid accepting that something extraordinary has occurred is to suppose that the shot was fired deliberately, but another is to suppose that there are many people about. So if the malicious-gunman hypothesis seems ruled out on other grounds (just as many find the designer of the Universe hypothesis hard to swallow) then the multiple-people hypothesis might seem a plausible alternative.

I suggest that anyone who is still inclined to think this way might like to put their money where their mouth is in the following simulation experiment (we can use paint-balls instead of bullets). You are blindfolded and ear-muffed in a large area, knowing that there is an n percent probability that you are in a large crowd, and that otherwise you are alone. (A ball is drawn from a hundred, n of which are red. A crowd is assembled just in case a red is drawn.) Clearly, if asked to bet that you are in a crowd, you should accept odds up to n:100–n. But now a paint-ball is fired randomly from a long distance and happens to hit you. Are you now more than n percent confident that you are part of a crowd? If so you should be willing to accept odds *higher* than n:100–n. And if so, I suggest we play the game repeatedly, with you betting at higher odds on each of the rare occasions that a bullet hits you. On this strategy I should win all your money in the long-run. For we will find that in only n percent of those occasions in which you are shot, you are part of a crowd. If we take reasonable betting odds as a guide to reasonable degrees of confidence, this experiment supports my claim that being shot gives you no reason to suppose that you are part of a crowd.

Conclusion

This example illustrates that removal of surprise need not involve confirmation. A hypothesis can be such that if we knew it to be true, it would make a certain event less surprising, yet the fact that it makes this event less surprising gives us no reason to suppose that the hypothesis is true.[18] We are now in a position to give a deeper analysis of the way in which the multiple-universe hypothesis reduces the surprisingness of the fine-tuning data. Assuming there is just one universe, the fact that it is life-permitting is surprising. For this otherwise extremely improbable outcome of the Big Bang is more probable on the assumption that there is a cosmic designer who might adjust the physical parameters to allow for the evolution of life. So the fine-tuning facts challenge us to question whether the Big Bang was merely an accident.

However, on the assumption that our universe is just one of very many, the existence of a designer does not raise the probability that our universe should be life-permitting. For while we might suppose that a designer would create some intelligent life somewhere, there is little reason to suppose it would be *here* rather than in one of the many other universes. It is only on the assumption that there are no other options that we should expect a designer to fine-tune *this* universe for life. Given the existence of many universes, it is already probable that some universe will be fine-tuned; the design hypothesis does not add to the probability that any particular universe will be fine-tuned. So the multiple-universe hypothesis screens off the probabilistic link between the design hypothesis and the fine-tuning data. Hence if we happened to know, on independent grounds, that there are many universes, the fine-tuning facts would give us little reason to question whether the Big Bang was an accident, and hence our knowledge of the existence of many universes would render the fine-tuning of our universe unsurprising. However, postulate as many other universes as you wish, they do not make it any more likely that ours should be life-permitting or that we should be here. So our good fortune to exist in a life-permitting universe gives us no reason to suppose that there are many universes.[19]

Postscript

Objection 1

Your case hinges on whether we know merely that there is a life-permitting universe, or more specifically that *this* one has life. But the issue is not so simple. We can of course refer to what we see as "this universe," or label it "α" and express our evidence as "α is life-permitting." But even if we learned merely that *some* universe had life, couldn't we just as easily label that universe "β" and then express our evidence as "β is life-permitting?"

Reply

I admit that there are difficult issues here in which I would rather not get entangled, and I regret putting the argument in these terms as I now think the crucial issue is independent of these matters (although it is still crucial that the multiple-universe hypothesis only raises the likelihood that some universe has life, not that any particular one does). We all agree that if I roll a pair of dice and get a double-six, this gives me no evidence that my colleagues are rolling dice in their offices (it makes no difference if we add an observational selection effect, say, that if they hadn't landed double-six I would have been shot before seeing them). Here is the relevant principle:

Observation principle: An observation I make gives me evidence for hypothesis H only if it is more likely given H that I would make that observation.

It is not enough for confirmation that if my colleagues are rolling dice, it is more likely that *someone* will see a double-six. If my observation is to provide *me* with evidence of these other rolls, they will have to make it more likely that *I* would observe this. This would be the case if, say, there was some mechanism such that any pair of dice landing double-six would cause me to be transported into that room to see them. The reasoning here is independent of the question of whether I know that *this* pair of dice landed double-six, on *this* roll, rather than just that there is a double-six. What we need is a probabilistic link between my experiences and the hypothesis in question. One way of establishing such a link in the present case is to suppose that I was once an unconscious soul waiting to be embodied in whichever universe produced a hospitable living organism. On this assumption the more universes there are, the more likely I am to observe one. This is not just a cheap shot. It is an illustration of the *kind* of story that we need to support the inference to multiple universes.

Objection 2

We need not appeal to disembodied souls. There are very many beings who could have been created other than me. And I'm no more likely to be born in this universe than in any other. The more universes there are, the more living creatures there are. So the more opportunities I had to be picked out of the pool of "possible beings," and hence the greater the likelihood that *I* should be observing anything.

Reply

The metaphysical picture behind this story is dubious. But, quite apart from that, we can see that something must be wrong with this line of reasoning.

The standard argument takes the fact that a universe must be extremely fine-tuned to support life, that a random Big Bang has a very slim chance of producing life, as crucial to the case for multiple universes. If the current objector's argument is cogent, then it should go through regardless of the need for fine-tuning for life. That is, even if a universe with just any set of fundamental constants is bound to produce life, we could still argue along these lines that the more universes there are the more opportunities I had for existing and observing, and hence that my observations provide evidence for multiple universes.

Indeed, if the objector's argument is sound, then the discovery that a universe must meet very tight constraints in order to support life should *diminish* the strength of the case for multiple universes. For if every universe is bound to produce life, then by increasing the number of universes we rapidly increase the number of conscious beings, whereas if each universe has a slim chance of producing life, then increasing the number of universes increases the number of conscious beings less rapidly, and hence (by the objector's argument) increases the likelihood of my existence less. I would be surprised if anyone wants to endorse an argument with these consequences, but, at any rate, it is not the standard one that takes the *fine-tuning* data to be crucial in the case for multiple universes.

Objection 3

Isn't there something mysteriously *indexical* about this observation principle to which you are appealing? I take it you grant that the multiple-universe hypothesis explains why there are observers. To expect a further explanation of why *I* am here to observe anything is misguided. Indeed, it is doubtful that anything could possibly explain why it is *me* that exists rather than someone else.

Reply

Let me make it clear that I do *not* claim that my existence requires an explanation or even that it could be given one. What I do claim is that it does not follow from the fact that M explains the existence of life, that we thereby have any evidence for M. If all my colleagues were rolling dice, this would explain why there is a double-six in one of the offices. Having explained this, it would be misguided to demand a further explanation of why *I* saw one rather than someone else. Nevertheless, my seeing a double-six gives me no reason to suppose that anyone else was rolling dice. This, I suggest, is because it does not raise the likelihood of my making any observation (a hypothesis may raise the likelihood of my making an observation, without thereby explaining why it was me that made it).

I don't see any difficulty in the use of the indexical "I." It does not matter, for instance, if I have forgotten which member of the department I am. All that matters is that, whoever I am, I know that I had the same likelihood of seeing a double-six regardless of whether anyone else was rolling dice. There are dozens of analogies of multiple-universe reasoning suggested in the literature involving a "multiple-Xs" hypothesis (e.g. multiple dice rolls, multiple people, multiple firing squads). In some cases the inference to multiple Xs seems sound but in others it doesn't. I think that in each case our judgments conform to the observation principle. If you doubt this principle, I would like to see another one that both accounts for the analogies and licenses the inference to multiple universes.

Objection 4

In challenging the inference to multiple universes, aren't you really trying to promote the design hypothesis?

Reply

Whatever I was really doing, it is appropriate in the context of this book to compare the design (D) and multiple-universe (M) hypotheses. According to the design hypothesis, an intelligent agent had the power to adjust those physical parameters on which life's existence depends. How plausible the design hypothesis is in the light of our data depends on its prior probability and the degree to which it raises the likelihood of the data. These are, of course, matters on which there is plenty of room for debate, but which I can't discuss here. My interest is in comparing the hypotheses D and M if it is granted, as many do, that the design hypothesis does significantly raise the likelihood of there being life, and is not too implausible to begin with.

It is worth noting a crucial difference between the two hypotheses. Of all the possible outcomes of a Big Bang, the design hypothesis may raise the likelihood of those with life-permitting parameters, while *lowering* the likelihood of others. But the multiple-universe hypothesis *shows no favoritism* among possible outcomes. The more universes there are, the more likely there will be one with life-permitting constants; but for *any* possible set of constants, M raises the likelihood of there being a universe with those constants.

This gives us another reason to be suspicious of arguments for multiple universes that appeal to the fact that our universe is life-permitting. No matter how the universe had been, the likelihood of there being a universe like that would be greater if there were other universes. But it can't be that no matter how the Universe was, it being that way would be evidence for the existence of other universes. By contrast, while the existence of a life-permitting universe may be more likely given the design hypothesis, the existence of some other kind, say one containing nothing but scattered hydrogen atoms, would be less likely. So while some possible outcomes support D, others disconfirm it.

Many who write on the subject suppose that M and D are more or less on a par; both can solve the puzzle of life's existence, so our preference for one over the other must be based on other grounds. If what I have argued is correct, this is a mistake. While both hypotheses, if known to be true, would render life's existence and indeed my existence unsurprising, only the design hypothesis is confirmed by the evidence. The multiple-universe hypothesis may indeed undermine the argument for design, but only to the extent that we have independent reasons to believe it.

Objection 5

Earlier you complained that while M raises the likelihood of someone's observing something, it doesn't raise the likelihood of *me* doing so. But isn't D in the same boat? We might expect a designer to create some living creatures, but there is no reason to suppose that he would want to create *us*. Indeed, I'm not sure it makes sense to suppose that the designer could choose not only what kind of beings, but *which* beings to create.

Reply

We don't need to suppose that the designer would choose to create me in order for D to raise the likelihood of my existence (this raising of likelihood does not amount to an *explanation* of why I exist). Let E = I exist and E' = someone exists. I take it that the objector wants to say that

$$(*)\quad P(E|E' \,\&\, D) = P(E|E' \,\&\, {\sim}D);$$

i.e. however likely my existence is given that someone exists, the design hypothesis makes my existence no more or less likely. Now the degree to which D raises the probability of E is given by the likelihood ratio

$$\begin{aligned} P(E|D)/P(E|{\sim}D) &= P(E \,\&\, E'|D)/P(E \,\&\, E'|{\sim}D) && \text{(since E entails E')} \\ &= P(E'|D) \times P(E|E' \,\&\, D)/P(E'|{\sim}D) \times P(E|E' \,\&\, {\sim}D) \\ &= P(E'|D)/P(E'|{\sim}D) && \text{(by (*))} \end{aligned}$$

So, on this assumption, D raises the likelihood of E to the same degree that it raises the likelihood of E'.

Objection 6

By reasoning parallel to that given above, it follows from your claims that $P(E'|{\sim}M) > P(E'|M)$ while $P(E|M) = P(E|{\sim}M)$, that

(**) P(E|E' & M) < P(E|E' & ~M).

However, one would have thought that P(E|E' & M) = P(E|E' & ~M), in which case M and D are on a par.

Reply

Given that there is life, if ours is the only universe, then there must be life in this universe, whereas if there are other universes, then there may be life in one of them but not in ours. So given E', M lowers the likelihood of there being life in our universe. So unless one thinks that I had the same chance of turning up in any universe, then, given there is life, M lowers the likelihood of my existence, i.e. (**) is correct.[20]

Notes

1 See also Clifton (1991), Leslie (1988), McGrath (1988), Smith (1986), and Whitaker (1988).

2 See Leslie (1989) for a summary of the fine-tuning data, and Barrow and Tipler (1986) for a detailed account.

3 See the above references for accounts of multiple-universe theories.

4 A partial exception is Hacking (1987), who as we will see agrees only in a special case. Earman (1987) expresses his doubts about the inference but does not argue the point at any length.

5 For convenience, I use "T_1" and the like sometimes as *names* for configurations, sometimes as *predicates*. The use should be clear from the context.

6 The name "*a*" is to be understood here as *rigidly designating* the universe that happens to be ours. Of course, in one sense, a universe cannot be *ours* unless it is life-permitting. But the universe that happens actually to be ours, namely *a*, might not have been ours, or anyone's. It had a slim chance of permitting life at all.

7 This can be seen briefly as follows: for any i, the probability that a particular universe is T_i is 1/n, so the probability that it is not is $1-(1/n)$, and the probability that each of k universes is not T_i is $(1-(1/n))^k$. Hence the probability that some universe is T_i, given that there are k universes, is $1-(1-(1/n))^k$.

8 The point is sometimes made in terms of explanation, where explanation is understood to involve the raising of probability. What is surprising, and needs explanation, the argument goes, is just that there is a life-permitting universe, not that there is *this* one. The multiple-universe hypothesis does explain the existence of a life-permitting universe by rendering it probable. Once this is explained, the specific question of why *this* universe is fine-tuned for life does not require an answer, since it is not surprising. The issue of surprisingness and the reduction of surprisingness is addressed in the sixth and seventh sections; *explanation* is briefly discussed in note 17.

9 Adapted from McGrath (1988: 265). Leslie (1988) considers an equivalent story in which a person is created *ex nihilo* upon a double-six. Whitaker's (1988) first story involves a two-month period during which a casino is allowed to open on a night only if a double-six is rolled in one go that night. We see a photo of the open casino in the gossip column and conclude that it was taken much later than the first night. In the second story, you send out researchers to knock on doors

until they find a particular unusual kind of family. When they return, you conclude that they were not successful at the first house, but at one much later. I believe my objections to McGrath's case are equally relevant to these cases.

10 Without this assumption Jane is not entitled to conclude that there have been *twenty-five* rolls, but she does have evidence that there have been multiple rolls. The crucial point here is that *she* will be awakened no matter which roll lands double-six. The problem that this raises will be discussed in relation to Cases B and B*.

11 Adapted from Hacking (1987: 399).

12 Both Whitaker and Hacking are mistaken on this point. Whitaker (1988: 264) claims that W_2 *follows* from W_1. Hacking (1987: 399) claims that "W_2 does not follow from, nor is it made probable by W_1." The correct view is that W_2 does not follow from but is made probable by W_1. The reasons why are slightly different here than in the dice case, since we do not know that we inhabit the *first* life-permitting universe in the sequence.

13 Adapted from McGrath (1988: 267). Whitaker adapts his story of the casino such that the rule applies only for one night, but to more than one casino. If there are several casinos, we should expect to see photos of one of them open, since the photographer will visit an open one. As before, I believe my objections apply equally to this case.

14 Some will object that the design hypothesis is so improbable given our background knowledge that it is not significantly confirmed by the fine-tuning data, and hence does not challenge our assumption that the outcome of the Big Bang was an accident. I disagree, but there is no need to argue the point here. The argument for multiple universes under consideration depends on the assumption that the life-permitting character of the Universe is *surprising*, which is the case only if there is some (not wildly improbable) hypothesis which renders the life-permitting character of the Universe far more probable than it is given that it was the result of chance. If the hypothesis is not one of intelligent design, I am not sure what it could be. If there is no such hypothesis, we should not be puzzled by the Universe containing life, but view it as just one of the many highly improbable possible outcomes of the Big Bang – in which case the motivation for multiple universes under consideration loses its force.

15 This is adapted from Leslie's (1988) version of the story. In the discussion that follows, it should be distinguished from Leslie's (1989) version, which is told from the point of view of the shooter.

16 On the assumption of D, E *disconfirms* M, for if the gunman is firing deliberately, he is less likely to shoot you if there are many equally interesting targets about.

17 One reason that it is tempting always to take a theory's ability to reduce the surprisingness of data as evidence in its favor is that it is plausible that a theory's ability to *explain* data is always evidence in its favor. And a central role of explanation is the reduction of surprisingness. I think that the example shows that reduction of surprisingness is not *sufficient* for explanation. It seems wrong to say that the multiple-people hypothesis *explains* your being shot, for at least three reasons. First, explanations should answer why-questions, but the answer to "Why were you shot?" is not "Because there were many people surrounding you." Second, the fact that you were part of a crowd is not causally relevant to your being shot. Third, your being in a crowd does not raise your chances of being shot. (Similarly, the answer to "Why is *α* life-permitting?" is not "Because there are lots of other universes." Nor is the existence of many universes causally or probabilistically relevant to *α* containing life.) If there is a sense of "explains"

in which your being in a crowd explains your being shot, this can only show that, in this sense of the term, explanation is not sufficient for confirmation.

18 Numerous examples illustrate this point. In Case B*, Jane has reason to be surprised when awakened if she thinks that she is the only sleeper (we can make it more surprising by using ten dice landing all sixes, instead of a pair), but not if she knows that there are many sleepers and dice rollers. But her being awake gives her no reason to suppose that there are other sleepers and dice rollers. Or consider one of Leslie's (1989) favorite analogies of fine-tuning and multiple universes. You stand before a firing squad, the guns go off, but you are still alive! Astonishing if you are alone in this situation; not so amazing if there are billions of people about before similar firing squads. Yet again, your surviving the firing squad gives you no reason to accept the multiple-firing squad hypothesis, even if this hypothesis is plausible to begin with.

19 My thoughts in the early sections of this chapter owe a great deal to numerous discussions with Phil Dowe. I must also thank William Alston, Adam Elga, Ned Hall, Neil A. Manson, Brent Mundy, Robert Stalnaker, Peter van Inwagen, and two anonymous referees for helpful discussions on this topic and/or comments on earlier drafts.

20 While I'm reluctant to attribute my specific formulations of these objections to anyone, the following people helped me think further about the arguments in the chapter: Nick Bostrom, Phil Bricker, Pete Graham, Ned Hall, Neil A. Manson, Calvin Normore, Derek Parfit, Josh Schecter, and Mike Thrush.

References

Barrow, J.D. and Tipler, F.J. (1986) *The Anthropic Cosmological Principle*, Oxford: Clarendon Press.

Clifton, R.K. (1991) "Critical notice of John Leslie's *Universes*," *Philosophical Quarterly* 41: 339–44.

Earman, J. (1987) "The SAP also rises: A critical examination of the anthropic principle," *American Philosophical Quarterly* 24: 307–17.

Hacking, I. (1987) "The Inverse Gambler's Fallacy: The argument from design. The anthropic principle applied to Wheeler universes," *Mind* 76: 331–40.

Horwich, P. (1982) *Probability and Evidence*, Cambridge: Cambridge University Press.

Kripke, S. (1980) *Naming and Necessity*, Cambridge, Massachusetts: Harvard University Press.

Leslie, J. (1989) *Universes*, London: Routledge.

—— (1988) "No Inverse Gambler's Fallacy in cosmology," *Mind* 97: 269–72.

McGrath, P.J. (1988) "The Inverse Gambler's Fallacy and cosmology – A reply to Hacking," *Mind* 97: 265–8.

Parfit, D. (1998) "Why anything? Why this?", *London Review of Books* 22: 24–7.

Smart, J.J.C. (1989) *Our Place in the Universe: A Metaphysical Discussion*, Oxford: Blackwell.

Smith, Q. (1986) "World ensemble explanations," *Pacific Philosophical Quarterly* 67: 73–86.

van Inwagen, P. (1993) *Metaphysics*, Boulder, Colorado: Westview Press.

Whitaker, M.A.B. (1988) "On Hacking's criticism of the Wheeler hypothesis," *Mind* 97: 259–64.

14

THE CHANCE OF THE GAPS

William Dembski

Probabilistic resources

Statistical reasoning must be capable of eliminating chance when the probability of events gets too small. If not, chance can be invoked to explain anything. Scientists rightly resist invoking the supernatural in scientific explanations for fear of committing a God-of-the-gaps fallacy (the fallacy of using God as a stopgap for ignorance). Yet without some restriction on the use of chance, scientists are in danger of committing a logically equivalent fallacy – one we may call the chance-of-the-gaps fallacy. Chance, like God, can become a stopgap for ignorance. For instance, in the movie *This is Spinal Tap*, one of the lead characters remarks that a former drummer in his band died by spontaneously combusting. Any one of us could instantly spontaneously combust if all the fast-moving air molecules in our vicinity suddenly converged on us. Such an event, however, is highly improbable, and we do not give it a second thought.

Even so, high improbability by itself is not enough to preclude chance. After all, highly improbable events happen all the time. Flip a coin a thousand times, and you will participate in a highly improbable event. Indeed, just about anything that happens is highly improbable once we factor in all the ways what did happen could have happened. Mere improbability therefore fails to rule out chance. In addition, improbability needs to be conjoined with an independently given pattern. An arrow shot randomly at a large blank wall will be highly unlikely to land at any one place on the wall. Yet land it must, and so some highly improbable event will be realized. But now fix a target on that wall and shoot the arrow. If the arrow lands in the target and the target is sufficiently small, then chance is no longer a reasonable explanation of the arrow's trajectory.

Highly improbable, independently patterned events are said to exhibit specified complexity. The term "specified complexity" has been around since 1973 when Leslie Orgel introduced it in connection with origins-of-life research: "Living organisms are distinguished by their specified complexity. Crystals such as granite fail to qualify as living because they lack complexity; mixtures of random polymers fail to qualify because they lack specificity"

(1973: 189). More recently, Paul Davies has also used the term in connection with the origin of life: "Living organisms are mysterious not for their complexity *per se*, but for their tightly specified complexity" (1999: 112). Events are specified if they exhibit an independently given pattern (cf. the target fixed on the wall). Events are complex to the degree that they are improbable. The identification of complexity with improbability here is straightforward. Imagine a combination lock. The more possibilities on the lock, the more complex the mechanism, and correspondingly the more improbable that it can be opened by chance. Note that the "complexity" in "specified complexity" has a particular probabilistic meaning and is not meant to exhaust the concept of complexity; Seth Lloyd, for instance, records dozens of types of complexity (Horgan 1996: 303, note 11).

The most controversial claim in my writings (Dembski 1998) is that specified complexity is a reliable empirical marker of intelligent agency. There are several ways in which to criticize this claim. Elliott Sober (Sober 1999; Fitelson *et al.* 1999) criticizes it for failing to meet Bayesian standards of probabilistic coherence. Robin Collins (2001) criticizes it for hinging on an ill-defined conception of specification. Taner Edis (2001) criticizes it for admitting a crucial counter-example – the Darwinian mechanism of natural selection and random variation is supposed to provide a naturalistic mechanism for generating specified complexity. None of these criticisms holds up under scrutiny (Dembski 2002). Nevertheless, a worry about small probability arguments remains: given an independently given pattern, or specification, what level of improbability must be attained before chance can legitimately be precluded? A wall so large that it cannot be missed and a target so large that it covers half the wall, for instance, are hardly sufficient to preclude chance (or "beginner's luck") as the reason for an archer's success in hitting the target. The target needs to be small to preclude hitting it by chance.

But how small is small enough? To answer this question we need the concept of a probabilistic resource. A probability is never small in isolation but only in relation to a set of probabilistic resources that describe the number of relevant ways an event might occur or be specified. There are thus two types of probabilistic resources, *replicational* and *specificational*. To see what is at stake, consider a wall so large that an archer cannot help but hit it. Next, let us say we learn that the archer hit some target fixed to the wall. We want to know whether the archer could reasonably have been expected to hit the target by chance. To determine this we need to know of any other targets at which the archer might have been aiming. Also, we need to know how many arrows were in the archer's quiver and might have been shot at the wall. The targets on the wall constitute the archer's specificational resources. The arrows in the quiver constitute the archer's replicational resources.

Note that to determine the probability of hitting some target with some arrow by chance, specificational and replicational resources multiply. Suppose the probability of hitting any given target with any one arrow has probability no more than p. Suppose further there are N such targets and M arrows in the quiver. Then the probability of hitting any one of these N targets, taken collectively, with a single arrow by chance is bounded by Np, and the probability of hitting any of these N targets with at least one of the M arrows by chance is bounded by MNp. Thus to preclude chance for a probability p means precluding chance for a probability MNp once M replicational and N specificational resources have been factored in. In practice it is enough that $MNp < 1/2$ or $p < 1/(2MN)$. The rationale here is that since factoring in all relevant probabilistic resources leaves us with an event of probability less than 1/2, that event is less probable than not, and consequently we should favor the opposite event, which is more probable than not and precludes it.[1]

To recap, probabilistic resources comprise the relevant ways an event can occur (replicational resources) and be specified (specificational resources). The important question therefore is not "What is the probability of the event in question?" but rather "What does its probability become after all the relevant probabilistic resources have been factored in?" Probabilities can never be considered in isolation, but must always be referred to a relevant reference class of possible replications and specifications. A seemingly improbable event can become quite probable when placed within the appropriate reference class of probabilistic resources. On the other hand, it may remain improbable even after all the relevant probabilistic resources have been factored in. If it remains improbable (and therefore complex) and if the event is also specified, then it exhibits specified complexity.

Universal probability bounds

In the observable universe, probabilistic resources come in very limited supplies. Within the known physical universe it is estimated that there are around 10^{80} elementary particles. Moreover, the properties of matter are such that transitions from one physical state to another cannot occur at a rate faster than 10^{45} times per second. This frequency corresponds to the Planck time, which constitutes the smallest physically meaningful unit of time (Halliday and Resnick 1988: 544).[2] Finally, the Universe itself is about a billion times younger than 10^{25} seconds (assuming the Universe is between 10 and 20 billion years old). If we now assume that any specification of an event within the known physical universe requires at least one elementary particle to specify it and cannot be generated any faster than the Planck time, then these cosmological constraints imply that the total number of specified events throughout cosmic history cannot exceed

$$10^{80} \times 10^{45} \times 10^{25} = 10^{150}.$$

It follows that any specified event of probability less than 1 in 10^{150} will remain improbable even after all conceivable probabilistic resources from the observable universe have been factored in. A probability of 1 in 10^{150} is therefore a *universal probability bound*.[3] A universal probability bound is impervious to all available probabilistic resources that may be brought against it. Indeed, all the probabilistic resources in the known physical world cannot conspire to render remotely probable an event whose probability is less than this universal probability bound. The universal probability bound of 1 in 10^{150} is the most conservative in the literature. The French mathematician Emile Borel proposed 1 in 10^{50} as a universal probability bound below which chance could definitively be precluded (i.e. any specified event as improbable as this could never be attributed to chance) (Borel 1962: 28; Knobloch 1987: 228). Cryptographers assess the security of cryptosystems in terms of a brute-force attack that employs as many probabilistic resources as are available in the Universe to break a cryptosystem by chance. In its report on the role of cryptography in securing the information society, the National Research Council set 1 in 10^{94} as its universal probability bound for ensuring the security of cryptosystems against chance-based attacks (Dam and Lin 1996: 380, note 17).[4] Such levels of improbability are easily attained by real physical systems. It follows that if such systems are also specified and if specified complexity is a reliable empirical marker of intelligence, then these systems are designed.

Implicit in a universal probability bound such as 10^{-150} is that the Universe is too small a place to generate specified complexity by sheer exhaustion of possibilities. Stuart Kauffman (2000) develops this theme at length in his book *Investigations*.[5] In one of his examples (and there are many like it throughout the book), he considers the number of possible proteins of length 200 (i.e. 20^{200} or approximately 10^{260}) and the maximum number of pairwise collisions of particles throughout the history of the Universe (he estimates 10^{193} total collisions supposing the reaction rate for collisions can be measured in femtoseconds). Kauffman concludes: "The known universe has not had time since the big bang to create all possible proteins of length 200 [even] once." To emphasize this point, he notes: "It would take at least 10 to the 67th times the current lifetime of the universe for the universe to manage to make all possible proteins of length 200 at least once" (Kaufmann 2000: 144).

Kauffman even has a name for numbers that are so big that they are beyond the reach of operations performable by and within the Universe – he refers to them as *transfinite*. For instance, in discussing a small discrete dynamical system whose dynamics are nonetheless so complicated that they cannot be computed, he writes: "There is a sense in which the computations are transfinite – not infinite, but so vastly large that they cannot be carried out by any computational system in the universe" (Kaufmann 2000: 138). He justifies such proscriptive claims in exactly the same terms that I justified the universal probability bound a moment ago. Thus, for justification he looks to the Planck time, the Planck length, the radius of the Universe, the number of particles in the Universe, and the rate at which particles can change states.[6] Kauffman's idea of transfinite numbers is insightful, but the actual term is infelicitous because it already has currency within mathematics, where transfinite numbers are by definition infinite (in fact, the transfinite numbers of transfinite arithmetic can assume any infinite cardinality whatsoever).[7] I therefore propose to call such numbers *hyperfinite numbers*.[8]

Kauffman often writes about the Universe being unable to exhaust some set of possibilities. Yet at other times he puts an adjective in front of the word 'universe', claiming it is the *known* universe that is unable to exhaust some set of possibilities.[9] Is there a difference between the Universe (no adjective in front) and the *known* or *observable* universe (adjective in front)? To be sure, there is no empirical difference. Our best scientific observations tell us that the world surrounding us appears quite limited. Indeed, the size, duration, and composition of the known universe are such that 10^{150} is a hyperfinite number. For instance, if the Universe were a giant computer, it could perform no more than this number of operations (quantum computation, by exploiting superposition of quantum states, enriches the operations performable by an ordinary computer but cannot change their number). If the Universe were devoted entirely to generating specifications, this number would set an upper bound. If cryptographers confine themselves to brute-force methods on ordinary computers to test cryptographic keys, the number of keys they can test will always be less than this number.

But what if the Universe is in fact much bigger than the known universe? What if the known universe is but an infinitesimal speck within the actual universe? Alternatively, what if the known universe is but one of many possible universes, each of which is as real as the known universe but causally inaccessible to it? If so, are not the probabilistic resources needed to eliminate chance vastly increased and is not the validity of 10^{-150} as a universal probability bound thrown into question? This line of reasoning has gained widespread currency among scientists and philosophers in recent years. In this chapter I will argue that this line of reasoning is fatally flawed. Indeed, I will argue that it is illegitimate to rescue chance by invoking probabilistic resources from outside the known universe. To do so artificially inflates one's probabilistic resources.

The inflationary fallacy

Only probabilistic resources from the known universe may legitimately be employed in testing chance hypotheses. In particular, probabilistic resources imported from outside the known universe are incapable of overturning the universal probability bound of 10^{-150}. My basic argument to support this claim is quite simple, though I need to tailor it to some of the specific proposals now current for inflating probabilistic resources. The basic argument is this: It is never enough to postulate probabilistic resources merely to prop up an otherwise failing chance hypothesis. Rather, one needs independent evidence of whether there really are enough probabilistic resources to render chance plausible.

Consider, for instance, two state lotteries, both of which have printed a million lottery tickets. Let us assume that each ticket has a one in a million probability of winning and that whether one ticket wins is probabilistically independent of whether another wins (multiple winners are therefore a possibility). Suppose now that one of these state lotteries sells the full 1 million tickets but that the other sells only two tickets. Ostensibly both lotteries have the same number of probabilistic resources – the same number of tickets were printed for each. Nevertheless, the probabilistic resources relevant for deciding whether the first lottery produced a winner by chance greatly exceed those of the second. Probabilistic resources are *opportunities* for an event to happen or be specified. To be relevant to an event, those opportunities need to be actual and not merely possible. Lottery tickets sitting on a shelf collecting dust might just as well never have been printed.

This much is uncontroversial. But let us now turn the situation around. Suppose we know nothing about the number of lottery tickets sold and are informed simply that the lottery had a winner. Suppose further that the probability of any lottery ticket producing a winner is extremely low. Now what can we conclude? Does it follow that many lottery tickets were sold? Hardly. We are entitled to this conclusion only if we have independent evidence that many lottery tickets were sold. Apart from such evidence we have no way of assessing how many tickets were sold, much less whether the lottery was conducted fairly and whether its outcome was due to chance. It is illegitimate to take an event, decide for whatever reason that it must be due to chance, and then propose numerous probabilistic resources because otherwise chance would be implausible. I call this the *inflationary fallacy* (Dembski 1998: section 6.6).

Stated thus, the inflationary fallacy is readily rejected as a bogus form of argument. Nevertheless, it can be nuanced so that the problem inherent in it is mitigated (though not eliminated). The problem inherent in the inflationary fallacy is always that it multiplies probabilistic resources in the absence of independent evidence that such resources exist. Typically, however, when probabilistic resources get inflated, the rationale for inflating them is not simply to render chance plausible when otherwise it would be implausible. Hardly anyone is so crass as to admit, "I didn't like the alternatives to chance so I simply decided to invent some probabilistic resources." The rationale for inflating probabilistic resources is always more subtle, seeking confirmation in general coherence or consilience considerations even though independent evidence is lacking.

The inflationary fallacy therefore has a crass and a nuanced form. The crass form can be expressed as follows:

(1) Alternatives to chance are for whatever reason unacceptable for explaining some event – call that event X.
(2) With the probabilistic resources available in the known universe, chance is not a reasonable explanation of X.
(3) If probabilistic resources could be expanded, then chance would be a reasonable explanation of X.
(4) Let there be more probabilistic resources.
(5) So, chance is now a reasonable explanation of X.

The problem with this argument is Premise 4 (the "fiat" premise), which creates probabilistic resources *ex nihilo* simply to ensure that chance becomes a reasonable explanation.

The more nuanced form of the inflationary fallacy is on the surface less objectionable; it can be expressed as follows:

(1) There is an important problem, call it Y, that admits a solution as soon as one is willing to posit some entity, process, or stuff outside the known universe. Call whatever this is that resides outside the known universe Z.
(2) Though not confirmed by any independent evidence, Z is also not inconsistent with any empirical data.
(3) With the probabilistic resources available in the known universe, chance is not a reasonable explanation of some event – call the event X.
(4) But when Z is added to the known universe, probabilistic resources are vastly increased and now suffice to account for X by chance.
(5) So, chance is now a reasonable explanation of X.

This nuanced form of the inflationary fallacy appears in various guises and has gained widespread currency. It purports to solve some problem of general interest and importance by introducing some factor Z, which we will call an *inflaton*.[10] By definition, an inflaton will be some entity, process, or stuff outside the known universe that in addition to solving some problem also has associated with it numerous probabilistic resources as a byproduct. These resources in turn help to shore up chance when otherwise chance would seem unreasonable in explaining some event.

Four widely discussed inflatons

I want next, therefore, to consider four inflatons that purport to resolve important problems and that have gained wide currency. The inflatons I will consider are these: the bubble universes of Alan Guth's inflationary cosmology (Guth 1997), the many worlds of Hugh Everett's interpretation of quantum mechanics (Everett 1957), the self-reproducing black holes of Lee Smolin's cosmological natural selection (Smolin 1997), and the possible worlds of David Lewis's extreme modal realist metaphysics (Lewis 1986). My choice of proposals, though selective, is representative of the forms that the inflationary fallacy takes. While I readily admit that these inflatons propose solutions to important problems, I will argue that the costs of these solutions outweigh their benefits. In general, inflatons that inflate probabilistic resources, so that what was unattributable to chance within the known universe now becomes attributable to chance after all, are highly problematic and create more difficulties than they solve.

Let us start with Alan Guth's inflationary cosmology. Inflationary cosmology posits a very brief period of hyper-rapid expansion of space just after the Big Bang. Though consistent with general relativity, such expansion is not required. What's more, the expansion has now stopped (at least as far as we can tell within the known universe). Guth introduced inflation to solve such problems in cosmology as the flatness, horizon, and magnetic monopole problems. In standard Big Bang cosmology the first two of these problems seem to require considerable fine-tuning of the initial conditions of the Universe whereas the third seems unresolvable if standard Big Bang cosmology is combined with Grand Unified Theories. Inflationary cosmology offers to resolve these problems in one fell swoop. In so doing, however, the known universe becomes a bubble universe within a vast sea of other bubble universes, and the actual universe then constitutes the sea that contains these bubble universes.

Next let us consider Hugh Everett's interpretation of quantum mechanics. Everett's many-worlds interpretation of quantum mechanics proposes a radical solution to what in quantum mechanics is known as the measurement problem. The state function of a quantum mechanical system corresponds to a probability distribution that upon measurement assumes a definite value. The problem is that any physical system whatsoever can be conceived as a quantum mechanical system described by a state function. Now what happens when the physical system in question is taken to be the entire Universe? Most physical systems one considers are proper subsets of the Universe and thus admit observers who are outside the system and who can therefore measure the system and, as it were, collapse the state function. But when the Universe as a whole is taken as the physical system in question, where is the observer to collapse the state function?[11] Everett's solution is to suppose that the state function does not collapse but rather splits into all different possible values that the state function could assume (mathematically this is very appealing – especially to quantum cosmologists – because it eliminates any break in dynamics resulting from state-function collapse). In effect, all possible quantum histories get lived out. Suppose, for instance, someone offers me a million dollars to play Quantum Russian Roulette (i.e. a quantum mechanical device is set up with six possibilities, each having probability one-sixth, and such that a bullet fires into my brain and kills me when exactly one of these possibilities occurs but leaves me unharmed otherwise). If I choose to play this game, then for every one quantum world in which I get a bullet to the head there are five in which I live happily ever after as a millionaire.

Next let us consider Lee Smolin's cosmological natural selection of self-reproducing black holes. Smolin's self-reproducing black holes constitute perhaps the most ambitious of the inflatons we will consider. Smolin characterizes his project as explaining how the laws of physics have come to take the form they do, but in fact he is presenting a full-blown cosmogony in which Darwinian selection becomes the mechanism by which universes are generated and flourish. According to Smolin, quantum effects preclude singularities at which time stops. Consequently, time does not stop in a black hole but rather "bounces" in a new direction, producing a region of space–time inaccessible to ours except at the moment of its origination. Moreover, Smolin contends that during a "bounce" the laws of nature change their parameters but not their general form. Consequently, the formation of black holes follows an evolutionary algorithm in which parameters get continually tightened to maximize the production of black holes. Within Smolin's scheme the known universe is but one among innumerable products of black holes that have formed by this process and that in turn generate other black holes. Cosmological natural selection accounts not only for the generation of universes but also for their fine-tuning and the possibility of such structures as life.

Finally, let us consider the possible worlds of David Lewis's extreme modal realist metaphysics. Lewis, unlike Guth, Everett, and Smolin, is not a scientist but a philosopher and in particular a metaphysician. For Lewis any logically possible world is as real as our world, which he calls the actual world. It is logically possible for a world to consist entirely of a giant tangerine. It is logically possible that the laws of physics might have been different, not only in their parameters but also in their basic form. It is logically possible that instead of turning to mathematics I might have become a rock-and-roll singer. For each of these logical possibilities Lewis contends that there are worlds as real as ours in which those possibilities are actualized. The only difference between those worlds and ours is that we happen to inhabit our world – that is what makes our world the actual world. Lewis's view is known as extreme modal realism. Modal realism asserts that logical possibilities are in some sense real (perhaps as abstractions in a mathematical space). *Extreme* modal realism emphasizes that logical possibilities are real in exactly the same way that the world we inhabit is real. Why does Lewis hold this view? According to him, possible worlds are indispensable for making sense of certain key philosophical problems, notably the analysis of counterfactual conditionals. What's more, he finds that all attempts to confer on possible worlds a status different from that of the actual world are incoherent (he refers to these disparagingly as *ersatz* possible worlds and finds them poor substitutes for his full-blown possible worlds).

I have provided only the briefest summary of the views of Alan Guth, Hugh Everett, Lee Smolin, and David Lewis. The problems these thinkers raise are important, and the solutions they propose need to be taken seriously. Moreover, except for David Lewis's possible worlds, which are purely metaphysical, the other three inflatons considered make contact with empirical data. Lee Smolin even contends that his theory of cosmological natural selection has testable consequences – he even runs through several possible tests. The unifying theme in Smolin's tests is that varying the parameters for the laws of physics should tend to decrease the rate at which black holes are formed in the known universe. It is a consequence of Smolin's theory that, for most universes generated by black holes, the parameters of the laws of physics should be optimally set to facilitate the formation of black holes. We ourselves are therefore highly likely to be in a universe where black hole formation is optimal. My own view is that our understanding of physics needs to proceed considerably further before we can establish convincingly that ours is a universe that optimally facilitates the formation of black holes. But even if this could be established now, it would not constitute independent evidence that a black hole is capable of generating a new universe. Smolin's theory, in positing that black holes generate universes, would explain why we are in a universe that optimally facilitates the formation of black holes. But it is not as though we would ever have independent evidence for Smolin's theory, say by looking inside a black hole and seeing whether there is a universe in it. Of all the objects in space (stars, planets, comets, etc.) black holes divulge the least amount of information about themselves.

Explanatory power and independent evidence

Each of the four inflatons considered here possesses explanatory power in the sense that each explains certain relevant data and thereby solves some problem of general interest and importance. These data are said to confirm or provide epistemic support for an inflaton in so far as it adequately explains the relevant data and does not conflict with other recognized data. What's more, in so far as an inflaton does not adequately explain the relevant data, it lacks explanatory power and is disconfirmed. In general, therefore, explanatory power entails testability in the weak sense that, if a claim fails adequately to explain certain relevant data, it is to be rejected (thus failing the test).

Nevertheless, even though the four inflatons considered here each possesses explanatory power, none of them possesses independent evidence for its existence. Independent evidence is by definition evidence that helps establish a claim apart from any appeal to the claim's explanatory power. The demand for independent evidence is neither frivolous nor tendentious. Instead, it is a necessary constraint on theory construction so that theory construction does not degenerate into total free-play of the mind.[12]

Consider for instance the "gnome theory of friction." Suppose a physicist claims that the reason objects do not slide endlessly across surfaces is because tiny invisible gnomes inhabit all surfaces and push back on any objects pushed along the surfaces. What's more, the rougher a surface, the more gnomes inhabit it, and consequently the greater the resistance to an object moving across the surface. Suitably formulated, the gnome theory of friction can explain how objects move across surfaces just as accurately as current physical theory. So why do we not take the gnome theory of friction seriously? One reason (though not the only reason – the gnome theory has many more problems than described here) is the absence of independent evidence for gnomes.

Independent evidence and explanatory power need to work in tandem, and for one to outpace the other typically leads to difficulties. In spinning out their theories, conspiracy theorists place all their emphasis on explanatory power but ignore the demand for independent evidence. In enumerating countless low-level facts, crude inductivists place all their emphasis on independent evidence. They miss the bold hypotheses and intuitive leaps that make for explanatory power and are thus capable of tying together their disparate facts. Independent evidence is the strict disciplinarian to explanatory power's carefree genius. Each is needed to balance the other. My favorite story illustrating the interplay between the two is due to John Leslie (1989: 10, 12). Suppose an arrow is fired at random into a forest and hits Mr Brown. To explain such a chance occurrence it would suffice for the forest to be full of people. The forest being full of people therefore possesses explanatory power. Even so, this explanation remains but a speculative possibility until it is supported by independent evidence of people other than Mr Brown in the forest.

The problem with the four inflatons considered above is that none of them admits independent evidence. The only thing that confirms them is their ability to explain certain data or resolve certain problems. With regard to inflationary cosmology, we have no direct experience of hyper-rapid inflation nor have we observed any process that could reasonably be extrapolated to hyper-rapid inflation. With regard to the many-worlds interpretation of quantum mechanics, we always experience exactly one world and have no direct access to alternate parallel worlds. If there is any access at all to these worlds, it is indirect and circumstantial. Indeed, to claim that quantum interference signals the influence of parallel worlds is to impose a highly speculative interpretation on the data of quantum mechanics that is far from compelling.[13] With regard to black hole formation, there is no way for anybody on the outside to get inside a black hole, determine that there actually is a universe inside there, and then emerge intact to report as much. With regard to possible worlds, they are completely causally separate from each other – other possible worlds never were and never can be accessible to us, either directly or indirectly.

The absence of independent evidence for these inflatons makes for them the problem of underdetermination especially acute. In general, when a hypothesis explains certain data, there are other hypotheses that also explain the data. In this way, data are said to underdetermine hypotheses. Nonetheless, it may be that one hypothesis explains the data better than the others so that it is possible to adjudicate among hypotheses simply on the basis of explanatory power. On the other hand, it may be that competing hypotheses exhibit identical explanatory power or that advocates of competing hypotheses claim that their preferred hypotheses exhibit the greater explanatory power. In either case, independent evidence will be required to adjudicate among the hypotheses. With the four inflatons here considered, no such independent evidence is forthcoming.

I want next, therefore, to examine these four inflatons in relation to design to see whether design might be amenable to independent evidence in a way that the four inflatons are not. As I defined it, an inflaton is some entity, process, or stuff outside the known universe that helps explain certain data and thereby resolve some problem. Notably absent from the inflatons described by Guth, Everett, Smolin, and Lewis is a designer. Their inflatons are fully compatible with naturalism and thoroughly non-teleological. Now the interesting thing is that a designer, especially when fleshed out into a full-blown theistic deity, can be employed to resolve the very problems that the four inflatons considered here were meant to resolve. The fine-tuning of the Universe and the form of the laws of physics that are central to Guth's and Smolin's concerns can be attributed to a divine act of creation. Moreover, such a deity could collapse the state function of the Universe and thereby resolve the measurement problem of quantum mechanics when this problem is applied to the Universe taken as a whole. And finally, such a deity, by being suitably omniscient and thus possessing what philosophers of religion call "middle knowledge," could provide a semantics for counterfactual conditionals and resolve many of the other problems for which David Lewis thinks he requires possible worlds.[14]

Now I want to stress that I am not advocating these theistic alternatives to the four inflatons considered above (I personally think there is something to the theistic fine-tuning arguments, but I am no fan of middle knowledge and have serious doubts about God's role as a state-function collapser). My point, rather, is this: Given that there are design-theoretic alternatives to the inflatons considered here and given that such alternatives immediately raise the problem of underdetermination, the only way to resolve this problem is via independent evidence. So let me pose this question: Is there independent evidence that would allow us to distinguish the four inflatons considered above from a design-theoretic alternative? We have already seen that there is no independent evidence that supports these four inflatons. But could there be independent evidence that supports a design-theoretic alternative and in so doing also disconfirms these four inflatons? I am going to argue that there is.

Arthur Rubinstein – consummate pianist or lucky poseur?

The four inflatons considered here allow for unlimited probabilistic resources. Now the problem with unlimited probabilistic resources is that they allow us to explain absolutely everything by reference to chance – not just natural objects that actually did result by chance and not just natural objects that look designed, but also all artificial objects that are in fact designed. In effect, unlimited probabilistic resources collapse the distinction between apparent design and actual design, and make it impossible to attribute anything with confidence to actual design. Was Arthur Rubinstein a great pianist or was it just that whenever he sat at the piano, he happened by chance to put his fingers on the right keys to produce beautiful music? It could happen by chance, and there is some possible world where everything is exactly as it is in this world except that the counterpart to Arthur Rubinstein cannot read music and happens to be incredibly lucky whenever he sits at the piano. Examples like this can be multiplied. There are possible worlds in which I cannot do arithmetic and yet sit down at my Macintosh computer and write probabilistic tracts about intelligent design. Perhaps Shakespeare was a genius. Perhaps Shakespeare was an imbecile who just by chance happened to string together a long sequence of apt phrases. Unlimited probabilistic resources ensure not only that we will never know, but also that we have no rational basis for preferring one to the other.

Given unlimited probabilistic resources, there is only one way to rebut this anti-inductive skepticism, and that is to admit that while unlimited probabilistic resources allow bizarre possibilities like this, these possibilities are nonetheless highly improbable in the little patch of reality that we inhabit. Unlimited probabilistic resources make bizarre possibilities unavoidable on a grand scale. The problem is how to mitigate the craziness that they entail, and the only way to do this once such bizarre possibilities are conceded is to render them improbable on a local scale. Thus, in the case of Arthur Rubinstein, there are worlds where someone named Arthur Rubinstein is a world-famous pianist and does not know the first thing about music. But it is vastly more probable that in worlds where someone named Arthur Rubinstein is a world-famous pianist, that person is a consummate musician. What's more, induction tells us that ours is such a world.

But can induction really tell us that? How do we know that we are not in one of those bizarre worlds where things happen by chance that we ordinarily attribute to design? Consider further the case of Arthur Rubinstein. Imagine it is January 1971 and you are at Orchestra Hall in Chicago listening to Arthur Rubinstein perform. As you listen to him perform Liszt's "Hungarian Rhapsody," you think to yourself, "I know the man I'm listening to right now is a wonderful musician. But there's an outside possibility that he doesn't know the first thing about music and is just banging away at the piano haphazardly. The fact that Liszt's 'Hungarian Rhapsody' is pouring forth would thus merely be a happy accident. Now if I take seriously the existence of other worlds, then there is some counterpart to me pondering these very same thoughts, only this time listening to the performance of someone named Arthur Rubinstein who is a complete musical ignoramus. How, then, do I know that I'm not that counterpart?"[15]

Indeed, how do you know that you are not that counterpart? First off, let us be clear that the Turing Test is not going to come to the rescue here by operationalizing the two Rubinsteins and rendering them operationally indistinguishable. According to the Turing Test (Turing 1950), if a computer can simulate human responses so that fellow humans cannot distinguish the computer's responses from an individual human's responses, then the computer passes the Turing Test and is adjudged intelligent. This operationalizing of intelligence has its own problems, but, even if we let them pass, success at passing the Turing Test is clearly not what is at stake in the Rubinstein example. The computer that passes the Turing Test presumably "knows" what it is doing (having been suitably programmed) whereas the Rubinstein who plays successful concerts by randomly positioning fingers on the keyboard does not have a clue. Think of it this way: Imagine a calculating machine whose construction guarantees that it performs arithmetic correctly and imagine another machine that operates purely by random processes. Suppose we pose the same arithmetic problems to both machines and out come identical answers. It would be inappropriate to assign arithmetic prowess to the random device, even - though it is providing the right answers, because that is not its proper function – it is simply by chance happening upon the right answers. On the other hand, it is entirely appropriate to attribute arithmetic prowess to the other machine because it is constructed to perform arithmetic calculations accurately – that is its proper function. Likewise, with the real Arthur Rubinstein and his chance-performing counterpart, the real Arthur Rubinstein's proper function is, if you will, to perform music with skill and expression whereas the counterpart is just a lucky poseur. When Turing operationalized intelligence, he clearly meant intelligence to be a proper function of a suitably programmed computer and not merely a happy accident.[16]

How, then, do you know that you are listening to Arthur Rubinstein the musical genius and not Arthur Rubinstein the lucky poseur? To answer this question, let us ask a prior question: How did you recognize in the first place that the man called Rubinstein performing in Orchestra Hall was a consummate musician? Reputation, formal attire, and famous concert hall are certainly giveaways, but they are neither necessary nor sufficient. Even so, a necessary condition for recognizing Rubinstein's musical skill (design) is that he was following a prespecified concert program, and in this instance that he was playing Liszt's "Hungarian Rhapsody" note for note (or largely so – Rubinstein was not immune to mistakes). In other words, you recognized that Rubinstein's performance exhibited specified complexity. Moreover, the degree of specified complexity exhibited enabled you to assess just how improbable it was that someone named Rubinstein was playing the "Hungarian Rhapsody" with éclat but did not have a clue about music. Granted, you may have lacked the technical background to describe the performance in these terms, but the recognition of specified complexity was there nonetheless, and without that recognition there would have been no way to attribute Rubinstein's playing to design rather than chance.

Independent evidence for a designer

Specified complexity is how we eliminate bizarre possibilities in which chance is made to account for things that we would ordinarily attribute to design. What's more, specified complexity is how we assess the improbability of those bizarre possibilities and therewith justify eliminating their chance occurrence. That being the case (and it certainly is the case for human artifacts), on what basis could we attribute chance to natural phenomena that exhibit specified complexity? Let us be clear that inflating probabilistic resources does not just diminish a universal probability bound and make it harder to attribute design – inflating probabilistic resources is not a matter of replacing one universal probability bound by another that is more stringent. Inflating probabilistic resources eliminates universal probability bounds entirely – the moment one posits unlimited probabilistic resources, anything of non-zero probability becomes certain (probabilistically this follows from the Strong Law of Large Numbers).[17] It seems, however, that in practical life we do allow for probability bounds to assess improbability and therewith specified complexity. A sentence or two repeated verbatim by another author can be enough to elicit the charge of plagiarism. It could happen by chance and given unlimited probabilistic resources there are patches of reality where it did happen by chance. But we do not buy it – at least not for our patch of reality. In practical life we tend not to be very conservative in setting probability bounds. They tend to be quite large, and certainly much larger than the universal probability bound of 10^{-150} that I have been advocating.

The difficulty confronting unlimited probabilistic resources can now be put quite simply: There is no principled way to discriminate between using unlimited probabilistic resources to retain chance and using specified complexity to eliminate chance. You can have one or the other, but you cannot have both. And the fact is, we already use specified complexity to eliminate chance. Let me stress that there is no *principled* way to make the discrimination. It is, for instance, possible to invoke naturalism as a philosophical presupposition and use it to discriminate between using probabilistic resources to retain chance when designers unacceptable to naturalism are implicated (e.g. God) and using specified complexity to eliminate chance when designers acceptable to naturalism are implicated (e.g. Francis Crick's space aliens who seed the Universe with life (Crick and Orgel 1973)). Thus, for artifactual objects exhibiting specified complexity and for which an embodied intelligence could plausibly have been involved, we would attribute design, but for natural objects exhibiting specified complexity and for which no embodied intelligence could plausibly have been involved, we would invoke unlimited probabilistic resources and thus attribute chance (or perhaps simply plead ignorance). But this is entirely arbitrary. Indeed, the problem of unlimited probabilistic resources throws naturalism itself into question, and it does no good to invoke naturalism to resolve the problem.

It is important to understand that I am not arguing that the inflation of probabilistic resources entails anti-inductive skepticism. Indeed, my argument here is not anti-inductive but pro-specified complexity. I did offer an anti-inductive argument in Chapter 6 of *The Design Inference* (1988). My focus there was on the set of all logically possible worlds, and thus on worlds that instantiate every possible set of natural laws. In that case, inflating probabilistic resources entails inductive skepticism since there are far more worlds that agree with our world up to the present and go haywire afterward than there are worlds that continue to obey the regularities observed thus far. My argument here, however, allows that the worlds that inflate probabilistic resources obey laws of the same form as the laws of our universe. In that case, the vast majority of worlds in which Rubinstein delivers an exquisite performance are worlds in which Rubinstein is a skilled musician rather than a lucky poseur. But to convince us for such worlds that Rubinstein is indeed a skilled musician rather than a lucky poseur requires specified complexity. Even with unlimited probabilistic resources, we need to distinguish design from non-design, and specified complexity is how we do it. Consequently, there is no principled way to discriminate between using unlimited probabilistic resources to retain chance and using specified complexity to eliminate chance. And since we already use specified complexity to eliminate chance, invoking unlimited probabilistic resources to retain chance is not a defensible option. I am not arguing that inflating probabilistic resources destroys induction. I am arguing that inflating probabilistic resources does not destroy specified complexity. In particular, probabilistic resources from outside the known universe are irrelevant to assessing specified complexity.[18]

We are now in a position to see why a designer outside the known universe could in principle be supported by independent evidence whereas the inflatons introduced by Guth, Everett, Smolin, and Lewis cannot. We already have experience of human and animal intelligences generating specified complexity. If we should ever discover evidence of extraterrestrial intelligence, a necessary feature of that evidence would be specified complexity. Thus, when we find evidence of specified complexity in nature for which no embodied, reified, or evolved intelligence could plausibly have been involved, it is a straightforward extrapolation to conclude that some unembodied intelligence must have been involved. Granted, this raises the question of how such an intelligence could coherently interact with the physical world.[19] But to deny this extrapolation merely because of a prior commitment to naturalism is not defensible. There is no principled way to distinguish between using specified complexity to eliminate chance in one instance and then in another invoking unlimited probabilistic resources to render chance plausible.

Design allows for the possibility of independent evidence whereas the inflatons of Guth, Everett, Smolin, and Lewis do not. Specified complexity can be a point of contact between the known universe and an intelligence outside it – designers within the Universe already generate specified complexity and a designer outside could potentially do the same. That is what allows for independent evidence to support unembodied designers. Provided nature supplies us with instances of specified complexity that cannot reasonably be attributed to any embodied intelligence, the inference to an unembodied intelligence becomes compelling and any instances of specified complexity used to support that inference can rightly be regarded as independent evidence (Dembski 2002: Ch. 5). By contrast, the inflatons of Guth, Everett, Smolin, and Lewis provide no such palpable connection with the known universe. Indeed, what in our actual experience can straightforwardly be extrapolated to hyper-rapid expansion of space, quantum many-worlds, cosmological natural selection, and causally inaccessible possible worlds? Is it, for instance, a straightforward extrapolation that takes us from biological natural selection of carbon-based life to cosmological natural selection of black holes? To be sure, there is an extrapolation here, but one where all meaningful analogies with actual experience break down.

Three crucial questions now face design:

(1) Is specified complexity exhibited in any natural systems where no embodied intelligence could plausibly have been involved?
(2) If so, does the design apparent in such systems match up meaningfully with known designs due to known embodied designers?
(3) Does a theory of design that treats specified complexity as a reliable marker of intelligence possess sufficient explanatory power to render it interesting and fruitful for science?

In *No Free Lunch* (Dembski 2002) I argue for an affirmative answer to each of these three questions.

Closing off quantum loopholes

In concluding this chapter, I want to address one possible worry that might remain. I have argued that it does no good to look outside the known universe to increase one's probabilistic resources. But what about looking inside the known universe for additional probabilistic resources? Take, for instance, quantum computation. Peter Shor (1994) has described an algorithm for quantum computers that is capable of factoring numbers vastly larger than can be factored with conventional computers (thus threatening cryptographic schemes that depend on factorization constituting a hard computational problem). David Deutsch therefore asks,

> When Shor's algorithm has factorized a number, using 10^{500} or so times the computational resources that can be seen to be present, where was the number factorized? There are only about 10^{80} atoms in the entire visible universe, an utterly minuscule number compared with 10^{500}. So if the visible universe were the extent of physical reality, physical reality would not even remotely contain the resources required to factorize such a large number. Who did factorize it, then? How, and where, was the computation performed?
>
> (1997: 217)

In raising these questions, Deutsch is advocating a many-worlds interpretation of quantum mechanics. This interpretation is not mandated. Indeed, interpretations of quantum mechanics abound and all of them, in so far as they are coherent and empirically adequate, are empirically indistinguishable. As Anthony Sudbery remarks,

> An interpretation of quantum mechanics is essentially an answer to the question "What is the state vector?" Different interpretations cannot be distinguished on scientific grounds – they do not have different experimental consequences; if they did they would constitute different *theories*.
>
> (1984: 212)

Yet if we resist the many-worlds interpretation of quantum mechanics and the unlimited probabilistic resources this interpretation provides, does not quantum mechanics, and quantum computation in particular, invite a huge number of probabilistic resources into our own known universe? I submit that it does not. True, quantum computation may alter the computational resources relevant to assessing the security of cryptosystems against brute-force attacks that enlist the entire Universe as a giant quantum computer. As a result, universal computation bounds will diverge from universal probability bounds – in the past they were largely identical because they were based on conventional computing whereas now they would diverge because of the increased computational resources due to quantum computing.

Even so, quantum computation provides no justification for altering the universal probability bound of 10^{-150}. To see this, let us pose a related but different question from the one raised by Shor. Shor asked how large a number could be factored with quantum computers as opposed to conventional computers. He found that quantum computers vastly increased the size of the numbers that could be factored. But now let us ask how many numbers could be factored with quantum computers as opposed to conventional computers. To factor a given number on either a conventional or a quantum computer means entering it as a specific sequence of bits or qubits, respectively, performing the relevant computation, and then identifying a specific output sequence as the answer. If we now ignore computation times, it follows that in terms of the sheer quantity of numbers that can be factored, quantum computation offers no advantage over conventional computation – specific numbers still have to be inputted and outputted. Input and output themselves take time, space, and material, and there are no more than 10^{150} specific numbers that computers, whether conventional or quantum, can ever input and output.

The lesson here is that specified complexity, precisely because it requires items of information to be specifically identified, provides no opening for quantum computation to exploit quantum parallelism or superposition and thereby generate specifications. We can imagine a quantum memory register of 1,000 qubits in a superposition of states representing every possible sequence of 0s and 1s of length 1,000. Nevertheless, this memory register is incapable of specifying even a single conventional bit string of length 1,000 until a measurement is taken and the superposition of states is projected onto an eigenstate.

Though quantum computation offers to dramatically boost computational power by allowing massively parallel computations, it does so by keeping computational states indeterminate until the very end of a computation. This indeterminateness of computationa lstate stakes the form of quantum superpositions, which are deliberately exploited in quantum computation to facilitate parallel computation. The problem with quantum superpositions, however, is that they are incapable of concretely realizing specifications. A quantum superposition is an indeterminate state. A specification is a determinate state. Measurement renders a quantum superposition determinate by producing an eigenstate, but, once it does, we are no longer dealing with a quantum superposition. Because quantum computation thrives precisely where it exploits superpositions and avoids specificity, it offers no means for boosting the number of specifications that can be concretely realized in the known universe.[20]

Is there any place else to look for additional probabilistic resources inside the known universe? According to Robin Collins, quantum mechanics offers still one other loophole for inflating probabilistic resources and thereby undercutting specified complexity as a reliable indicator of design. Collins says that the state function of a quantum mechanical system can take continuous values and thus assume infinitely many possible states. From this he draws the following conclusion: "This means that in Dembski's scheme one could only absolutely eliminate chance for events of zero probability!" (Collins 2001: 336, note 7). Presumably he thinks that because quantum systems can produce infinitely many possible events, this means that quantum systems also induce infinitely many probabilistic resources. And since infinitely many probabilistic resources coincide with a probability threshold of zero, my scheme could therefore only eliminate chance for events of probability zero. The problem here is that Collins fails to distinguish between the *range* of possible events that might occur and the *opportunities* for a given event to occur or be specified. A reference class of possibilities may well be infinite (as in the case of certain quantum mechanical systems). But the opportunities for sampling from such a reference class and thereby inducing information are always finite and extremely limited. Probabilistic resources always refer to the opportunities for sampling from a range of possible events. The range of possible events itself might well be infinite. But this has no bearing on the probabilistic resources associated with a given event in that range.

It appears, then, that we are back to our own known little universe, with its very limited number of probabilistic resources but therewith also its increased possibilities for detecting design. This is one instance where less is more, where having fewer probabilistic resources opens possibilities for knowledge and discovery that would otherwise be closed. Limited probabilistic resources enrich our knowledge of the world by enabling us to detect design where otherwise it would elude us. At the same time, limited probabilistic resources protect us from the unwarranted confidence in natural causes that unlimited probabilistic resources invariably seem to engender. In short, limited probabilistic resources eliminate the chance of the gaps.

Notes

1 Full details for this rationale are given in Chapter 6 of *The Design Inference* (Dembski 1998), and specifically in section 6.3 entitled "The magic number 1/2."

2 Note that universal time-bounds for electronic computers have clock speeds that are between ten and twenty magnitudes slower than the Planck time (Wegener 1987: 2).

3 For the details justifying this universal probability bound, see Dembski (1998: section 6.5).

4 See also Singh (1999), which is full of arguments that tacitly appeal to universal probability bounds. For instance, Singh quotes William Crowell, Deputy Director of the National Security Agency: "If all the personal computers in the world – approximately 260 million computers – were to be put to work on a single PGP encrypted message, it would take on average an estimated 12 million times the age of the universe to break a single message" (317).

5 Although Kauffman does not explicitly mention the phrase "specified complexity," his emphasis throughout this book is on the complexity of biological systems, and the type of complexity he is trying to explain is in fact specified complexity.

6 Kaufmann (2000: 137–8, 144, 162, 167).
7 See Hallett (1984: 55–6).
8 Peter Rüst refers to such numbers as "transastronomical" (Rüst 1992: 80). Emile Borel referred to the reciprocal of such numbers as "probabilities which are negligible on the supercosmic scale." See Borel (1962: 28–30).
9 See Kauffman (2000: 144) where he switches indiscriminately between referring to "the known universe" and simply "the universe."
10 Within inflationary cosmology, inflatons are fields that drive inflation. I am using the term in a more general sense.
11 Strictly speaking an observer is not necessary. All that is necessary for quantum measurement is that to each eigenstate for a subsystem there corresponds a unique relative state for the remainder of the whole system. If the subsystem is the whole Universe, however, then there is no remainder and nothing (apparently) to do the measuring. Everett's solution is to deny that state functions collapse to eigenstates and assert instead that all possible eigenstates are realized. Simon Saunders (1993) thinks that sense can be made of Everett's solution without postulating many worlds.
12 The need for independent evidence to confirm a scientific theory has frequently been noted in connection with intelligent design. For instance, citing Leibniz, Philip Kitcher (1982: 138) describes the need for "independent criteria of design" before design can be taken seriously in science. In *No Free Lunch* (Dembski 1998) I attempt to answer Kitcher's challenge for the case of intelligent design. Nevertheless, it is a challenge that all scientific theories must at some point face, the inflatons considered here being a case in point.
13 David Deutsch would reject my claim that the many-worlds interpretation lacks independent evidence. Describing the double-slit experiment, Deutsch writes:

> A real, tangible photon behaves differently according to what paths are open, elsewhere in the apparatus, for something to travel along and eventually intercept the tangible photon. Something does travel along those paths, and to refuse to call it "real" is merely to play with words. "The possible" cannot interact with the real: non-existent entities cannot deflect real ones from their paths. If a photon is deflected, it must have been deflected by something, and I have called that thing a "shadow photon".
>
> (1997: 49)

For Deutsch, shadow photons reside in universes different from our own and yet causally interact with our universe by, for instance, deflecting photons. In fact, to read Deutsch one would think that the many-worlds, or as he calls it the "multiverse," interpretation of quantum mechanics is the only one that is coherent and experimentally supported. As he writes, "I have merely described some physical phenomena and drawn inescapable conclusions....Quantum theory describes a multiverse" (1997: 50). Or:

> The quantum theory of parallel universes is not the problem, it is the solution. It is not some troublesome, optional interpretation emerging from arcane theoretical considerations. It is the explanation – the only one that is tenable – of a remarkable and counter-intuitive reality.
>
> (1997: 51)

But, in fact, one can interpret the double-slit experiment and other quantum mechanical results without multiple worlds and do so coherently – i.e. without internal contradiction and without contradicting any empirical data. And there are plenty of such interpretations. The uniting feature of these different interpretations is that they are empirically equivalent – if not, there would be multiple quantum theories. As it is, there is only one quantum theory and many interpretations. See Sudbery (1984: 212–25).

Deutsch sees the deflection of photons in a double-slit experiment as sure evidence of parallel universes interacting with our own. Deutsch's very reference to "deflected photons" is a throwback to metaphors of classical physics that have no proper place in quantum mechanics. To invoke them as independent evidence of the many-worlds interpretation of quantum mechanics is to confuse what needs to be explained with what adjudicates among competing explanations or interpretations. The behavior of photons passing through two slits and exhibiting an interference pattern on a screen needs to be explained, but that behavior does not single out the many-worlds interpretation as, to quote Deutsch, "the only one that is tenable." Deutsch's uncompromising advocacy of the many-worlds interpretation of quantum mechanics is as dogmatic as it is unfounded.

14 For a sampling of theistic solutions to such problems consult the essays in Craig and Moreland (2000) and in Murray (1999).

15 Note that I am not wedded to any particular metaphysical position about counterparts. My argument here treats counterparts as separate individuals and thus not as a single transworld individual. But for my argument to work it is enough that separate persons with similar cognitive faculties and background beliefs exist in separate worlds and be listening to separate Rubinsteins, the one real and the other fake. Transworld identity is therefore not required nor is a theory of counterparts. For David Lewis's theory of counterpart relations and his critique of transworld identity in modal metaphysics see Lewis (1986: 9–13 and 210–20, respectively). For Alvin Plantinga's indexical account of transworld identity and his critique of Lewis's counterpart theory see Plantinga (1974: 88–101 and 102–20, respectively).

16 For more on proper function see Plantinga (1993).

17 For the Strong Law of Large Numbers see Bauer (1981: 172).

18 I am grateful to Rob Koons for pressing me to clarify this point.

19 See Dembski (2002: section 6.5).

20 For an overview of quantum computation see Williams and Clearwater (1998) and Hey (1999).

References

Bauer, H. (1981) *Probability Theory and Elements of Measure Theory*, trans. R.B. Burckel, 2nd English edn, New York: Academic Press.

Borel, E. (1962) *Probabilities and Life*, trans. M. Baudin, New York: Dover.

Collins, R. (2001) "An evaluation of William A. Dembski's *The Design Inference*: A review essay," *Christian Scholar's Review* 30(3): 329–41.

Craig, W.L. and Moreland, J.P. (eds) (2000) *Naturalism: A Critical Analysis*, London: Routledge.

Crick, F. and Orgel, L. (1973) "Directed panspermia," *Icarus* 19: 341–6.

Dam, K.W. and Lin, H.S. (eds) (1996) *Cryptography's Role in Securing the Information Society*, Washington, DC: National Academy Press.

Davies, P. (1999) *The Fifth Miracle*, New York: Simon & Schuster.

Dembski, W. (2002) *No Free Lunch: Why Specified Complexity Cannot be Purchased without Intelligence*, Lanham, Maryland: Rowman & Littlefield.

—— (1998) *The Design Inference: Eliminating Chance through Small Probabilities*, Cambridge: Cambridge University Press.

Deutsch, D. (1997) *The Fabric of Reality: The Science of Parallel Universes – and Its Implications*, New York: Penguin.

Edis, T. (2001) "Darwin in mind: 'Intelligent design' meets artificial intelligence," *Skeptical Inquirer* 25(2): 35–9.

Everett, H. (1957) "'Relative state' formulation of quantum mechanics," *Reviews of Modern Physics* 29: 454–62.

Fitelson, B., Stephens, C., and Sober, E. (1999) "How not to detect design – critical notice: William A. Dembski, *The Design Inference*," *Philosophy of Science* 66: 472–88.

Guth, A. (1997) *The Inflationary Universe: The Quest for a New Theory of Cosmic Origins*, Reading, Massachusetts: Addison-Wesley.

Hallett, M. (1984) *Cantorian Set Theory and Limitation of Size*, Oxford: Oxford University Press.

Halliday, D. and Resnick, R. (1988) *Fundamentals of Physics*, 3rd edn extended, New York: Wiley.

Hey, A.J.G. (ed.) (1999) *Feynman and Computation: Exploring the Limits of Computers*, Reading, Massachusetts: Perseus.

Horgan, J. (1996) *The End of Science*, New York: Broadway Books.

Kauffman, S. (2000) *Investigations*, New York: Oxford University Press.

Kitcher, P. (1982) *Abusing Science: The Case against Creationism*, Cambridge, Massachusetts: MIT Press.

Knobloch, E. (1987) "Emile Borel as a probabilist," in L. Krüger, L.J. Daston, and M. Heidelberger (eds) *The Probabilistic Revolution*, vol. 1, Cambridge, Massachusetts: MIT Press.

Leslie, J. (1989) *Universes*, London: Routledge.

Lewis, D. (1986) *On the Plurality of Worlds*, Oxford: Basil Blackwell.

Murray, M.J. (ed.) (1999) *Reason for the Hope Within*, Grand Rapids, Michigan: Eerdmans.

Orgel, L. (1973) *The Origins of Life*, New York: Wiley.

Plantinga, A. (1993) *Warrant and Proper Function*, Oxford: Oxford University Press.

—— (1974) *The Nature of Necessity*, Oxford: Clarendon Press.

Rüst, P. (1992) "How has life and its diversity been produced?" *Perspectives on Science and Christian Faith* 44(2): 80–94.

Saunders, S. (1993) "Decoherence, relative states, and evolutionary adaptation," *Foundations of Physics* 23: 1,553–95.

Shor, P. (1994) "Algorithms for quantum computation: Discrete logarithms and factoring," *Proceedings of the 35th Annual Symposium on Foundations of Computer Science*, pp. 124–34.

Singh, S. (1999) *The Code Book*, New York: Doubleday.

Smolin, L. (1997) *The Life of the Cosmos*, Oxford: Oxford University Press.

Sober, E. (1999) "Testability," *Proceedings and Addresses of the American Philosophical Association* 73(2): 47–76.

Sudbery, A. (1984) *Quantum Mechanics and the Particles of Nature*, Cambridge: Cambridge University Press.

Turing, A. (1950) "Computing machinery and intelligence," *Mind* 59: 434–60.
Wegener, I. (1987) *The Complexity of Boolean Functions*, Stuttgart: Wiley-Teubner.
Williams, C.P. and Clearwater, S.H. (1998) *Explorations in Quantum Computing*, New York: Springer-Verlag.

Part IV

BIOLOGY

15

THE MODERN INTELLIGENT DESIGN HYPOTHESIS

Breaking rules[1]

Michael Behe

Differences from Paley

In this chapter I will argue that some biological systems at the molecular level appear to be the result of deliberate intelligent design (ID). In doing so I am well aware that arguments for design in biology have been made before, most notably by William Paley in the nineteenth century. So I think it is important right at the beginning to clearly distinguish modern arguments for ID from earlier versions. The most important difference is that my argument is limited to design itself; I strongly emphasize that it is not an argument for the existence of a benevolent God, as Paley's was. I hasten to add that I myself do believe in a benevolent God, and I recognize that philosophy and theology may be able to extend the argument. But a scientific argument for design in biology does not reach that far. Thus, while I argue for design, the question of the identity of the designer is left open. Possible candidates for the role of designer include: the God of Christianity; an angel – fallen or not; Plato's demiurge; some mystical new-age force; space aliens from Alpha Centauri; time travelers; or some utterly unknown intelligent being. Of course, some of these possibilities may seem more plausible than others based on information from fields other than science. Nonetheless, as regards the identity of the designer, modern ID theory happily echoes Isaac Newton's phrase, *hypothesis non fingo*.

The fact that modern ID theory is a minimalist argument for design itself, not an argument for the existence of God, relieves it of much of the baggage that weighed down Paley's argument. First of all, it is immune to the argument from evil. It matters not a whit to the scientific case whether the designer is good or bad, interested in us or uninterested. It only matters whether an explanation of design appears to be consistent with the biological examples I point to. Second, questions about whether the designer is omnipotent, or even especially competent, do not arise in my case, as they did in Paley's. Perhaps the designer isn't omnipotent or very competent. More to the point, perhaps the designer was not interested in every detail of biology, as Paley thought, so that, while some features were indeed designed, others were left to the vagaries of nature. Thus the modern argument for design need only show that intelligent agency appears to be a good explanation for some biological features.

Thus, compared to William Paley's argument, modern ID theory is very restricted in scope. However, what it lacks in scope, it makes up for in resilience. Paley conjoined a number of separable ideas in his argument – design, omnipotence, benevolence, and so on – that made his overall position quite brittle. For example, arguments against the perceived benevolence of the design became arguments against the very existence of design. Thus one got the seeming *non sequitur* stating that because biological feature A appears malevolent, therefore all biological features arose by natural selection or some other unintelligent process. With the much more modest claims of modern ID theory, such a move is not possible. Attention is kept focused on the basic question of whether unintelligent processes could have produced the complex structures of biology, or whether intelligence was indeed required.

Another important point to emphasize right at the beginning is that mine is indeed a scientific argument, not a philosophical or theological argument. Let me explain what I mean by that without getting entangled in trying to define those elusive terms. By calling the argument scientific I mean first that it does not rest on any tenet of any particular creed, nor is it a deductive argument from first principles. Rather, it depends critically on physical evidence found in nature. Second, because it depends on physical evidence it can potentially be falsified by other physical evidence. Thus it is tentative, only claiming that it currently seems to be the best explanation given the information we have available to us right now.

I do acknowledge that the scientific argument for design may have theological implications, but that does not change its status as a scientific idea. I would like to draw a parallel between the modern argument for design in biology and the Big Bang theory in physics. The Big Bang theory strikes many people as having theological implications, as shown by those who do not welcome those implications. For example, in 1989, John Maddox, the editor of *Nature*, the world's leading science journal, published a very peculiar editorial, entitled "Down with the Big Bang." He wrote:

> Apart from being philosophically unacceptable, the Big Bang is an over-simple view of how the Universe began, and it is unlikely to survive the decade ahead....Creationists...seeking support for their opinions have ample justification in the Big Bang.
>
> (Maddox 1989: 425)

Nonetheless, despite its theological implications, the Big Bang theory is a completely scientific one, which justifies itself by physical data, not by appeals to holy books. I think a theory of ID in biology fits into the same category: while it may have theological implications it justifies itself by physical data. Furthermore, just as the Big Bang theory could be overturned tomorrow by new evidence, so could ID theory. Both are tentative.

With these preliminary remarks in mind, I now turn to considering the scientific case for ID in biology. I will proceed as follows. First, I will briefly make the case for design. Second, I will then address several specific scientific objections put forward by critics of design. Finally, I will discuss the question of falsifiability.

Darwinism and design

In 1859 Charles Darwin published his great work *On the Origin of Species*, in which he proposed to explain how the great variety and complexity of the natural world might have been produced solely by the action of blind physical processes. His proposed mechanism was, of course, natural selection working on random variation. In a nutshell, Darwin reasoned that the members of a species whose chance variation gave them an edge in the struggle to survive would tend to survive and reproduce. If the variation could be inherited, then over time the characteristics of the species would change. And over great periods of time, perhaps great changes would occur.

It was a very elegant idea. Nonetheless, Darwin knew his proposed mechanism could not explain everything, and in *Origin* he gave us a criterion by which to judge his theory. He wrote:

> If it could be demonstrated that any complex organ existed which could not possibly have been formed by numerous, successive, slight modifications, my theory would absolutely break down.
>
> (Darwin 1999 [1859]: 154)

He added, however, that he could "find out no such case." Darwin of course was justifiably interested in protecting his fledgling theory from easy dismissal, and so he threw the burden of proof – to prove a negative, to "demonstrate" that something "could not possibly" have happened – onto his opponents, which is essentially impossible to do in science. Nonetheless, let's ask what might at least potentially meet Darwin's criterion? What sort of organ or system seems unlikely to be formed by "numerous, successive, slight modifications"? A good place to start is with one that is irreducibly complex. In *Darwin's Black Box: The Biochemical Challenge to Evolution*, I defined an irreducibly complex system as:

> [A] single system which is composed of several well-matched, interacting parts that contribute to the basic function, and where the removal of any one of the parts causes the system to effectively cease functioning.
>
> (Behe 1996: 39)

A good illustration of an irreducibly complex system from our everyday world is a simple mechanical mousetrap. A common mousetrap has several parts, including a wooden platform, a spring with extended ends, a hammer, holding bar, and catch. Now, if the mousetrap is missing the spring, or hammer, or platform, it doesn't catch mice half as well as it used to, or a quarter as well. It simply doesn't catch mice at all. Therefore it is irreducibly complex. It turns out that irreducibly complex systems are headaches for Darwinian theory, because they are resistant to being produced in the gradual, step-by-step manner that Darwin envisioned.

As biology has progressed with dazzling speed in the past half-century, we have discovered many systems in the cell, at the very foundation of life, which, like a mousetrap, are irreducibly complex. Time permits me to mention only one example here – the bacterial flagellum. The flagellum is quite literally an outboard motor that some bacteria use to swim. It is a rotary device that, like a boat's motor, turns a propeller to push against liquid, moving the bacterium forwards in the process. It consists of a number of parts, including a long tail that acts as a propeller, the hook region that attaches the propeller to the drive shaft, the motor that uses a flow of acid from the outside of the bacterium to the inside to power the turning, a stator that keeps the structure stationary in the plane of the membrane while the propeller turns, and bushing material to allow the drive shaft to poke up through the bacterial membrane. In the absence of the hook, or the motor, or the propeller, or the drive shaft, or most of the forty different types of proteins that genetic studies have shown to be necessary for the activity or construction of the flagellum, one doesn't get a flagellum that spins half as fast as it used to, or a quarter as fast. Either the flagellum doesn't work, or it doesn't even get constructed in the cell. Like a mousetrap, the flagellum is irreducibly complex. And again, like the mousetrap, its evolutionary development by "numerous, successive, slight modifications" is quite difficult to envision. In fact, if one examines the scientific literature, one quickly sees that no one has ever proposed a serious, detailed model for how the flagellum might have arisen in a Darwinian manner, let alone conducted experiments to test such a model. Thus in a flagellum we seem to have a serious candidate to meet Darwin's criterion. We have a system that seems very unlikely to have been produced by "numerous, successive, slight modifications."

Is there an alternative explanation for the origin of the flagellum? I think there is, and it's really pretty easy to see. But in order to see it, we have to do something a bit unusual: we have to break a rule. The rule is rarely stated explicitly. But it was set forth candidly by the Nobel laureate Christian De Duve in his 1995 book, *Vital Dust*, in which he speculated about the expansive history of life. He wrote:

> A warning: All through this book, I have tried to conform to the overriding rule that life be treated as a natural process, its origin,

> evolution, and manifestations, up to and including the human species, as governed by the same laws as nonliving processes.
>
> (De Duve 1995: xiv)

In science journals the rule is always obeyed, at least in letter, yet sometimes it is violated in spirit. For example, several years ago David DeRosier, professor of biology at Brandeis University, published a review article on the bacterial flagellum in which he remarked:

> More so than other motors, the flagellum resembles a machine designed by a human.
>
> (DeRosier 1998)

That same year the journal *Cell* published a special issue (92(3)) on the topic of "Macromolecular machines." On the cover of the journal was a painting of a stylized protein apparently in the shape of an animal, with what seems to be a watch in the foreground (perhaps William Paley's watch). Articles in the journal had titles such as "The cell as a collection of protein machines"; "Polymerases and the replisome: Machines within machines"; and "Mechanical devices of the spliceosome: Motors, clocks, springs and things." By way of introduction, on the contents page was written:

> Like the machines invented by humans to deal efficiently with the macroscopic world, protein assemblies contain highly coordinated moving parts.
>
> (*Cell* 6 February 1998)

Well, if the flagellum and other biochemical systems strike scientists as looking like "machines" that were "designed by a human" or "invented by humans," then why don't we actively entertain the idea that perhaps they were indeed designed by an intelligent being? We don't do that, of course, because it would violate the rule. But sometimes, when a fellow is feeling frisky, he throws caution to the wind and breaks a few rules. In fact, this is just what I did in *Darwin's Black Box*: I proposed that, rather than Darwinian evolution, a more compelling explanation for the irreducibly complex molecular machines discovered in the cell is that they were indeed designed, as David DeRosier and the editors of *Cell* apprehended – purposefully designed by an intelligent agent. This proposal has attracted a bit of attention. Some of my critics have asserted that the proposal of ID is a religious idea, not a scientific one. I disagree. I think the conclusion of ID in these cases is completely empirical. That is, it's based entirely on the physical evidence, along with an appreciation for how we come to a conclusion of design. Every day of our lives we decide, consciously or not, that some things were designed, others not. How do we do that? How do we come to a conclusion of design?

To help see how we conclude design, imagine that you are walking with a friend in the woods. Suddenly your friend is pulled up by the ankle by a vine and left dangling in the air. After you cut him down you reconstruct the situation. You see that the vine was tied to a tree limb that was bent down and held by a stake in the ground. The vine was covered by leaves so that you wouldn't notice it, and so on. From the way the parts were arranged you would quickly conclude that this was no accident – it was a designed trap. This is not a religious conclusion, but one based firmly in the physical evidence.

Although I think that ID is a rather obvious hypothesis, nonetheless my book seems to have caught a number of people by surprise, and so it has been reviewed pretty widely. The *New York Times*, the *Washington Post*, the *Allentown Morning Call* – all the major media have taken a look at it. Unexpectedly, not everyone agreed with me. In fact, in response to my argument, several scientists have pointed to experimental results that, they claim, either cast much doubt over the claim of ID, or falsify it outright. In the remainder of the chapter I will discuss these counter-examples. I hope to show why I think they not only fail to support Darwinism, but why they actually fit much better with a theory of ID. After that, I will discuss the issue of falsifiability.

An "evolved" operon

Kenneth Miller, a professor of cell biology at Brown University, has written a book recently, entitled *Finding Darwin's God*, in which he defends Darwinism from a variety of critics, including myself. In a chapter devoted to rebutting *Darwin's Black Box*, he quite correctly states that "a true acid test" of the ability of Darwinism to deal with irreducible complexity would be to "[use] the tools of molecular genetics to wipe out an existing multipart system and then see if evolution can come to the rescue with a system to replace it" (Miller 1999: 145). He then cites the careful work over the past twenty-five years of Barry Hall of the University of Rochester on the experimental evolution of a lactose-utilizing system in *E. coli*.

Here is a brief description of how the system, called the *lac* operon, functions. The *lac* operon of *E. coli* contains genes coding for several proteins that are involved in the metabolism of a type of sugar called lactose. One protein of the *lac* operon, called a permease, imports lactose through the otherwise impermeable cell membrane. Another protein is an enzyme called galactosidase, which can break down lactose to its two constituent monosaccharides, galactose and glucose, which the cell can then process further. Because lactose is rarely available in the environment, the bacterial cell switches off the genes until lactose is available. The switch is controlled by another protein called a repressor, whose gene is located next to the operon. Ordinarily the repressor binds to the *lac* operon, shutting it off by physically interfering with the operon. However, in the presence of the natural "inducer" allolactose or the artificial chemical inducer IPTG, the repressor binds to the inducer and releases the operon, allowing the *lac* operon enzymes to be synthesized by the cell.

After giving his interpretation of Barry Hall's experiments, Kenneth Miller excitedly remarks:

> Think for a moment – if we were to happen upon the interlocking biochemical complexity of the reevolved lactose system, wouldn't we be impressed by the intelligence of its design? Lactose triggers a regulatory sequence that switches on the synthesis of an enzyme that then metabolizes lactose itself. The products of that successful lactose metabolism then activate the gene for the lac permease, which ensures a steady supply of lactose entering the cell. Irreducible complexity. What good would the permease be without the galactosidase?...No good, of course.
>
> By the very same logic applied by Michael Behe to other systems, therefore, we could conclude that the system had been designed. Except we *know* that it was *not* designed. We know it evolved because we watched it happen right in the laboratory! No doubt about it – the evolution of biochemical systems, even complex multipart ones, is explicable in terms of evolution. Behe is wrong.
>
> (Miller 1999: 146–7)

For the next few minutes I will try to show that the picture Miller paints is greatly exaggerated. In fact, far from being a difficulty for design, the very same work that Miller points to as an example of Darwinian prowess I would cite as showing the limits of Darwinism and the need for design.

So what did Barry Hall actually do? To study bacterial evolution in the laboratory, in the mid-1970s Hall produced a strain of *E. coli* in which the gene for *just* the galactosidase of the *lac* operon was deleted. He later wrote:

> All of the other functions for lactose metabolism, including lactose permease and the pathways for metabolism of glucose and galactose, the products of lactose hydrolysis, remain intact, thus re-acquisition of lactose utilization requires only the evolution of a new ß-galactosidase function.
>
> (Hall 1999: 2)

Thus, contrary to Miller's own criterion for "a true acid test," a multipart system was not "wiped out" – only one component of a multipart system was deleted. The *lac* permease and repressor remained intact. What's more, as we shall see, the artificial inducer IPTG was added to the bacterial culture, and an alternate, cryptic galactosidase was left intact.

Without galactosidase, Hall's cells would not grow when cultured on a medium containing only lactose as a food source. However, when grown on a plate that also included alternative nutrients, bacterial colonies were established. When the other nutrients were exhausted the colonies stopped growing. However, Hall noticed that after several days to several weeks, hyphae grew on some of the colonies. Upon isolating cells from the hyphae, Hall saw that they frequently had two mutations, one of which was in a gene for a protein he called "evolved ß-galactosidase" ("*ebg*"), that allowed it to metabolize lactose efficiently. The *ebg* gene is located in another operon, distant from the *lac* operon, and is under the control of its own repressor protein. The second mutation Hall found was always in the gene for the *ebg* repressor protein, which caused the repressor to bind lactose with sufficient strength to de-repress the *ebg* operon.

The fact that there were two separate mutations in different genes – neither of which by itself allowed cell growth (Hall 1982a) – startled Hall, who knew that the odds against the mutations appearing randomly and independently were prohibitive (Hall 1982b). Hall's results and similar results from other laboratories led to research in a new area dubbed "adaptive mutations" (Cairns 1998; Foster 1999; Hall 1998; McFadden and Al Khalili 1999; Shapiro 1997). As Hall later wrote:

> Adaptive mutations are mutations that occur in nondividing or slowly dividing cells during prolonged nonlethal selection, and that appear to be specific to the challenge of the selection in the sense that the only mutations that arise are those that provide a growth advantage to the cell. The issue of the specificity has been controversial because it violates our most basic assumptions about the randomness of mutations with respect to their effect on the cell.
>
> (Hall 1997: 39)

The mechanism(s) of adaptive mutation are currently unknown. While they are being sorted out, it seems unwise to cite results of processes which "violate our most basic assumptions about the randomness of mutations" to argue for Darwinian evolution, as Miller does.

The nature of adaptive mutation aside, a strong reason to consider Barry Hall's results to be quite modest is that the *ebg* proteins – both the repressor and galactosidase – are homologous to the *E. coli lac* proteins and overlap the proteins in activity. Both of the unmutated *ebg* proteins already bind lactose. Binding of lactose even to the unmutated *ebg* repressor induces a hundred-fold increase in synthesis of the *ebg* operon (Hall 1982a). Even the unmutated *ebg* galactosidase can hydrolyze lactose at a level of about 10 percent that of a "Class II" mutant galactosidase that supports cell growth (Hall 1999). These activities are not sufficient to permit growth of *E. coli* on lactose, but they are already present. The mutations reported by Hall simply enhance pre-existing activities of the proteins. In a recent paper (Hall 1999) Professor Hall pointed out that both the *lac* and *ebg* galactosidase enzymes are part of a family of highly conserved galactosidases, identical at thirteen of fifteen active site amino acid residues, which apparently diverged by gene duplication more than 2 billion years ago. The two mutations in *ebg* galactosidase that increase its ability to hydrolyze lactose change two non-identical residues back to those of other galactosidases, so that their active sites are identical. Thus – before any experiments were done – the *ebg* active site was already a near-duplicate of other galactosidases, and only became more active by becoming a complete duplicate. Significantly, by phylogenetic analysis Hall concluded that those two mutations are the *only* ones in *E. coli* that confer the ability to hydrolyze lactose – that is, no other protein, no other mutation in *E. coli* will work. Hall wrote:

> The phylogenetic evidence indicates that either Asp-92 and Cys/Trp-977 are the only acceptable amino acids at those positions, or that all of the single base substitutions that might be on the pathway to other amino acid replacements at those sites are so deleterious that they constitute a deep selective valley that has not been traversed in the two billion years since those proteins diverged from a common ancestor.
>
> (Hall 1999: 6–7)

To my mind, such results hardly support extravagant claims for the creativeness of Darwinian processes.

Another critical caveat not mentioned by Kenneth Miller is that the mutants that were initially isolated would be unable to use lactose in the wild – they required the artificial inducer IPTG to be present in the growth medium. As Barry Hall states clearly, in the absence of IPTG, no viable mutants are seen. The reason for this is that a permease is required to bring lactose into the cell. However, *ebg* only has a galactosidase activity, not a permease activity, so the experimental system had to rely on the pre-existing *lac* permease. Since the *lac* operon is repressed in the absence of either allolactose or IPTG, Hall decided to include the artificial inducer in all media up to this point so that the cells could grow. Thus the system was being artificially supported by intelligent intervention.

The prose in Miller's book obscures the facts that most of the lactose system was already in place when the experiments began, that the system was carried through non-viable states by inclusion of IPTG, and that the system will not function without pre-existing components. From a skeptical perspective, the admirably careful work of Barry Hall involved a series of micromutations stitched together by intelligent intervention. He showed that the activity of a deleted enzyme could be replaced only by mutations to a second, homologous protein with a nearly-identical active site; and only if the second repressor already bound lactose; and only if the system were also artificially induced by IPTG; and only if the system were also allowed to use a pre-existing permease. In my view, such results are entirely in line with the expectations of irreducible complexity requiring intelligent intervention, and of limited capabilities for Darwinian processes.

Blood clotting

A second putative counter-example to ID concerns the blood-clotting system. Blood clotting is a very intricate biochemical process, requiring many protein parts. I had devoted a chapter of *Darwin's Black Box* to the blood-clotting cascade, claiming that it is irreducibly complex and so does not fit well within a Darwinian framework. However, Russell Doolittle, a prominent biochemist, member of the National Academy of Sciences, and expert on blood clotting, disagreed. While discussing the similarity of the proteins of the blood-clotting cascade to each other in an essay in *Boston Review* in 1997, he remarked that "the genes for new proteins come from the genes for old ones by gene duplication" (Doolittle 1997: 28). Doolittle's invocation of gene duplication has been repeated by many scientists reviewing my book, but it reflects a common confusion. Genes with similar sequences only suggest common descent – they do not speak to the mechanism of evolution. This point is critical to my argument and bears emphasis: *evidence of common descent is not evidence of natural selection.* Similarities among either organisms or proteins are the evidence for descent with modification, that is, for evolution. Natural selection, however, is a proposed explanation for how evolution might take place – its mechanism – and so it must be supported by other evidence if the question is not to be begged.

Doolittle then cited a paper (Bugge *et al.* 1996a) entitled "Loss of fibrinogen rescues mice from the pleiotropic effects of plasminogen deficiency." (By way of explanation, fibrinogen is the precursor of the clot material; plasminogen is a protein that degrades blood clots.) He commented:

> Recently the gene for plaminogen [*sic*] was knocked out of mice, and, predictably, those mice had thrombotic complications because fibrin clots could not be cleared away. Not long after that, the same workers knocked out the gene for fibrinogen in another line of mice. Again, predictably, these mice were ailing, although in this case hemorrhage was the problem. And what do you think happened when these two lines of mice were crossed? For all practical purposes, the mice lacking both genes were normal! Contrary to claims about irreducible complexity, the entire ensemble of proteins is not needed. Music and harmony can arise from a smaller orchestra.
>
> (Doolittle 1997: 29)

The implied argument seems to be that the modern clotting system is actually not irreducibly complex, and so a simpler clotting cascade might be missing factors such as plasminogen and fibrinogen, and perhaps it could be expanded into the modern clotting system by gene duplication. However, that interpretation does not stand up to a careful reading of Bugge *et al.*

In their paper, Bugge *et al.* (1996a) note that the lack of plasminogen in mice results in many problems, such as high mortality, ulcers, severe thrombosis, and delayed wound healing. On the other hand, lack of fibrinogen results in failure to clot, frequent hemorrhage, and death of females during pregnancy. The point of Bugge *et al.* (1996a) was that if one crosses the two knockout strains, producing plasminogen-plus-fibrinogen deficiency in individual mice, the mice do not suffer the many problems that afflict mice lacking plasminogen alone. Since the title of the paper emphasized that mice were "rescued" from some ill effects, one might be misled into thinking that the double-knockout mice were normal. They are not. As Bugge *et al.* state in their abstract, "Mice deficient in plasminogen and fibrinogen are phenotypically indistinguishable from fibrinogen-deficient mice" (1996a: 709). In other words, the double-knockouts have all the problems that mice lacking only fibrinogen have: they do not form clots, they hemorrhage, and the females die if they become pregnant. They are definitely not promising evolutionary intermediates.

The probable explanation is straightforward. The pathological symptoms of mice missing just plasminogen apparently are caused by uncleared clots. But fibrinogen-deficient mice cannot form clots in the first place. So problems due to uncleared clots don't arise either in fibrinogen-deficient mice or in mice that lack both plasminogen and fibrinogen. Nonetheless, the severe problems that attend lack of clotting in fibrinogen-deficient mice continue in the double knockouts. Pregnant females still perish.

Most important for the issue of irreducible complexity, however, is that the double-knockout mice do not merely have a less sophisticated but still functional clotting system. They have no functional clotting system at all. They are not evidence for the Darwinian evolution of blood clotting. Therefore my argument, that the system is irreducibly complex, is unaffected by this example.

Other work from the same laboratory is consistent with the view that the blood-clotting cascade is irreducibly complex. Experiments with "knockout" mice in which the genes for other clotting components, called tissue factor and prothrombin, have been deleted separately, show that those components are required for clotting, and in their absence the organism suffers severely (Bugge *et al.* 1996b; Sun *et al.* 1998).

In ending this section let me just make explicit the point that two very competent scientists, Professors Miller and Doolittle, both of whom are highly motivated to discredit claims of ID, and both of whom are quite capable of surveying the entire biomolecular literature for experimental counter-examples, both came up with examples that, when looked at skeptically, actually buttress the case for irreducible complexity, rather than weaken it. Of course, this does not prove that claims of irreducible complexity are true, or that ID is correct. But it does show, I think, that scientists really don't have a handle on irreducible complexity, and that the idea of ID is considerably stronger than its detractors would have us believe. It also shows the need to treat Darwinian scenarios, such as those Miller and Doolittle offered, with a hermeneutic of suspicion. Some scientists believe so strongly in Darwinism that their critical judgments are affected, and they will unconsciously overlook pretty obvious problems with Darwinian scenarios, or confidently assert things that are objectively untrue.

Falsifiability

Let us now consider the issue of falsifiability. Let me say up front that I know most philosophers of science do not regard falsifiability as a necessary trait of a successful scientific theory. Nonetheless, falsifiabilty is still an important factor to consider since it is nice to know whether or not one's theory can be shown to be wrong by contact with the real world.

A frequent charge made against ID is that it is unfalsifiable, or untestable. For example, in its recent booklet *Science and Creationism* the National Academy of Sciences writes:

> [I]ntelligent design…[is] not science because [it is] not testable by the methods of science.
>
> (National Academy of Sciences 1999: 25)

Yet that claim seems to be at odds with the criticisms I have just summarized. Clearly, both Russell Doolittle and Kenneth Miller advanced scientific arguments aimed at falsifying ID. If the results of Bugge *et al.* (1996a) had been as Doolittle first thought, or if Barry Hall's work had indeed shown what Miller implied, then they correctly believed that my claims about irreducible complexity would have suffered quite a blow.

Now, one can't have it both ways. One can't say both that ID is unfalsifiable (or untestable) and that there is evidence against it. Either it is unfalsifiable and floats serenely beyond experimental reproach, or it can be criticized on the basis of our observations and is therefore testable. The fact that critical reviewers advance scientific arguments against ID (whether successfully or not) shows that they think ID is indeed falsifiable. What's more, it is wide open to falsification by a series of rather straightforward laboratory experiments such as those that Miller and Doolittle pointed to, which is exactly why they pointed to them.

Now let's turn the tables by asking the following question: how could one falsify the claim that a particular biochemical system was produced by a

Darwinian process? Kenneth Miller announced an "acid test" for the ability of natural selection to produce irreducible complexity. He then decided that the test was passed, and unhesitatingly proclaimed ID to be falsified ("Behe is wrong") (Miller 1999: 147). But if, as it certainly seems to me, *E. coli* actually fails the lactose-system "acid test," would Miller consider Darwinism to be falsified? Almost certainly not. He would surely say that Barry Hall started with the wrong bacterial species, or used the wrong selective pressure, and so on. So it turns out that his "acid test" was not a test of Darwinism; it tested only ID.

The same one-way testing was employed by Russell Doolittle. He pointed to the results of Bugge *et al.* to argue against ID. But when the results turned out to be the opposite of what he had originally thought, Professor Doolittle did not abandon Darwinism.

It seems then, perhaps counter-intuitively to some, that ID is quite susceptible to falsification, at least on the points under discussion. Darwinism, on the other hand, seems quite impervious to falsification. The reason for this can be seen when we examine the basic claims of the two ideas with regard to a particular biochemical system like, say, the bacterial flagellum. The claim of ID is that "*No* unintelligent process could produce this system." The claim of Darwinism is that "*Some* unintelligent process could produce this system." To falsify the first claim, one need only show that at least one unintelligent process could produce the system. To falsify the second claim, one would have to show the system could not have been formed by any of a potentially infinite number of possible unintelligent processes, which is effectively impossible to do.

The danger of accepting an effectively unfalsifiable hypothesis is that science has no way to determine if the belief corresponds to reality. In the history of science, the scientific community has believed in any number of things that were in fact not true, not real – for example, the universal ether. If there were no way to test those beliefs, the progress of science might be substantially and negatively affected. If, in the present case, the expansive claims of Darwinism are in reality not true, then its unfalsifiability will cause science to bog down in these areas, as I believe it has.

So, what can be done? I don't think that the answer is to never investigate a theory that is unfalsifiable. After all, although it is unfalsifiable, Darwinism's claims are potentially positively demonstrable. For example, if some scientist conducted an experiment showing the production of a flagellum (or some equally complex system) by Darwinian processes, then the Darwinian claim would be affirmed. The question only arises in the face of negative results.

I think several steps can be prescribed. First of all, one has to be aware – raise one's consciousness – about when a theory is unfalsifiable. Second, as far as possible, an advocate of an unfalsifiable theory should try as diligently as possible to demonstrate positively the claims of the hypothesis. Third, one needs to relax Darwin's criterion from this:

> If it could be demonstrated that any complex organ existed which could not possibly have been formed by numerous, successive, slight modifications, my theory would absolutely break down.
>
> (Darwin 1999 [1859]: 154)

to something like this:

> If a complex organ exists which seems very unlikely to have been produced by numerous, successive, slight modifications, and if no experiments have shown that it or comparable structures can be so produced, then maybe we're barking up the wrong tree. So…
>
> *Let's break some rules!*

Of course, people will differ on the point at which they decide to break rules. But at least with the realistic criterion there could be evidence against the unfalsifiable. At least then people like Doolittle and Miller would run a risk when they cite an experiment that shows the opposite of what they had thought. At least then science would have a way to escape from the rut of unfalsifiability and think new thoughts.

Notes

1 This paper was delivered on 28 May 2000 to a plenary session of the Gifford Bequest International Conference, "Natural Theology: Problems and Prospects," held at the University of Aberdeen, Scotland.

References

Behe, M.J. (1996) *Darwin's Black Box: The Biochemical Challenge to Evolution*, New York: The Free Press.

Bugge, T.H., Kombrinck, K.W., Flick, M.J., Daugherty, C.C., Danton, M.J., and Degen, J.L. (1996a) "Loss of fibrinogen rescues mice from the pleiotropic effects of plasminogen deficiency," *Cell* 87: 709–19.

Bugge, T.H., Xiao, Q., Kombrinck, K.W., Flick, M.J., Holmback, K., Danton, M.J., Colbert, M.C., Witte, D.P., Fujikawa, K., Davie, E.W., and Degen, J.L. (1996b) "Fatal embryonic bleeding events in mice lacking tissue factor, the cell-associated initiator of blood coagulation," *Proceedings of the National Academy of Sciences of the United States of America* 93: 6,258–63.

Cairns, J. (1998) "Mutation and cancer: The antecedents to our studies of adaptive mutation," *Genetics* 148: 1,433–40.

Darwin, C. (1999 [1859]) *The Origin of Species*, New York: Bantam Books.

De Duve, C. (1995) *Vital Dust: Life as a Cosmic Imperative*, New York: Basic Books.

DeRosier, D.J. (1998) "The turn of the screw: The bacterial flagellar motor," *Cell* 93: 17–20.

Doolittle, R.F. (1997) "A delicate balance," *Boston Review* February/March: 28–9.

Foster, P.L. (1999) "Mechanisms of stationary phase mutation: A decade of adaptive mutation," *Annual Review of Genetics* 33: 57–88.

Hall, B.G. (1999) "Experimental evolution of Ebg enzyme provides clues about the evolution of catalysis and to evolutionary potential," *FEMS Microbiology Letters* 174: 1–8.

—— (1998) "Adaptive mutagenesis: A process that generates almost exclusively beneficial mutations," *Genetica* 102–3: 109–25.

—— (1997) "On the specificity of adaptive mutations," *Genetics* 145: 39–44.

—— (1982a) "Evolution of a regulated operon in the laboratory," *Genetics* 101: 335–44.

—— (1982b) "Evolution on a Petri dish: The evolved ß-galactosidase system as a model for studying acquisitive evolution in the laboratory," in M.K. Hecht, B. Wallace, and G.T. Prance (eds) *Evolutionary Biology*, New York: Plenum Press, pp. 85–150.

McFadden, J. and Al Khalili, J. (1999) "A quantum mechanical model of adaptive mutation," *Biosystems* 50: 203–11.

Maddox, J. (1989) "Down with the Big Bang," *Nature* 340: 425.

Miller, K.R. (1999) *Finding Darwin's God: A Scientist's Search for Common Ground between God and Evolution*, New York: Cliff Street Books.

National Academy of Sciences (1999) *Science and Creationism: A View from the National Academy of Sciences*, Washington, DC: National Academy Press.

Shapiro, J.A. (1997) "Genome organization, natural genetic engineering and adaptive mutation," *Trends in Genetics* 13: 98–104.

Sun, W.Y., Witte, D.P., Degen, J.L., Colbert, M.C., Burkart, M.C., Holmback, K., Xiao, Q., Bugge, T.H., and Degen, S.J. (1998) "Prothrombin deficiency results in embryonic and neonatal lethality in mice," *Proceedings of the National Academy of Sciences of the United States of America* 95: 7,597–602.

16

ANSWERING THE BIOCHEMICAL ARGUMENT FROM DESIGN

Kenneth R. Miller

One of the things that makes science such an exhilarating activity is its revolutionary character. As science advances, there is always the possibility that some investigator, working in the field or at a laboratory bench, will produce a discovery or experimental result that will completely transform our understanding of nature. The history of science includes so many examples of such discoveries that in many respects the practice of science has a built-in bias in favor of the little guy, the individual investigator who just might hold the key to our next fundamental scientific advance. Indeed, if there is one dogma in science, it should be that science has no dogma.

What this means, as a practical matter, is that everything in science is open to question. Can we be sure that the speed of light isn't an absolute upper limit? Is it possible that genetic information can be carried by proteins, rather than DNA? Was Einstein correct in his formulation of the theory of general relativity? It is never easy to upset the scientific apple cart, but the practice of science requires, as an absolute, that everything in science be open to question. Everything.

In 1996, Michael Behe took a bold step in this scientific tradition by challenging one of the most useful, productive, and fundamental concepts in all of biology – Charles Darwin's theory of evolution. Behe's provocative claim, carefully laid out in his book, *Darwin's Black Box*, was that whatever else Darwinian evolution can explain successfully, it cannot account for the biochemical complexity of the living cell. As Behe put it: "for the Darwinian theory of evolution to be true, it has to account for the molecular structure of life. It is the purpose of this book to show that it does not" (1996a: 24–5).

As we will see, Behe's argument is crafted around the existence of complex molecular machines found in all living cells. Such machines, he argues, could not have been produced by evolution, and therefore must be the products of intelligent design, a point of view he has articulated elsewhere in this volume. This argument has been picked up by a variety of anti-evolution groups around the USA, and has become a focal point for those who would argue that "intelligent design" theory (ID) deserves a place in the science classroom as a scientific alternative to Darwin. What I propose to do in this brief review is to put this line of reasoning to the test. I will examine both the scientific evidence for this claim and the logical structure of the biochemical argument from design, and will pose the most fundamental question one can ask of any scientific hypothesis – does it fit the facts?

An exceptional claim

For nearly more than a century and a half, one of the classic ways to argue against evolution has been to point to an exceptionally complex and intricate structure and then to challenge an evolutionist to "evolve this!" Examples of such challenges have included everything from the optical marvels of the human eye to the chemical defenses of the bombardier beetle. At first glance, Behe's examples seem to fit this tradition. As examples of cellular machinery for which no evolutionary explanations exist he cites the cilia and flagella that produce cell movement, the cascade of blood-clotting proteins, the systems that target proteins to specific sites within the cell, the production of antibodies by the immune system, and the intricacies of biosynthetic pathways.

As he realizes, however, the mere existence of structures and pathways that have not yet been given step-by-step Darwinian explanation does not make much of a case against evolution. Critics of evolution have laid down such challenges before, only to see them backfire when new scientific work provided exactly the evidence they had demanded. Behe himself once made a similar claim when he challenged evolutionists to produce transitional fossils linking the first fossil whales with their supposed land-based ancestors (Behe 1994: 61). Ironically, not one, not two, but *three* transitional species between whales and land-dwelling Eocene mammals had been discovered by the end of 1994 when his challenge was published (Gould 1994: 8–15).

Given that the business of science is to provide and test explanations, the fact that there are a few things that have, as yet, no published evolutionary explanations is not much of an argument against Darwin. Rather, it means that the field is still active, vital, and filled with scientific challenges. Behe realizes this, and therefore his principal claim for design is quite different. He observes, quite correctly, that science has not explained the evolution of the bacterial flagellum, but then he goes one step further. No such explanation is even *possible*, according to Behe. Why? Because the flagellum has a characteristic that Behe calls "irreducible complexity":

> By irreducibly complex I mean a single system composed of several well-matched, interacting parts that contribute to the basic function, wherein the removal of any one of the parts causes the system to effectively cease functioning.
>
> (Behe 1996a: 39)

Irreducible complexity is the key to Behe's argument against Darwin. Why? Because it opens a chain of reasoning that allows the critic of evolution to reach the conclusion of design. It alone allows one to state that the notion of an evolutionary origin for any complex biochemical structure can be ruled out in principle. To make his point perfectly clear, Behe uses a common mechanical device, the mousetrap, as an example of irreducible complexity:

> A good example of such a system is a mechanical mousetrap.... The mousetrap depends critically on the presence of all five of its components; if there were no spring, the mouse would not be pinned to the base; if there were no platform, the other pieces would fall apart; and so on. The function of the mousetrap requires all the pieces: you cannot catch a few mice with just a platform, add a spring and catch a few more mice, add a holding bar and catch a few more. All of the components have to be in place before any mice are caught. Thus the mousetrap is irreducibly complex.
>
> (Behe 1998: 178)

Since every part of the mousetrap must be in place before it is functional, this means that partial mousetraps, ones that are missing one or two parts, are useless – you cannot catch mice with them. Extending the analogy to irreducibly complex biochemical machines, they also are without function until all of their parts are assembled. What this means, of course, is that natural selection could not produce such machines gradually, one part at a time. They would be non-functional until all of their parts were assembled, and natural selection, which can only select functioning systems, would have nothing to work with.

Behe has made this exact point quite clear:

> An irreducibly complex system cannot be produced directly by numerous, successive, slight modifications of a precursor system, because any precursor to an irreducibly complex system that is missing a part is by definition nonfunctional. Since natural selection can only choose systems that are already working, then if a biological system cannot be produced gradually it would have to arise as an integrated unit, in one fell swoop, for natural selection to have anything to act on.
>
> (1996b: 39)

In Behe's view, this observation, in and of itself, makes the case for design. If the biochemical machinery of the cell cannot be produced by natural selection, then there is only one reasonable alternative – design by an intelligent agent. Lest anyone doubt his claim for the absolute impossibility that evolution might have produced such machinery, Behe assures his readers that the immense scientific literature on evolution contains not a single example to the contrary:

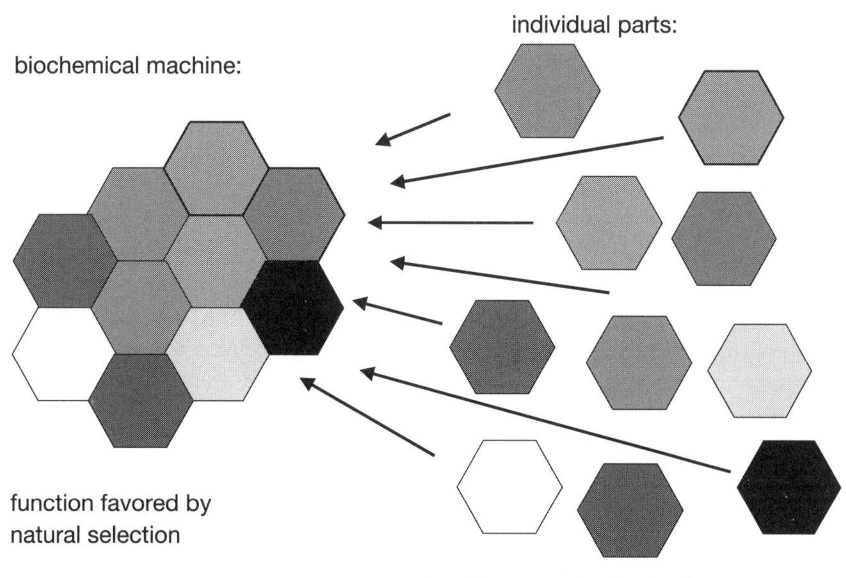

Figure 16.1 The Biochemical Argument from Design

Note: According to the biochemical argument from design, natural selection could not produce an irreducibly complex biochemical machine because its individual parts are, by definition, without any selectable function.

> There is no publication in the scientific literature – in prestigious journals, specialty journals, or books – that describes how the molecular evolution of any real, complex, biochemical system either did occur or even might have occurred.
>
> (1996a: 185)

Powerful stuff. The great power of Behe's argument is that it claims to have discovered, in the biochemical machinery of the living cell, a new property (irreducible complexity) that makes it possible to rule out, even in principle, any possibility that evolution could have produced it. The next question we should ask is simple – is he right?

Mr Darwin's workshop

If Behe's arguments have a familiar ring to them, they should. They mirror the classic "Argument from Design," articulated so well by William Paley nearly 200 years ago in his book *Natural Theology*. Darwin was well aware of the argument, so much so that he devoted special care to answering it when he wrote *On the Origin of Species*. Darwin's answer, in essence, was that evolution produces complex organs in a series of fully functional intermediate stages. If each of the intermediate stages can be favored by natural selection, then so can the whole pathway. Is there something different about biochemistry, a reason why Darwin's answer would not apply to the molecular systems that Behe cites?

In a word, no.

In 1998, Siegfried Musser and Sunney Chan described the evolutionary development of the cytochrome c oxidase proton pump, a complex, multipart molecular machine that plays a key role in energy transformation by the cell. In human cells, the pump consists of six proteins, each of which is necessary for the pump to function properly. It would seem to be a perfect example of irreducible complexity. Take one part away from the pump, and it no longer works. And yet these authors were able to produce, in impressive detail, "an evolutionary tree constructed using the notion that respiratory complexity and efficiency progressively increased throughout the evolutionary process" (Musser and Chan 1998: 517).

How is this possible? If you believed Michael Behe's assertion that biochemical machines were irreducibly complex, you might never bother to check, and this is the real scientific danger of his ideas. Musser and Chan did check, and found that two of the six proteins in the proton pump were quite similar to a bacteria enzyme known as the cytochrome bo_3 complex. Could this mean that part of the proton pump evolved from a working cytochrome bo_3 complex? Certainly.

An ancestral two-part cytochrome bo_3 complex would have been fully functional, albeit in a different context, but that context would indeed have allowed natural selection to favor its evolution. How can we be sure that this "half" of the pump would be any good? By reference to modern organisms that have full, working versions of the cytochrome bo_3 complex. Can we make the same argument for the rest of the pump? Well, it turns out that each of the pump's major parts is closely related to working protein complexes found in micro-organisms. Evolution assembles complex biochemical machines, as Musser and Chan proposed, from smaller working assemblies that are adapted to fit novel functions. The multiple parts of complex biochemical machines are themselves assembled from smaller, working machines developed by natural selection, as shown in Figure 16.2.

What of the statement that there is no publication anywhere describing how the "molecular evolution of any real, complex, biochemical system either did occur or even might have occurred?" Simply put, that statement is not correct.

In 1996, Enrique Meléndez-Hevia and his colleagues published, in the *Journal of Molecular Evolution*, a paper entitled "The puzzle of the Krebs citric acid cycle: Assembling the pieces of chemically feasible reactions, and

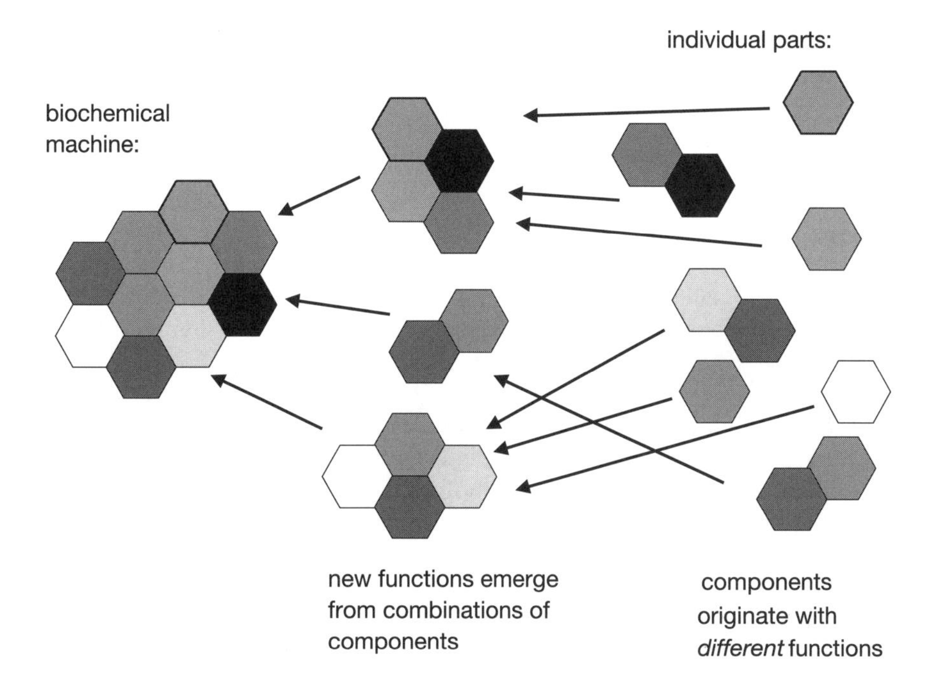

Figure 16.2 The Evolution of Biochemical Machines

Note: A Darwinian view of the evolution of complex biochemical machines requires that their individual parts and components have selectable functions.

opportunism in the design of metabolic pathways during evolution" (Meléndez-Hevia *et al.* 1996). The Krebs cycle is *real, complex, and biochemical*, and this paper does exactly what Behe says cannot be done, even in principle – it presents a feasible proposal for its evolution from simpler biochemical systems. This paper, as well as a subsequent review of the Krebs cycle by other authors (Huynen *et al.* 1999), shows that the scheme indicated in Figure 16.2 is a perfectly adequate model to account for biochemical complexity.

These are not isolated examples. Recently Martino Rizzotti published a series of detailed, step-by-step hypotheses for the evolution of a wide variety of cellular structures, including the bacterial flagellum and the eukaryotic cilium (Rizzotti 2000). I do not claim, even for a moment, that every one of Rizzotti's explanations represents the final word on the evolution of these structures. Nonetheless, any validity one might have attached to the claim that the literature lacks such explanations vanishes upon inspection.

What all of this means, of course, is that two principal claims of the ID movement are disproved, namely that it is impossible to present a Darwinian explanation for the evolution of a complex biochemical system, and that no such papers appear in the scientific literature. It is possible, and such papers do exist.

Getting to the heart of the matter

To fully explore the scientific basis of the biochemical argument from design, we should investigate the details of some of the very structures used in Behe's book as examples of irreducibly complex systems. One of these is the eukaryotic cilium, an intricate whip-like structure that produces movement in cells as diverse as green algae and human sperm. And,

> Just as a mousetrap does not work unless all of its constituent parts are present, ciliary motion simply does not exist in the absence of microtubules, connectors, and motors. Therefore we can conclude that the cilium is irreducibly complex.
>
> (Behe 1996a: 65)

Remember Behe's statement that the removal of any one of the parts of an irreducibly complex system effectively causes the system to stop working? The cilium provides us with a perfect opportunity to test that assertion. If it is correct, then we should be unable to find examples of functional cilia anywhere in nature that lack the cilium's basic parts. Unfortunately for the argument, that is not the case. Nature presents many examples of fully functional cilia that are missing key parts. One of the most compelling is the eel sperm flagellum (Figure 16.3), which lacks at least three important parts normally found in the cilium: the central doublet, central spokes, and the dynein outer arm (Wooley 1997).

This leaves us with two points to consider. First, a wide variety of motile systems exist that are missing parts of this supposedly irreducibly complex structure. Second, biologists have known for years that each of the major components of the cilium, including proteins tubulin, dynein, and actin, have distinct functions elsewhere in the cell that are unrelated to ciliary motion.

Given these facts, what is one to make of the core argument of biochemical design – namely, that the parts of an irreducibly complex structure have no functions on their own? The key element of the claim was that "any precursor to an irreducibly complex system that is missing a part is by definition nonfunctional." But the individual parts of the cilium, including tubulin, the motor protein dynein, and the contractile protein actin, are fully functional elsewhere in the cell. What this means, of course, is that a selectable function exists for each of the major parts of the cilium, and therefore that the argument is wrong.

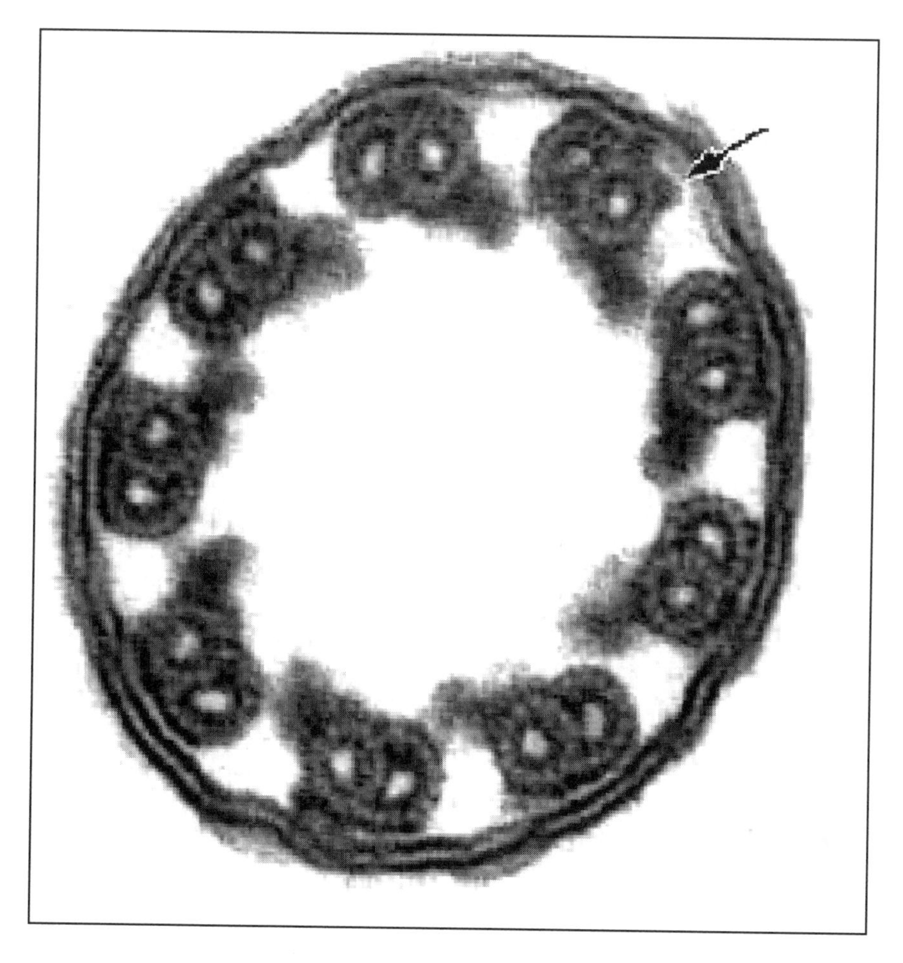

Figure 16.3 A Living Contradiction to "Irreducible Complexity"

Note: A cross-section of an eel sperm flagellum. In other organisms, this "irreducibly complex" structure includes a central pair of microtubules, spokes linking the central pair to the outer doublets, and dynein outer arms linking the doublets. Each of these structures is missing in the eel sperm flagellum (the arrow shows the location of one of these missing dynein arms), and yet the structure is fully functional. From Woolley (1997: 91).

Whips and syringes

In many ways, the "poster child" for irreducible complexity has been the bacterial flagellum. The well-matched parts of this ion-driven rotary engine pose, in the view of many critics, an insurmountable challenge to Darwinian evolution. Once again, however, a close examination of this remarkable biochemical machine tells a quite different story.

To begin with, there is more than one type of "bacterial flagellum." Flagella found in the archaebacteria are clearly not irreducibly complex.

Recent research has shown that the flagellar proteins of these organisms are closely related to a group of cell surface proteins known as the Class IV pilins (Jarrel *et al.* 1996). Since these proteins have a well-defined function that is not related to motility, the archael flagella fail the test of irreducible complexity.

Clearly, when he speaks of the bacterial flagellum, Behe refers to flagella found in the eubacteria. Representations of eubacterial flagella appear in *Darwin's Black Box* (Behe 1996a: 71) and have been used by Dr Behe in a number of public presentations. Surely these structures must fit the test of irreducible complexity? Ironically, they don't.

In 1998 the flagella of eubacteria were discovered to be closely related to a non-motile cell membrane complex known as the Type III secretory apparatus (Hueck 1998). These complexes play a deadly role in the cytotoxic (cell-killing) activities of bacteria such as *Yersinia pestis*, the bacterium that causes bubonic plague. When these bacteria infect an organism, bacteria cells bind to host cells, and then pump toxins directly through the secretory apparatus into the host cytoplasm. Efforts to understand the deadly effects of these bacteria on their hosts led to molecular studies of the proteins in the Type III apparatus, and it quickly became apparent that at least ten of them are homologous to proteins that form part of the base of the bacterial flagellum (Hueck 1998: 410).

This means that a portion of the whip-like bacterial flagellum functions as the "syringe" that makes up the Type III secretory apparatus. In other words, a subset of the proteins of the flagellum is fully functional in a completely different context – not motility, but the deadly delivery of toxins to a host cell. This observation falsifies the central claim of the biochemical argument from design – namely, that a subset of the parts of an irreducibly complex structure must be "by definition nonfunctional." Here are ten proteins from the flagellum that are missing not just one part but more than forty, and yet they are fully functional in the Type III apparatus.

Disproving design

If the biochemical argument from design is a scientific hypothesis, as its proponents claim, then it should make specific predictions that are testable in scientific terms. The most important prediction of the hypothesis of irreducible complexity is shown in Figure 16.4, and it is that components of irreducibly complex structures should not have functions that can be favored by natural selection.

As we have seen, a subset of the proteins from the flagellum does indeed have a selectable function in Type III secretion. However, we can make a more general statement about many of the components of the eubacterial flagellum (see Figure 16.5). Proteins that make up the flagellum itself are

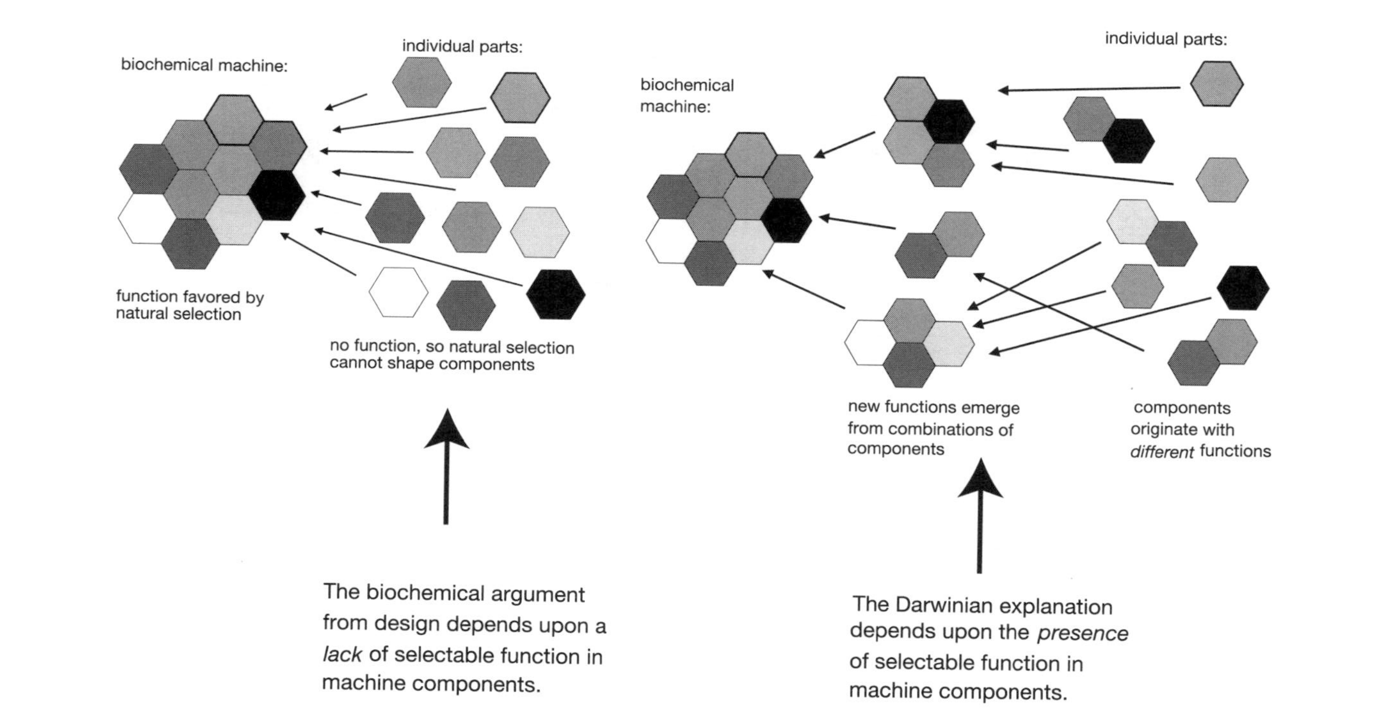

Figure 16.4 Putting Design to the Test

Note: The biochemical argument from design makes a specific, testable prediction about the components of "irreducibly complex" structures. That prediction is that individual portions of such machines should not have selectable functions. The Darwinian explanation for such structures makes a contrary prediction – namely, that components of the machine should have such functions. The scientific literature provides more than enough evidence to distinguish between the two alternatives.

closely related to a variety of cell surface proteins, including the pilins found in a variety of bacteria. A portion of the flagellum functions as an ion channel, and ion channels are found in all bacterial cell membranes. Part of the flagellar base is functional in protein secretion, and, once again, all bacteria possess membrane-bound protein secretory systems. Finally, the heart of the flagellum is an ion-driven rotary motor, a remarkable piece of protein machinery that converts ion movement into rotary movement that makes flagellar movement possible. Surely this part of the flagellum must be unique? Not at all. All bacteria possess a membrane protein complex known as the ATP synthase that uses ion movements to produce ATP. How does the synthase work? It uses the energy of ion movements to produce rotary motion. In short, at least four key elements of the eubacterial flagellum have other selectable functions in the cell that are unrelated to motility.

These facts demonstrate that the one system most widely cited as the premier example of irreducible complexity contains individual parts that have selectable functions. What this means, in scientific terms, is that the hypothesis of irreducible complexity is falsified. The Darwinian explanation of complex systems, however, is supported by the same facts.

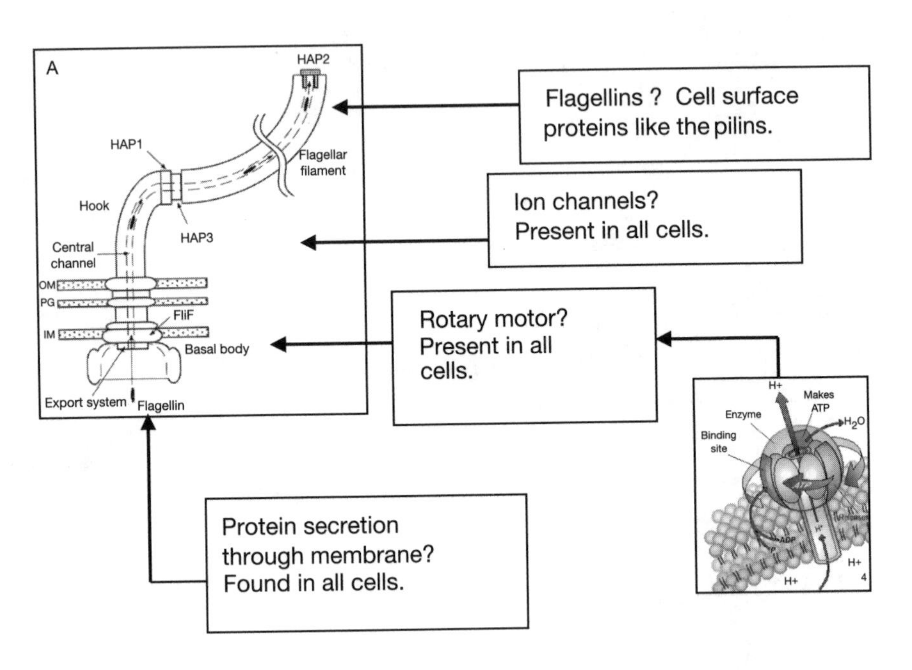

Figure 16.5 A Reducible Flagellum

Note: At least four components of the eubacterial flagellum have selectable functions that are unrelated to motility.

One might, of course, raise the objection that I have not provided a detailed, step-by-step explanation of the evolution of the flagellum. Isn't such an explanation required to dispose of the biochemical argument from design?

In a word, no. Not unless the argument has allowed itself to be reduced to a mere observation that an evolutionary explanation of the eubacterial flagellum has yet to be written. I would certainly agree with such a statement. However, the contention made by Behe is quite different from this – it is that evolution cannot explain the flagellum *in principle* (because its multiple components have no selectable function). By demonstrating the existence of such functions, even in just a handful of components, we have invalidated the argument.

Caught in the mousetrap

Why does the biochemical argument from design collapse so quickly upon close inspection? I would suggest that this is because the logic of the argument itself is flawed. Consider, for example, the mechanical mousetrap as an analogy of irreducibly complex systems. Behe has written that a mousetrap does not work if even one of its five parts is removed. However, with a little ingenuity, it turns out to be remarkably easy to construct a working mousetrap *after* removing one of its parts, leaving just four. In fact, Professor John McDonald of the University of Delaware has taken this several steps further, posting drawings on a website that show how a mousetrap may be constructed with just three, two, or even just one part. McDonald's mousetrap plans are available at:*http://udel.edu/~mcdonald/mousetrap.html* .

Behe has responded to these simpler mousetraps by pointing out, quite correctly, that human intervention and ingenuity are needed to construct the simpler mousetraps, and therefore they do not present anything approaching a model for the "evolution" of the five-part trap. However, his response overlooks the crucial question: are subsets of the five-part trap useful (selectable) in different contexts? Considering the following examples: for my personal use I sometimes wear a tie clip consisting of just three parts (platform, spring, and hammer) and use a key chain consisting of just two (platform and hammer). It is possible, in fact, to imagine a host of uses for parts of the "irreducibly complex" mousetrap, some of which are listed in Figure 16.6.

The meaning of this should be clear. If portions of a supposedly irreducibly complex mechanical structure are fully functional in different contexts, then the central claim built upon this concept is incorrect. If bits and pieces of a machine are useful for different functions, it means that natural selection could indeed produce elements of a biochemical machine for different purposes. The mousetrap example provides, unintentionally, a perfect analogy for the way in which natural selection builds complex structures.

Individual parts of a supposedly irreducibly complex machine are fully functional for different purposes. Examples:

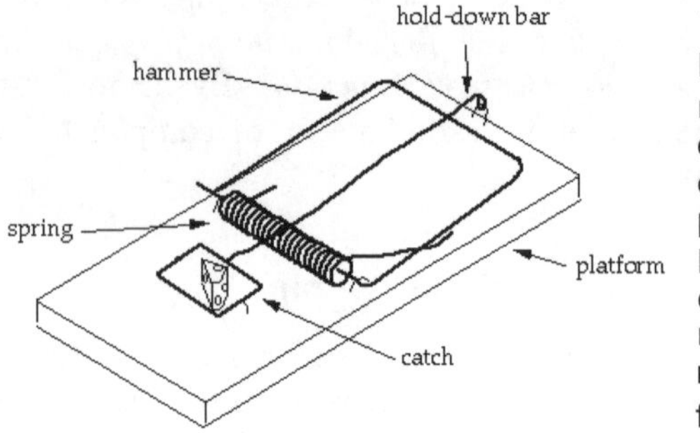

tie clip (3)
key ring (2)
refrigerator clip (3 +1)
clipboard holder (2)
door knocker (3)
paperweight (1)
kindling block (1)
catapult (4)
nutcracker (3)
nose ring (2)
fish hook (1)
toothpick (1)

Figure 16.6 A Reducible Mousetrap

Note: There are alternate, selectable functions for partial assemblies of the five parts of a standard mousetrap. The numbers in parentheses indicate the number of parts required for each function. For example, just one part (the hold-down bar) is required for a toothpick. Three parts are needed for a tie-clip (base, spring, and hammer). A refrigerator clip can be fashioned from the same three parts by adding one additional part (a magnet).

Breaking the chain

Critics of evolution are fond of claiming that they have "discovered" evidence of intelligent design in biochemical systems, suggesting that they have found positive evidence of the work of the designer. Behe himself uses such language when he writes:

> The result of these cumulative efforts to investigate the cell – to investigate life at the molecular level – is a loud, clear, piercing cry of "design!" The result is so unambiguous and so significant that it must be ranked as one of the greatest achievements in the history of science. The discovery rivals those of Newton and Einstein, Lavoisier and Schrödinger, Pasteur, and Darwin.
>
> (1996a: 232–3)

What, exactly, is the source of this "loud, piercing cry"? It turns out not to be any direct evidence, but rather a chain of reasoning – begining with the observation of "irreducible complexity" and leading, step by step, to a conclusion of design (see below) – that is well-removed from experimental evidence:

What is the "evidence" for design?

What follows is the logical chain of reasoning leading from the observation of biochemical complexity to the conclusion of intelligent design.

1 *Observation*: the cell contains biochemical machines in which the loss of a single component may abolish function. *Definition*: such machines are therefore said to be "irreducibly complex."

⇓

2 *Assertion*: any irreducibly complex structure that is missing a part is by definition non-functional, leaving natural selection with nothing to select for.

⇓

3 *Conclusion*: therefore, irreducibly complex structures *could not* have been produced by natural selection.

⇓

4 *Secondary conclusion*: therefore, such structures must have been produced by another mechanism. Since the only credible alternate mechanism is intelligent design, the very existence of such structures must be evidence of intelligent design.

When the reasoning behind the biochemical argument from design is laid out in this way, it becomes easy to spot the logical flaw in the argument. The first statement is true – the cell does indeed contain any number of complex molecular machines in which the loss of a single part may affect function. However, the second statement, the assertion of non-functionality, is demonstrably false. As we have seen, the individual parts of many such machines do indeed have well-defined functions within the cell. Once this is realized, the logic of the argument collapses. If the assertion in the second statement is shown to be false, the chain of reasoning is broken and both conclusions are falsified.

The cell does not contain biochemical evidence of design.

Paley's ghost

Paley's twenty-first century followers claim that the ID movement is based upon new discoveries in molecular biology, and represents a novel scientific movement that is worthy of scientific and educational attention. Couched in the modern language of biochemistry, Behe's formulation of Paley represents the best hope of the movement to establish its views as scientifically legitimate. As we have seen in this brief review, however, it is remarkably easy to answer each of his principal claims.

This analysis shows that the "evidence" used by modern advocates of intelligent design to resurrect Paley's early nineteenth-century arguments is neither novel nor new. Indeed, their only remaining claim against Darwin is that *they* cannot imagine how evolution might have produced such systems. Time and time again, other scientists, unpersuaded by such self-serving pessimism, have shown (and published) explanations to the contrary. When closely examined, even the particular molecular machines employed by the movement as examples of "irreducible complexity" turn out to be incorrect. Finally, the logic of the argument itself turns out to have an obvious and fatal flaw.

Behe argues that anti-religious bias is the reason the scientific community resists the explanation of design for his observations:

> Why does the scientific community not greedily embrace its startling discovery? Why is the observation of design handled with intellectual gloves? The dilemma is that while one side of the elephant is labeled intelligent design, the other side might be labeled God.
> (1996a: 232)

I would suggest that the actual reason is much simpler. The scientific community has not embraced the explanation of design because it is quite clear, on the basis of the evidence, that it is wrong.

References

Behe, M. (1998) "Intelligent design theory as a tool for analyzing biochemical systems," in W. Dembski (ed.) *Mere Creation*, Downers Grove, Illinois: InterVarsity Press.

—— (1996a) *Darwin's Black Box*, New York: The Free Press.

—— (1996b) "Evidence for intelligent design from biochemistry," from a speech delivered at the Discovery Institute's "God and Culture" conference, 10 August 1996. Available online at the Discovery Institute website: *www.discovery.org/crsc*.

—— (1994) "Experimental support for regarding functional classes of proteins to be highly isolated from each other," in J. Buell and V. Hearn (eds) *Darwinism: Science or Philosophy?,* Houston, Texas: The Foundation for Thought and Ethics.

Gould, S.J. (1994) "Hooking Leviathan by its past," *Natural History* May: 8–15.

Hueck, C.J. (1998) "Type III protein secretion systems in bacterial pathogens of animals and plants," *Microbiology and Molecular Biology Review* 62: 379–433.

Huynen, M.A., Dandekar, T., and Bork, P. (1999) "Variation and evolution of the citric acid cycle: A genomic perspective," *Trends in Microbiology* 7: 281–91.

Jarrel, K.F., Bayley, D.P., and Kostyukova, A.S. (1996) "The archael flagellum: A unique motility structure," *Journal of Bacteriology* 178: 5,057–64.

Meléndez-Hevia, E., Waddell, T.G., and Cascante, M. (1996) "The puzzle of the Krebs citric acid cycle: Assembling the pieces of chemically feasible reactions, and opportunism in the design of metabolic pathways during evolution," *Journal of Molecular Evolution* 43: 293–303.

Musser, S.M. and Chan, S.I. (1998) "Evolution of the cytochrome c oxidase proton pump," *Journal of Molecular Evolution* 46: 508–20.

Rizzotti, M. (2000) *Early Evolution: From the Appearance of the First Cell to the First Modern Organisms*, Basel: Birkhauser.

Wooley, D.M. (1997) "Studies on the eel sperm flagellum," *Journal of Cell Science* 110: 85–94.

17

MODERN BIOLOGISTS AND THE ARGUMENT FROM DESIGN

Michael Ruse

> If the God of the Bible is the creator of the universe, then it is not possible to understand fully or even appropriately the processes of nature without any reference to that God. If, on the contrary, nature can be appropriately understood without reference to the God of the Bible, then that God cannot be the creator of the universe, and consequently he cannot be truly God and be trusted as a source of moral teaching either.
>
> (Pannenberg 1993: 16)

Charles Darwin's *Origin of Species* (1859) was a watershed in the history of the argument from design – the argument that traditionally focuses on organic adaptation, concluding that the intricate complexity and functioning of such adaptation is inexplicable save through the invocation of a creative intelligence (Ruse 1979). The argument had had its critics before Darwin, notably David Hume (1947 [1779]), but as the noted atheist Richard Dawkins (1986) has conceded, until the publication of *Origin* it was difficult if not impossible to deny that there might be something to the argument, somewhere, somehow. Then, natural selection did what God supposedly had wrought. No longer was there the compulsion to introduce an intelligence at work on the manufacture of the world, especially the living world. The hand and the eye had natural causes.

Of course, one might argue that Darwin did not make impossible a designer. It was just that he made appeal to a designer no longer compelling. One could go on believing that God did have a hand in the making of the eye. One no longer had to agree that God *had* to have had a hand in the making of the eye. However, my sense is that after Darwin (especially in Protestant circles) the argument from design lost much of its appeal (Moore 1979; Numbers 1998). People would no doubt have agreed that, even if traditional "natural theology" failed – it is no longer plausible to use the world and reason to argue for God – it is still open to the Christian to accept what the German theologian Wolfhart Pannenberg (1993) has called a "theology of nature." One can see in the world the work of God, even if the chief source for belief lies now in faith rather than reason and the senses. And people would certainly have agreed that one can see the marks of ongoing providence – especially inasmuch as evolution seems to lead upwards to the appearance of humankind. But, generally, most people simply dropped the whole design way of thought. For instance, with an exception to be noted later, one searches in vain through the published Gifford Lectures – lectures whose subject is explicitly that of natural theology – for the obsession with God's handiwork that one associates with Archdeacon William Paley (1819 [1802]) or with the authors of the 1830s *Bridgewater Treatises* (Gillespie 1950). Basically it is just not there.

Let me qualify somewhat what I have just said. It is true that today, for many who look at the science/religion interface, there is renewed interest in design. But the real point of excitement lies not in the traditional issues – the hand and the eye. It comes instead from the physical sciences, and particularly from the so-called "anthropic principle." More than a few trained in physics and chemistry now assure us that the basic constants of the Universe are so fine-tuned for the possibility of intelligent life that some purpose must lie behind everything (Barrow and Tipler 1986; Polkinghorne 1989). Admittedly, this line of argument has severe critics – and not just in the camps of the atheists, notably from Nobel Prize-winner Steven Weinberg (1994) – but it does persist and attract interest and excitement.

My concern here, however, is not with the anthropic principle – true or false – but with two interesting exceptions to the indifference to the traditional design argument – the design argument that focuses on organisms, specifically on organic adaptations. At opposite ends of the pole though they may be, the exceptions are back to back in their insistence that Paley was right. They claim that there is something interesting and significant about the way the organic world is put together, and that from a theological perspective we should sit up and take notice. It is with these people – evangelical Christians at one end and atheists at the other – that this essay is concerned. I take them in turn, showing why I think neither succeeds in what they respectively attempt. But I conclude with praise for what I think is their sensitivity to important issues, too much ignored after Darwin.

Intelligent design

We begin with people who think that Darwinism is ineffective, at least inasmuch as it claims to make superfluous or unnecessary a direct appeal to a designer of some sort. These are people who think that a full understanding of the organic world demands the invocation of some force beyond nature, a force that is purposeful or at least purpose-creating: an intelligent designer. There are two parts to this approach: an empirical and a philosophical. Let us take them in turn.

He who has most fully articulated the empirical case for a designer is the Lehigh University biochemist Michael Behe. Focusing on something that he calls "irreducible complexity," Behe writes:

> By irreducibly complex I mean a single system composed of several well-matched, interacting parts that contribute to the basic function, wherein the removal of any one of the parts causes the system to effectively cease functioning. An irreducibly complex system cannot be produced directly (that is, by continuously improving the initial function, which continues to work by the same mechanism) by slight, successive modifications of a precursor system, because any precursor to an irreducibly complex system that is missing a part is by definition nonfunctional.
>
> (1996: 39)

Behe adds, surely truly, that any

> irreducibly complex biological system, if there is such a thing, would be a powerful challenge to Darwinian evolution. Since natural selection can only choose systems that are already working, then if a biological system cannot be produced gradually it would have to arise as an integrated unit, in one fell swoop, for natural selection to have anything to act on.
>
> (1996: 39)

As an example of something irreducibly complex, Behe instances the common mousetrap. This has various parts (five in a standard model – spring, base, and so forth) that are put together to produce a fatal snapping motion when the trigger is activated by a small rodent attempting to take the bait. The point is that this mousetrap is an all-or-nothing phenomenon. Take away any one part and you have something that simply does not function at all. You could not have a mousetrap with four and a half parts nor even could you have a trap with only four parts. "If the hammer were gone, the mouse could dance all night on the platform without becoming pinned to the wooden base. If there were no spring, the hammer and platform would jangle loosely, and again the rodent would be unimpeded" (Behe 1996: 42) – and so on, through the various parts. The mousetrap functions only when it is up and running, entire. It could not have come about through gradual development – and of course, as we know, it did not. It came through the conscious intent and actions of a human designer. It was planned and fabricated.

Now turn to the world of biology, and in particular turn to the microworld of the cell and of mechanisms (or "mechanisms") that we find at that level. Take bacteria that use a flagellum, driven by a kind of rotary motor, to move around. Every part is incredibly complex, and so are the various parts, combined. The external filament of the flagellum (called a "flagellin"), for instance, is a single protein that makes a kind of paddle surface contacting the liquid during swimming. Near the surface of the cell, just as needed, is a thickening, so that the filament can be connected to the rotor drive. This naturally requires a connector, known as a "hook protein." There is no motor in the filament, so that has to be somewhere else. "Experiments have demonstrated that it is located at the base of the flagellum, where electron microscopy shows several ring structures occur" (Behe 1996: 70). All of these components are, Behe alleges, way too complex to have come into being in a gradual fashion. Only a one-step process will do, and this one-step process must involve some sort of designing cause. Behe is careful not to identify this designer with the Christian God, but the implication is that it is a force from without the normal course of nature.

Darwinism is ruled out and we must look for another explanation. This can be done only by postulating a "hopeful monster" that luckily gets all of the proteins of the right nature in the right order at once, or by the guidance of an intelligent agent (Behe 1996: 96). There is only one possible answer. Irreducible complexity spells design.

The explanatory filter

Backing the empirical argument is a conceptual argument due to the philosopher-mathematician William Dembski. His aim is two-fold: first, to give us the criteria by which we distinguish something that we would label "designed" rather than otherwise; and, second, to put this into context by showing how we distinguish something designed from something produced naturally by law or something we would put down to chance. As far as inferring design is concerned, there are three notions of importance: contingency, complexity, and specification. Contingency is the idea of something happening, but not being ascribable simply to blind law. My being hanged is contingent. My falling according to Galileo's laws of motion is not.

> In practice, to establish the contingency of an object, event, or structure, one must establish that it is compatible with the regularities involved in its production, but that these regularities also permit any number of alternatives to it.
>
> (Dembski 2002: 8)

For example, a crystal of salt results from forces of chemical necessity that can be described by the laws of chemistry. By contrast, a setting of silverware is not.The laws of nature do not determine that the fork should be on the left and the knife on the right. This setting is therefore contingent in a way that the structure of the crystal is not. The structure is the result of physical necessity.

Design has to be something that is contingent. The example that Dembski uses is the message from outer space received in the movie *Contact*. The series of dots and dashes, zeros and ones, can not be deduced from the laws of physics. But does it show evidence of design? Suppose we can interpret the series in a binary fashion, and the initial yield is the number group, 2, 3, 5. As it happens, these are the beginning of the prime-number series, but with so small a yield that no one is going to get very excited. It could just be chance. So no one is going to insist on design yet. But suppose now you keep going on with the series, and it turns out that it yields in exact and precise order the prime numbers up to 101. Now you will start to think that something is up, because the situation seems just too complex to be mere chance. It is highly improbable.

> Complexity as I am describing it here is a form of probability. To see the connection between complexity and probability, consider a combination lock. The more possible combinations of the lock, the more complex the mechanism and correspondingly the more improbable that the mechanism can be opened by chance. A combination lock whose dial is numbered from 0 to 39 and which must be turned in three alternating directions will have 64,000 (= 40 × 40 × 40) possible combinations and thus a 1/64,000 probability of being opened by chance. A more complicated combination lock whose dial is numbered from 0 to 99 and which must be turned in five alternating directions will have 10,000,000,000 (= 100 × 100 × 100 × 100 × 100) possible combinations and thus a 1/10,000,000,000 probability of being opened by chance. Complexity and probability therefore vary inversely: the greater the complexity, the smaller the probability. Thus to determine whether something is sufficiently complex to warrant a design inference is to determine whether it has sufficiently small probability.
>
> (Dembski 2002: 9)

But although you are probably happy now to conclude (on the basis of the prime-number sequence) that there are extraterrestrials out there, in fact there is another thing needed.

> If I flip a coin 1000 times, I will participate in a highly complex (i.e. highly improbable) event. Indeed, the sequence I end up flipping will be one in a trillion trillion trillion…where the ellipsis needs twenty-two more "trillions". This sequence of coin tosses will not, however, trigger a design inference. Though complex, this sequence will not exhibit a suitable pattern.

Here, then, we have a contrast with the prime-number sequence from 2 to 101: "Not only is this sequence complex, but it also embodies a suitable pattern. The SETI researcher who in the movie *Contact* discovered this sequence put it this way: 'This isn't noise, this has structure"' (Dembski 2002: 9).

What is going on here? You recognize in design something that is not just arbitrary or chance or which is given status only after the experiment or discovery, but rather something that was or could be in some way specified, insisted upon, before you set out. You know or could work out the sequence of prime numbers at any time before or after the contact from space. The random sequence of penny tosses will come only after the event. Likewise, suppose (say) you shoot an arrow at a target. And suppose that only after the shooting do you then draw bull's-eyes around the landed arrow. No one will be impressed with your skills. But if you specify where to find the bull's-eyes before you set out, then your abilities will be applauded. What is it about the pre-drawn bull's-eye that leads one to think of design, whereas the post-drawn bull's-eye does not? The key concept is that of independence. Dembski defines a specification as a match between an event and an independently given pattern. Events that are both highly complex and specified (that is, that match an independently given pattern) indicate design.

Dembski points out that if an archer hits a target that was already in place when he fired, the pattern is independent of the event. However, if the archer draws a circle around the arrow, the pattern is not independent of the event of shooting. This type of non-independent pattern he calls fabrication. More generally, Dembski divides patterns into two types: those in which complexity bears an inference to design and those in which complexity does not bear such an inference:

> The first type of pattern I call a *specification*, the second a *fabrication*. Specifications are the non-*ad hoc* patterns that can legitimately be used to eliminate chance and warrant a design inference. In contrast, fabrications are the *ad hoc* patterns that cannot legitimately be used to warrant a design inference.
>
> (Dembski 2002: 12).

Dembski is now in a position to move on to the second part of his argument where we actually detect design. Here we have what he calls an "explanatory filter." We have a particular phenomenon. The question is, what caused it? Is it something that might not have happened, given the laws of nature? Is it contingent? Or was it necessitated? The Moon goes endlessly round the Earth. We know that it does this because of Newton's laws. End of discussion. No design here. However, now we have some rather strange new phenomenon, the causal origin of which is a puzzle. Suppose we have a mutation such that, although we can quantify its appearance over large numbers, we cannot predict its appearance at an individual level. There is no immediate subsumption beneath law, and therefore there is no reason to think that at this level it was necessary. Let us say, as supposedly happened in the extended royal family of Europe, there was a mutation to a gene responsible for hemophilia. Is it complex? Obviously not, for it leads to breakdown rather than otherwise. Hence it is appropriate to talk now of chance. There is no design. The hemophilia mutation was just an accident.

Suppose now that we do have complexity. I would think that a rather intricate mineral pattern in the rocks might qualify here. Suppose we have veins of precious metals set in other materials, the whole being intricate and varied – certainly not a pattern you could simply deduce from the laws of physics or chemistry or geology or whatever. Nor would one think of it as being a breakdown mess, as one might a malmutation. Is this now design? Almost certainly not, for there is no way that one might prespecify such a pattern. It is all a bit *ad hoc*, and not something that comes across as the result of conscious intention. And then finally there are phenomena that are complex and specified. One presumes that the microscopical biological apparatuses and processes discussed by Behe would qualify here. They are contingent, for they are irreducibly complex. They are design-like, for they do what is needed for the organism in which they are to be found. That is to say, they are of prespecified form. And so, having survived the explanatory filter, they are properly considered the product of real design.

Of course, the nature of this designer is another matter. Because one accepts intelligent design, one is certainly not thereby necessarily pushed into a crude literalistic reading of the early chapters of Genesis – six days, 6,000 years, universal flood. Dembski is explicit in saying that he thinks it rules out evolution, even a guided "theistic" evolution. But is the intelligent designer the Christian God? Dembski does not want to assert this absolutely – although as a matter of fact he himself does accept the Christian God – but he certainly gives the green light to one who, having accepted an intelligent designer, would understand or interpret this in the light of the Christian faith. Bemoaning the fact that so many of today's theologians refuse to see clear evidence of God's actions in the world (more on this point later), Dembski himself opts for a God of the Gospels who is also the God of biology:

> How then do we determine whether God has so arranged the physical world that our natural intellect can discover reliable evidence of him? The answer is obvious: put our natural intellect to the task and see whether indeed it produces conclusive evidence of design in nature. Doing so poses no threat to the Christian faith. It challenges neither the cross, the tomb, the resurrection on the third day, the ascension into heaven, the sitting at the right hand of the Father nor the second coming of Christ. Indeed, nature is silent about the revelation of Christ in Scripture. On the other hand, nothing prevents nature from independently testifying to the God

revealed in the Scripture. Now intelligent design does just this – it puts our natural intellect to work and thereby confirms that a designer of remarkable talents is responsible for the natural world. How this designer connects with the God of Scripture is then for theology to determine.

(Dembski and Richards 2001: 228–9)

Intelligent design criticized

Irreducible complexity is supposedly something that could not have come through unbroken law, and especially not through the agency of natural selection. Critics point out that Behe's example of a mousetrap is somewhat unfortunate, for it is simply not the case that the trap will work only with all five pieces in place. For a start, one could reduce the number to four by removing the base and fixing the trap to the floor. It may be better if you could move it around, but selection never made a claim to perfection – simply to functioning better than any alternative. In fact, as can be seen from an online cartoon (*www.udel.edu/~mcdonald/mousetrap.html*), it may even be possible to reduce the number of components down to one! Not a great trap, but a trap nevertheless. Rather more significant, though, is the fact that the Behe mousetrap example shows a misunderstanding of the very nature and workings of natural selection. No one is denying that in natural processes there may well be parts that, if removed, would lead at once to the non-functioning of the systems in which they occur. The point, however, is not whether the parts now in place could not be removed without collapse, but whether they could have been put in place by natural selection. To counter Behe's artifactual analogy, it is not difficult to think of other artifactual analogies that do show precisely how the apparently impossible (if such it be) could be achieved. Consider an arched bridge, made from cut stone, without cement, held in place only by the force of the stones against each other. If you tried to build the bridge from scratch, upwards and then inwards, you would fail – the stones would keep falling to the ground, as indeed the whole bridge would collapse were you to remove the center keystone or any surrounding it. Rather, what you must do is first build a supporting structure (possibly an earthen embankment), on which you will lay the stones of the bridge, until they are all in place. At this point you can remove the structure, for it is no longer needed, and in fact is in the way. Likewise, one can imagine a biochemical sequential process with several stages, on the parts of which other processes piggyback, as it were. Then the hitherto non-sequential parasitic processes link up and start functioning independently, the original sequence finally being removed by natural selection as redundant or inconveniently draining of resources.

Of course, this is all pretend, as is the mousetrap. But Darwinian evolutionists have hardly ignored the matter of complex processes. Indeed, it is discussed in detail by Darwin in his *Origin*, where he refers to that most puzzling of all adaptations, the eye. At the biochemical level, today's Darwinians have many examples of the most complex of processes that have been put in place by selection. Take that staple of the body's biochemistry, the process where energy from food is converted into a form that can be used by the cells. Rightly does a standard textbook refer to this vital organic system, the so-called "Krebs cycle," as something that "undergoes a very complicated series of reactions" (Hollum 1987: 408). This process, which occurs in the cell parts known as mitochondria, involves the production of ATP (adenosine triphosphate), a complex molecule that is energy-rich and which is degraded by the body as needed (say in muscle action) into another less rich molecule, ADP (adenosine diphosphate). The Krebs cycle remakes ATP from other energy sources – an adult human male needs nearly 200 kg a day – and, by any measure, the cycle is enormously involved and intricate. For a start, nearly a dozen enzymes (substances that facilitate chemical processes) are required, as one subprocess leads on to another.

Yet the cycle did not come out of nowhere. It was cobbled together out of other cellular processes that do other things. It was a "bricolage." Each one of the bits and pieces of the cycle exists for other purposes and has been coopted for the new end. The scientists who have made this connection could not have made a stronger case against Behe's irreducible complexity than if they had had him in mind from the first. In fact, they set up the problem virtually in Behe's terms:

> The Krebs cycle has been frequently quoted as a key problem in the evolution of living cells, hard to explain by Darwin's natural selection: How could natural selection explain the building of a complicated structure in toto, when the intermediate stages have no obvious fitness functionality?
>
> (Meléndez-Hevia *et al.* 1996: 302)

What these workers do not offer is a Behe-type answer. First, they brush away a false lead. Could it be that we have something like the evolution of the mammalian eye, where primitive existent eyes in other organisms suggest that selection can and does work on proto-models (as it were), refining features that have the same function (if not as efficient) as more sophisticated models? Probably not, for there is no evidence of anything like this. But then we are put on a more promising track:

> In the Krebs cycle problem the intermediary stages were also useful, but for different purposes, and, therefore, its complete design was a very clear case of opportunism. The building of the eye was really a

> creative process in order to make a new thing specifically, but the Krebs cycle was built through the process that Jacob (1977) called "evolution by molecular tinkering," stating that evolution does not produce novelties from scratch: it works on what already exists. The most novel result of our analysis is seeing how, with minimal new material, evolution created the most important pathway of metabolism, achieving the best chemically possible design. In this case, a chemical engineer who was looking for the best design of the process could not have found a better design than the cycle which works in living cells.
>
> (Meléndez-Hevia *et al.* 1996: 302)

Rounding off the response to Behe, let us note that, if his arguments are well taken, then in some respects we are in bigger trouble than otherwise! His position seems simply not viable given what we know of the nature of mutation and the stability of biological systems over time. When exactly is the intelligent designer supposed to strike and to do its work? Unlike Dembski, Behe seems not entirely opposed to some kind of evolution, and in *Darwin's Black Box* the suggestion is made that everything might have been done long ago and then left to its own devices:

> The irreducibly complex biochemical systems that I have discussed...did not have to be produced recently. It is entirely possible, based simply on an examination of the systems themselves, that they were designed billions of years ago and that they have been passed down to the present by the normal processes of cellular reproduction.
>
> (Behe 1996: 227–8)

Unfortunately, as Behe's most doughty biological critic retorts, we cannot ignore the history of the preformed genes from the point between their origin (when they would not have been needed) and today when they are in full use.

> As any student of biology will tell you, because those genes are not expressed, natural selection would not be able to weed out genetic mistakes. Mutations would accumulate in these genes at breathtaking rates, rendering them hopelessly changed and inoperative hundreds of millions of years before Behe says that they will be needed.

There is a mass of experimental evidence to show that this is the case. Behe's idea of a designer doing everything long ago and then leaving matters to their natural fate is "pure and simple fantasy" (Miller 1999: 162–3).

What is the alternative? Presumably that the designer is at work all of the time, producing mechanisms as and when needed. So I take it that, if we are lucky, we might expect to see some produced in our lifetime. Indeed, there must be a sense of disappointment among biologists that no such creative acts have so far been reported. Are these acts to be fused into what exists already, in which case the designer seems less and less necessary as it builds on or alters what already is working quite nicely, or are they (as one presumes Dembski thinks) to be creative clearings of the deck and the provision of entirely new organisms? And if the latter, then why do we have so many similarities (homologies) at the biochemical level between different organisms, giving at least an appearance of evolution? The similarities are not necessary from a functional perspective, any more than, say, the similarities between the bones of the wing of a bird and those of a human are necessary from a functional perspective. The whole thing is, simply and bluntly speaking, a mess.

More than this, as we turn from science towards theology, there are serious problems. Most obviously, what about malmutations? If the designer is needed and available for complex engineering problems, why could not the designer take some time on the simple matters, specifically those simple matters that, if unfixed, lead to absolutely horrendous problems? Some of the worst genetic diseases are caused by one little alteration in one little part of the DNA (sickle-cell anemia, for example). If the designer is able and willing to do the very complex because it is very good, why does it not do the very simple because the alternative is very bad? Behe speaks of this as being part of the problem of evil, which is true, but not very helpful. Given that the opportunity and ability to do good was so obvious and yet not taken, we need to know the reason why.

Explanatory filters

It is here that Dembski would come to the rescue. Refer back to the explanatory filter. A malmutation would surely get caught by the filter half-way down. It would be siphoned off to the side as chance, if not indeed simply put down as necessity. It certainly would not pass the specification test. This would mean, for example, that sickle-cell anemia would not be the fault of the designer, whereas the blood-clotting cascade another of Behe's examples would be to the designer's credit. Dembski stresses that these are mutually exclusive alternatives:

> To attribute an event to design is to say that it cannot plausibly be referred to either law or chance. In characterizing design as the set-theoretic complement of the disjunction law-or-chance, one therefore guarantees that these three modes of explanation will be mutually exclusive and exhaustive.
>
> (Dembski 1998: 36)

One's response to this is that one can of course define things as one will, and if one stipulates that design and law and chance are mutually exclusive, then so be it. But the downside is that one now has a stipulative definition and not necessarily a lexical definition – that is, one that accords to general use. Suppose that something is put down to chance. Does this mean that law is ruled out? Surely not! If I argue that a Mendelian mutation is chance, what I mean is that it is chance with respect to that particular theory, but I may well believe (I surely will believe) that the mutation came about by normal regular causes and that, if these were all known, then it would no longer be chance at all but necessity. The point is that chance is, in this case, a confession of ignorance, not (as one might well think would be the case in the quantum world) an assertion about the way that things are. That is, claims about chance are not necessarily ontological assertions, as presumably claims about designers must be. More than this, one might well argue that the designer always works through law. This takes us back to deism, or to the kind of position endorsed by the pre-Darwinian Baden Powell. The designer may prefer to have things put in motion in such a way that his/her/its intentions unfurl and reveal themselves as time goes by. The pattern in a piece of cloth made by machine is as much an object of design as the pattern from cloth produced by a hand loom. In other words, in a sense that would conform to the normal usage of the terms, one might want to say of something that it is produced by laws, is chance with respect to our knowledge or theory, and fits into an overall context of design by the great orderer or creator of things. One finds, indeed, real people who have made precisely this kind of claim! Sir Ronald A. Fisher for a start. He argued that his fundamental theorem of natural selection made for ongoing progress in evolution. The theorem worked on the multiplicity of chance mutations that occur in every natural population. And he wrapped the whole picture up and saw it as the manifestation of the actions and intentions of a good God, in fact the Anglican God Fisher worshiped all of his life.

Dembski's filter does not let Behe's designer off the hook. If the designer can make – and rightfully takes credit for – the very complex and good, then the designer could prevent – and by its failure is properly criticized for – the very simple and awful. The problems in theology are as grim as are those in science.

The blind watchmaker

Turning now to the side of active Darwinism, it is Richard Dawkins who has written most eloquently and fervently on these issues. What he has to say is extremely negative as regards the Christian perspective. He will have no nonsense with neo-orthodox separations of science and religion. They make claims and they clash. One side is right and one side is wrong. Dawkins gives and expects no quarter. "A cowardly flabbiness of the intellect afflicts otherwise rational people confronted with long-established religions" (Dawkins 1997: 397). As a Darwinian, Dawkins is not about to deny that the Christians got the question right. It is their answer that is the problem! Speaking of Archdeacon William Paley with respect, if not reverence, Dawkins willingly allows that *Natural Theology* articulates the teleological argument "more clearly and convincingly than anyone had before" (Dawkins 1986: 4). The trouble is that it is completely and absolutely wrong! "The analogy between telescope and eye, between watch and living organism, is false. All appearances to the contrary, the only watchmaker in nature is the blind forces of physics, albeit deployed in a very special way." A genuine watchmaker plans everything and puts his purposes into action. Natural selection simply acts. It has no purposes.

> It has no mind and no mind's eye. It does not plan for the future. It has no vision, no foresight, no sight at all. If it can be said to play the role of watchmaker in nature, it is the *blind* watchmaker.
>
> (Dawkins 1986: 5)

Now, obviously not all Christians are going to be upset by this (Ruse 2001). Darwin showed that the route to creation must lie through evolution by natural selection. But why should not God have worked this way rather than in some other fashion? Darwin himself thought this when he wrote *Origin of Species*. It is true that, by this time, Darwin was a deist rather than a theist, but the philosopher of science and Catholic priest Ernan McMullin (1985) has pointed out that there is theological warrant for the Christian God having created in this evolutionary fashion. Saint Augustine, the greatest of the Church Fathers and even to this day a major influence on the nature of Christianity, claimed that God is outside time. For Him, therefore, the thought of creation, the act of creation, and the end of creation are as one. According to Augustine, God put in place the seeds of future development, the potentiality of unfurling, as an essential and integral part of the very existence of organisms. You cannot separate Becoming and Being. Although Augustine himself was certainly no evolutionist, he was laying the scientific groundwork for precisely such a scientific theory.

Dawkins has a number of back-up arguments, trying to stop just such moves as this. One, which clearly resonates strongly with him, is that such an appeal to God is redundant. He (Dawkins) has no need of that hypothesis:

> We cannot disprove beliefs like these, especially if it is assumed that God took care that his interventions always closely mimicked what would be expected from evolution by natural selection. All that we can say about such beliefs is, firstly, that they are superfluous and, secondly, that they assume the existence of the main thing we want to explain, namely organized complexity.

Dawkins goes on to say that the "one thing that makes evolution such a neat theory is that it explains how organized complexity can arise out of primeval simplicity" (Dawkins 1986: 316). In any case, any God capable of creating such complexity must itself have been even more complex in the first place. Why bother to get into this, postulating such a God? If you are not going to argue properly, then simply claim that life as we know it exists, and leave it at that.

Another of Dawkins's arguments is that the world is not so perfectly designed that one must necessarily invoke a designer. There are many things that work well enough, without being perfect. You do not have to go beyond unbroken material law. And this of course is quite apart from the things that do not work very well at all, and which you might not expect on the basis of a designer. But one senses that this is really not a line of argument with much intuitive appeal to Dawkins, for he is much more concerned to stress how good is the design in the living world. Much more powerful to him as a counter-argument against God is the traditional problem of pain and evil, something he clearly thinks is exacerbated by the Darwinian approach. Natural selection presupposes a struggle for existence, and the struggle on many, many occasions is downright nasty. Using the notion of "reverse engineering" for the process of picking backwards to try to work out something's purpose, and of a "utility function" for the end purpose being intended, Dawkins draws attention to the cheetah/antelope interaction, and asks: "What was God's utility function?" Cheetahs seem wonderfully designed to kill antelopes. "The teeth, claws, eyes, nose, leg muscles, backbone and brain of a cheetah are all precisely what we should expect if God's purpose in designing cheetahs was to maximize deaths among antelopes." But conversely, "we find equally impressive evidence of design for precisely the opposite end: the survival of antelopes and starvation among cheetahs." It is almost as though we had two Gods, making the different animals, and then competing. If there is only one God who made the two animals, then what on Earth is going on? What kind of God is this? "Is He a sadist who enjoys spectator blood sports? Is He trying to avoid overpopulation in the mammals of Africa? Is He maneuvering to maximize David Attenborough's television ratings?" The whole thing is ludicrous (Dawkins 1995: 105). Truly, concludes Dawkins, there are no ultimate purposes to life, no deep religious meanings. "The universe we observe has precisely the properties we should expect if there is, at bottom, no design, no purpose, no evil and no good, nothing but blind, pitiless indifference" (Dawkins 1995: 133).

Pain

It is at this point that one regrets how little sympathy there is for forthright Darwinism in today's scientific/religious community. With few exceptions, the nigh universal reaction to Dawkins's polemics is to retreat. There seems to be a general agreement that Dawkins has a good point – Darwinism does show that life is mean and nasty, without any purpose. The only way to avoid so unpleasant a conclusion is to abandon Darwinism. No one wants to give up evolution *per se*, but the scramble is on to find an evolutionism with a warmer, friendlier face. There has to be a way to get what you want – adaptation possibly, humans definitely – without all of the nastiness. To this end, reminiscent of the end of the nineteenth century, many would introduce fairly large one-step changes, or to argue that all of the really significant changes come from the way the laws of nature function, quite irrespective of selection. This latter move acts only in a minor way, fine-tuning or removing the detritus.

Paul Davies argues in this mode:

> My own opinion is that emergent laws of complexity offer reasonable hope for a better understanding not only of biogenesis, but of biological evolution too. Such laws might differ from the familiar laws of physics in a fundamental and important respect. Whereas the laws of physics merely shuffle information around, a complexity law might actually create information, or at least wrest it from the environment and etch it onto a material structure.
>
> (Davies 1999: 259)

Apparently Darwinism is to be given some role, but a minor one. In Davies's evolutionary world,

> relatively small replicator molecules form by chance and start to evolve by Darwinian means, but the process is sometimes aided, and even overridden, by organizational principles that confer specificity and information. These organizational principles serve to amplify greatly the selectivity of the evolutionary process, and lead to sudden jumps in complexity rather than the incremental advance expected from Darwinian evolution acting alone.
>
> (Davies 1999: 260)

Significantly, the jackpot for Davies is that his new approach will explain complexity and progress towards the human mind. The adaptations of Paley, Darwin, and Dawkins do not rate much mention. The struggle for existence is pushed to the edges.

Total hogwash, I am afraid. And completely unneeded. When physicists start taking biology as seriously as they expect the rest of us to take their theories and ideas, we might perhaps have some possibility of constructive interaction. Until then, to those who will listen, one can simply state flatly and truly that Darwinism – adaptation brought about by natural selection – is an active and forward-looking science. In Kuhn's language, it is a highly successful paradigm. Whether you like it or not, we are stuck with it. The Darwinian

Revolution is over. Darwin won. So, if there is to be any satisfactory response to Dawkins, it must be on his terms – adaptation, selection, pain, and all. And once this is recognized, things start to fall into place. Dawkins is absolutely right. Neo-orthodoxy is no response. Darwinism does talk about origins and the theologically inclined must take note. Pure Paley is no longer possible. Natural selection does rule out the necessity of an appeal to an intervening God. This leaves only the third option, Pannenberg's theology of nature. The only possible response to Dawkins is that, say what you may, Darwin or not, as a Christian you simply must agree that our understanding of nature, of living things, is changed and illuminated and made complete by our acceptance of the existence and creative power and sustaining nature of God.

Where stands this response now in the light of the Darwinians' attacks? What about the problem of evil? The critics want to claim that the coming of Darwin made the God hypothesis impossible. After Darwin we see that the world is simply not how it would be given an all-loving, all-powerful creator (recall the cheetah and antelope argument). Nothing like this could possibly occur given a loving God. Everything like this is expected, given blind, purposeless law. However, as the appeal to blind unbroken law – things not working perfectly and pain and strife being commonplaces – is a traditional argument against the possibility of the Christian God, so the appeal to blind unbroken law – things not working perfectly and pain and strife being commonplaces – yields a traditional argument protecting the possibility of the Christian God. One must and can properly invoke some version of the argument used by Leibniz, namely that God's powers only extend to doing the possible. Having once committed Himself to creation by law, everything else follows as a matter of course. It is true that Leibniz's argument was parodied by Voltaire in *Candide* – everything in this world happens for the best of all possible reasons – but just as that counter fails in the world of religion, so also it fails in the world of science.

Imagine what kind of wholesale change would be needed to pain-proof processes. Fire could no longer burn, for fear that children and others might get trapped in smoke-filled apartments. But if fire did not burn, how could I warm myself through the winter and how could I cook my food and so much more? One change by God would require another and another and another, until everything was altered. And could this be possible? Where would it end, and where could it end in a satisfactory manner?

> Human beings are sentient creatures of nature. As physiological beings they interact with Nature; they cause natural events and in turn are affected by natural events. Hence, insofar as humans are natural, sentient beings, constructed of the same substance as Nature and interacting with it, they will be affected in any natural system by lawful natural events.
>
> (Reichenbach 1982: 111–12)

You start altering things around, and there is no end to it – except that you will have to change humans so that they are no longer truly human. And even now, who dare say we humans would be better situated? "Whether humans would have evolved but no infectious virus or bacilli, or whether there would have resulted humans with worse and more painful diseases, or whether there would have been no conscious, moral beings at all, cannot be discerned" (Reichenbach 1982: 113). The world is a package deal, and we simply have no right or authority to say that God could have created it in such a way as to prevent such physical evil as there is.

In fact, paradoxically and somewhat humorously, one can appeal to Dawkins's own writings to drive home this line of argument. As a good Darwinian, he is insistent that adaptive complexity is the mark of the living. This and this alone picks out the quick from the dead, or rather from the never quick. But how is one to get this adaptive complexity? Dawkins insists that it is through and only through natural selection. Alternative mechanisms like Lamarckism or hopeful monsters (mutationism) simply will not work. "Wherever in the universe adaptive complexity shall be found, it will have come into being gradually through a series of small alterations, never through large and sudden increments in adaptive complexity" (Dawkins 1983: 412). The point is that physical processes do not suddenly and spontaneously bring about adaptive complexity. The only sudden changes are those that destroy or degrade. They are never creative. Boeing 747s crash into the ground and in an instant they are no more. Boeing 747s do not lie in pieces around the junkyard or on the ocean bottom and then in an instant form fully functioning flying machines. In the case of organisms, there is no known physical rival to the slow, creative, adaptive-complexity-forming process of natural selection. So it is selection or nothing.

> However diverse evolutionary mechanisms may be, if there is no other generalization that can be made about life all around the Universe, I am betting that it will always be recognizable as Darwinian life. The Darwinian Law...may be as universal as the great laws of physics.
>
> (Dawkins 1983: 423)

You cannot get adaptive complexity without natural selection.

In other words, if God's process of creation is through unbroken law, then He had to do it as He did – natural selection, pain and agony, imperfection, and all. One might still argue that God should not have created in the first place (it is an essential part of Christian theology that the act of Creation was one done freely from love by God), but that is a somewhat different point. If one objects, as did the novelist Dostoevsky, that the pain in the world could never be trumped by the happy end result – no amount of bliss in heaven for me or Mother Teresa could ever balance the agony of the child at Auschwitz – one is not necessarily implicating the coming of Darwinism. The pain would happen irrespective of natural selection. So, ultimately, effective though this line of argument may or may not be, it is irrelevant to our main line of discussion. Darwinism as such does not make irresistible the argument from natural evil. It may even make the solution easier to grasp.

Redundancy?

After Darwin, perhaps it is not even the problem of pain that is the most serious for the believer. It is the problem of redundancy. Do we need the God hypothesis at all? Is it not far simpler to go with natural selection and nothing more? I will ignore Dawkins's rather odd argument that the complex can be explained only by the more complex – I always thought the strength of really great theories is their simplicity – and I will move to the main issue. Does Darwinism dispel the need for God? And at once I will point out that the one thing that the theist need not and should not be is unduly constrained at this point. When the Darwinian is attacked by (say) Richard Lewontin on some particular adaptive claim, it is legitimate to appeal to the whole of evolutionary understanding to make reasonable a conjecture that, on its own, may be somewhat unsupported. Likewise for the Christian believer. For the theist, it is not just the science/religion interface that is pertinent here, but the whole of human experience. It is revelation and faith, and also human relationships – can one (as the Jewish philosopher Martin Buber would have said) understand the "I–thou" relationship except in a God-given sense? And can one explain the horrors of Auschwitz without at some level acknowledging and invoking original sin? Is the grotesque, warped banality of Hitler simply to be reduced to psychology? For Pannenberg, even if one conceded the redundancy of God in the post-Darwinian biological world, this would not deny the existence of God on other grounds, which then makes it proper (and necessary) to bring God into an understanding of the biological world.

This being so, it is surely open to share in the position taken in the mid-twentieth century by Canon Charles Raven, Anglican priest and voluminous writer on biology and its history. Although no Darwinian, he is one exception I have found to the indifference of the Gifford Lecturers to the design argument. Raven spoke of how much time he had spent and sheer pleasure he had derived from following and studying butterflies all over England and Scotland:

> Every specimen differed from the rest, in detail from those of its own group, in total effect from those of others. Each was in itself a perfect design, satisfying in whole and parts, inviting one to concentrate one's whole attention upon it. To move from one to another, to

> sense the difference of impact, to work out the quality of this difference in the detailed modifications of the general pattern, this was a profoundly moving experience.
>
> (Raven 1953: 112)

For Raven, this was the real edge of the science/religion encounter. This is what makes it all meaningful to the believer. Not proof, but simply flooding, overwhelming experience that could not be denied.

This is the answer to Dawkins. Not an answer of logic or proof. Not an answer strong enough for the person who cannot be convinced. Not an answer needed for the person who is convinced. Simply the way things are. The theist can and does rejoice in nature and feel (with William Whewell) awed by the wonderful processes that God uses to produce us and the world around us. Evolution – Darwinian evolution – speaks to His greatness. Logically, the God hypothesis may be redundant. Emotionally and religiously, the God hypothesis has never been stronger. To quote one of the more attractive writers on these matters, the late-Victorian, Oxonian, Anglo-Catholic theologian, Aubrey Moore, "Darwinism appeared, and, under the guise of a foe, did the work of a friend." It shows us that either God is everywhere or He is nowhere.

> We must frankly return to the Christian view of direct Divine agency, the immanence of Divine power from end to end, the belief in a God in Whom not only we, but all things have their being, or we must banish him altogether.
>
> (Moore 1891: 73)

Conclusion

The argument from design is not resuscitated by the intelligent design theorists. Irreducible complexity does not compel and explanatory filters do not sift salt from sand. The argument from design is not abolished by the Darwinian atheists. It may not be logically compelling but it is not pathetically worthless. These are negative conclusions, but this is not a negative discussion. Too many Christian writers on the science/religion relationship, people mentioned at the beginning of the chapter, have less sense or appreciation of the traditional argument from design. Even when they accept some form of the argument (particularly that based on the anthropic principle), for them, as physicists or non-scientists, adaptation is simply not a pressing issue. They want and expect little insight into God and His nature as revealed by the organic world, by the intricacies of the hand and the eye. For this reason, I celebrate and applaud the fact that this is not true of people such as Michael Behe and Richard Dawkins. Whatever you might think of their theology, their passionate love of the living world comes through in every line that they write. Dawkins on the eye or Behe on the cell are people for whom the creation (take that in whatever sense you like) is a real and wonderful part of experience and their lives. I am not now claiming that Dawkins is a crypto-Christian. He is not. I am not now claiming that Behe is a crypto-materialist. He is not. I am saying that they stand in the tradition of the great naturalists, including Darwin, in responding joyously to the argument from design.

I conclude with the words of Charles Raven:

> Here is beauty – whatever the philosophers and art critics who have never looked at a moth may say – beauty that rejoices and humbles, beauty remote from all that is meant by words like random or purposeless, utilitarian or materialistic, beauty in its impact and effects akin to the authentic encounter with God.
>
> (1953: 112–13)

References

Barrow, J.D. and Tipler, F.J. (1986) *The Anthropic Cosmological Principle*, Oxford: Clarendon Press.

Behe, M. (1996) *Darwin's Black Box: The Biochemical Challenge to Evolution*, New York: Free Press.

Darwin, C. (1859) *On the Origin of Species*, London: John Murray.

Davies, P. (1999) *The Fifth Miracle: The Search for the Origin and Meaning of Life*, New York: Simon & Schuster.

Dawkins, R. (1997) "Obscurantism to the rescue," *Quarterly Review of Biology* 72: 397–9.

—— (1995) *A River out of Eden*, New York: Basic Books.

—— (1986) *The Blind Watchmaker*, New York: Norton.

—— (1983) "Universal Darwinism," in D.S. Bendall (ed.) *Molecules to Men*, Cambridge: Cambridge University Press, pp. 403–25.

Dembski, W. (2002) *No Free Lunch: Why Specified Complexity Cannot be Purchased without Intelligence*, Lanham, Maryland: Rowman and Littlefield.

—— (1998) *The Design Inference: Eliminating Chance through Small Probabilities*, Cambridge: Cambridge University Press.

Dembski, W.A. and Richards, J.W. (eds) (2001) *Unapologetic Apologetics: Meeting the Challenges of Theological Studies*, Downers Grove, Illinois: InterVarsity Press.

Gillespie, C.C. (1950) *Genesis and Geology*, Cambridge, Massachusetts: Harvard University Press.

Hollum, J.R. (1987) *Elements of General and Biological Chemistry*, New York: Wiley.

Hume, D. (1947 [1779]) *Dialogues Concerning Natural Religion*, ed. N.K. Smith, Indianapolis: Bobbs-Merrill Co.

Jacob, F. (1977) "Evolution and tinkering," *Science* 196: 1,161–6.

McMullin, E. (ed.) (1985) *Evolution and Creation*, Notre Dame, Indiana: University of Notre Dame Press.

Meléndez-Hevia, E., Waddell, T.G., and Cascante, M. (1996) "The puzzle of the Krebs citric acid cycle: Assembling the pieces of chemically feasible reactions, and

opportunism in the design of metabolic pathways during evolution," *Journal of Molecular Evolution* 43: 293–303.
Miller, K. (1999) *Finding Darwin's God*, New York: Harper & Row.
Moore, A. (1891) "The Christian doctrine of God," in C. Gore (ed.) *Lux Mundi* (12th edition), London: John Murray, pp. 41–81.
Moore, J. (1979) *The Post-Darwinian Controversies: A Study of the Protestant Struggle to Come to Terms with Darwin in Great Britain and America, 1870–1900*, Cambridge: Cambridge University Press.
Numbers, R. (1998) *Darwinism Comes to America*, Cambridge, Massachusetts: Harvard University Press.
Paley, W. (1819 [1802]) *Natural Theology (Collected Works: IV)*, London: Rivington.
Pannenberg, W. (1993) *Towards a Theology of Nature*, Louisville, Kentucky: Westminster/John Knox Press.
Polkinghorne, J. (1989) *Science and Providence: God's Interaction with the World*, Boston: Shambhala.
Raven, C.R. (1953) *Natural Religion and Christian Theology. The Gifford Lectures 1952, Second Series: Experience and Interpretation*, Cambridge: Cambridge University Press.
Reichenbach, B.R. (1982) *Evil and a Good God*, New York: Fordham University Press.
Ruse, M. (2001) *Can a Darwinian be a Christian? The Relationship between Science and Religion*, Cambridge: Cambridge University Press.
—— (1979) *The Darwinian Revolution: Science Red in Tooth and Claw*, Chicago: University of Chicago Press.
Weinberg, S. (1994) "The emergence of life," *Scientific American* October: 46–9.

18

THE PARADOXES OF EVOLUTION

Inevitable humans in a lonely universe?

Simon Conway Morris

Science, like the proverbial activity of travel, is meant to broaden the mind. The scientific enterprise can be likened to a journey, as it provides a renewed invitation to reflect on the strangeness of our world. In such a manner we may ask deceptively simple questions such as: "How did we come to be here?" and "Are humans unique?" Both of these questions can be answered in various ways, but it would be strange if the respondent failed to make some mention of organic evolution. So far as the second question is concerned, the notion of human uniqueness has taken many hard knocks, irrespective of such achievements as landing a rover vehicle on the surface of Mars, or scanning the cosmos with telescopes, let alone performing Robert Wylkynson's mysteriously beautiful *Jesus autem transiens* – a work, in the words of John Milsom, of "harmonious chaos...through [which] the medieval listener could more vividly imagine part of the divine order that rested on those pillars of eternity." Despite our obsession with technologies, for many it is music such as Wylkynson's, master of the choristers at Eton College in the early sixteenth century, which serves to define our place in the natural order.

Yet the realities of evolution and the discoveries of astronomy serve to define a paradox. Locally, that is within the Solar System, it is quite likely that life is restricted to the Earth. So far as the rest of the Galaxy, and beyond, is concerned, here too we have no direct evidence for extraterrestrial life, yet there is an almost universal confidence that such life must, so to speak, be "out there." Most people would probably concede that the majority of inhabited planets only reach the stage of "pond scum" or its equivalents (e.g. Ward and Brownlee 2000) and that sentient life-forms are much less common. As for the likelihood of any extraterrestrial civilizations being built by some sort of sentient biped, this would be regarded as quite fanciful (e.g. Simpson 1964). These are some of the lessons of Copernican astronomy and the principle of mediocrity: there is nothing special about us, and whatever other beings there might be living on remote planets, they will probably be different, perhaps so different as to elude communication and perhaps even recognition. Here I will argue for effectively the opposite: from first principles life ought to be a universal principle but actually is exceedingly rare, yet paradoxically where it does occur then something like a human is very much on the cards. In brief, humans are inevitable, but we live in a very lonely universe. If this is true then its religious implications will be self-evident.

What is best?

Irrespective of whether there are or are not planets teeming with extraterrestrial life, we can still inquire whether life will share some universal properties. Such an inquiry will be seized upon by those skeptics who might invite us to consider, for example, silicon-based life-forms. Indeed, such imaginative excursions have gone far beyond considering mere analogues of Earthly life, to conjuring organisms disporting themselves in oceans composed of ammonia, and even to speculating on the inhabitants of neutron stars (Shapiro and Feinberg 1995). Such conjectures are always to be encouraged lest the mundane view cripple expectations. Yet the general consensus of what delimits the ranges of life is rather conservative. Can an organism, for example, be anything but carbon- and water-based? In terms of the necessary molecular "backbone," the only other terrestrial contender seems to be silicon. So far as life is concerned, however, both its strength of bonding (think of quartz) and, more importantly, the chemistry of its compounds seem to make this element distinctly discouraging as an alternative to carbon.

So here, maybe, is one constraint. There may be no hand to shake, but whatever prehensile organ is extended in greeting by the extraterrestrial most probably will rely on carbon. If the alien offers us a cup of water (or something stronger) then it is likely that the ice floating on the surface will seem too commonplace to mention. But a world in which ice behaves like nearly all other solids as they crystallize out from the liquid phase would, as has often been pointed out, be very different. In this counter-intuitive but physically unremarkable world, whenever water froze it would quietly sink out of sight and carpet the sea floor in increasingly thick layers of ice. Such a strange planet might harbor life, but as has been repeatedly emphasized (e.g. Henderson 1913; Denton 1998) the many other physical oddities of water not only make it peculiarly well-suited to life, but make it difficult to envisage a credible alternative.

Life, or at least planetary life, will therefore be almost certainly carbon-based and require water. Beyond that, few biologists would probably be willing to impose any further restrictions, but there may be more specific constraints. George Wald, for example, has drawn attention (e.g. Wald 1974) to the curious mismatch between the spectral absorbance of chlorophyll (of which there are several different types, e.g. chlorophyll a and b) and the energy that pours out from the Sun in the visible part of the light spectrum. Chlorophyll, of course, is responsible for trapping the quanta of light energy that come from the Sun and thus this molecule underpins the process of photosynthesis. Such is the premium on intercepting the maximum levels of energy that one would predict that, in a Darwinian fashion, the absorption spectra of the plant's chlorophyll would evolve to match as best as possible that of the Sun. In fact, there is a surprising discrepancy and a significant part of the Sun's output remains untapped. This leads Wald to suggest that not only is this the best chlorophyll can manage, but also that there can be no alternatives to chlorophyll, for otherwise they would have evolved to replace chlorophyll. To paraphrase Wald, wherever there are planets with plants, there too will be chlorophyll.

Wald proposed that not only was chlorophyll effectively universal, but so too was the biochemistry familiar to us on Earth. As he once remarked, "Learn your biochemistry here and you will be able to pass examinations on Arcturus" (Wald 1973: 16). Here too I suspect that very few biologists would be inclined to take this literally, but Wald was actually making an important point (see also Pace 2001). Are there biological systems that do not use either the coding system of DNA or amino acids to build proteins? Are there genuinely alien biochemistries? With a sample of one – the terrestrial biosphere – any answer will be speculative. An inability to conceive of an alien biochemistry may again be a failure of our imagination, while the notion that the familiar biochemical pathways are the norm elsewhere in the Universe would seem to verge on the preposterous. Yet perhaps the view that terrestrial biochemistry has a universal relevance has some merit. Consider DNA. For all of its iconic, if not totemic, significance it is not widely appreciated that the molecular construction of DNA is really very peculiar. In principle, other replication systems should exist. The team led by Albert Eschenmoser has undertaken some remarkable work on what he calls the "etiology of DNA" (e.g. Eschenmoser 1999). This is effectively an exploration of DNA "space" whereby various components that make up DNA (effectively the base pairs, e.g. adenine, and the sugar ribose) are substituted and the properties of the alternative, so-called homo-DNA are investigated. Such experiments might indicate both how DNA came to have its present form and how viable the alternatives are. For the most part, although the different types of homo-DNA have interesting properties, they are unlikely to have arisen by natural processes. Homo-DNAs are, of course, variants on the theme of a sugar–nucleic acid replication system and do not directly address the question of whether another organic molecule could encode genetic information in a reliable fashion. Could it be the case that the ribonucleic acids – by which I mean both DNA and its likely precursor RNA (via the "RNA world") – are uniquely (and universally) suitable?

Not only is DNA a remarkable molecule, but so too is the so-called genetic code. The "codons" are the combinations of three base pairs (e.g.

adenine–adenine–guanine, or AAG) – known therefore as triplets – which are responsible for the coding of the twenty different amino acids (thus AAG codes for lysine), which are then assembled via polypeptides into the proteins. The efficiency of the genetic code has often been emphasized, with an in-built redundancy that in general ensures that if a triplet contains an error, say AAA instead of AAG, then either the same amino acid (in this case it is still lysine) or one with similar chemical properties will be transcribed. The system is not foolproof, but it is remarkably efficient. But how good is good? In a series of investigations that effectively randomize the code and see if a better alternative can be found, Freeland and Hurst (e.g. 1998) came to the remarkable conclusion that the genetic code is not just good, it is astonishingly good. As they themselves noted, "It is one in a million." Could it even be the very best? This question cannot be directly answered because the number of possible combinations of triplets is astronomically large, but by making certain assumptions about the related biochemistry, which reduces the total range of possible combinations to approximately 200 million, Freeland *et al.* (2000) conclude that the genetic code is one of the best, if not the best. How very curious.

If life is so constrained that it can use only chlorophyll for photosynthesis and DNA for the genetic basis of life, if life ends up with a genetic code that is eerily efficient and has a biochemistry that is as familiar on Arcturus as it is to us (or at least biochemists and their Arcturan equivalents), then this would be in marked contrast to the now-popular view of evolution as being little more than a contingent muddle. Thus, we are asked to admire an endless coruscation of life, with no species in possession of any privileged position, no overall trends, and emphatically no progress. One of the chief exponents of this view was S.J. Gould. It seems fairly clear that he found this view of evolution attractive both because it demotes the position of humans to evolutionary insignificance – after all, they are just another evolutionary accident – and makes the discovery of any moral dimension solely a byproduct of this fortuitous event. One implication of this view is that moral precepts (authority is too strong a word) may be tailored by us to suit our circumstances – no doubt in the pursuit of freedom and goodness rather than state terror and genocide. It is difficult, in these naturalistic circumstances, to explain the widely shared admiration for such men and women – and let me just mention two heroic Germans – as Dietrich Bonhoeffer and Edith Stein.

Hallmarks of creation?

It was G.K. Chesterton who rightly reminded us that we should be careful when we employ the natural world in a theological context. As he wrote, "We may accept Nature as a messenger from Heaven; but certainly not as a plenipotentiary ambassador" (Chesterton 1999: 265). There are, of course, such theologians as Karl Barth who regard any reference to nature as a mere distraction from the business at hand, but this seems to be difficult to square with the claim of religion, or at least Christianity, to have a universal significance, notably in the Pauline and Johannine traditions. In the Christian context of eschatology the future of the natural world is debatable, but as ever what we expect probably will not happen; rather, we should be prepared for a surprise. It seems to me also that even from a scientific perspective there is an argument for taking the created world that we inhabit with some seriousness (and happiness) by treating it as the product of the Creator. Michael Ruse (2001) has argued forcibly that holding to the Christian faith is compatible with being an evolutionary biologist. Moreover, such a view gives some accommodation to what we perceive as natural evils, but Ruse also emphasizes that his argument for the compatibility of Christianity and Darwinism makes less sense in the absence of the concept of evolutionary progress. This notion has certainly enjoyed a rough ride, but a survey even of sensory modalities in the natural order (vision, hearing, olfaction, echolocation, vocalization, etc.) makes it difficult to dismiss the idea of evolutionary progress. Can we not say the modalities in this progression are more complex, more sophisticated, and, by leading to a richer world, perhaps even better?

When I claim that the world around us bears the hallmarks of creation, I may need to stress that this chapter is not meant to offer any comfort to those who wish to find the action of God in some case of irreducible complexity, nor am I willing to give a jot of support to the so-called scientific creationists. Much has been made of the anthropic principle, and especially the remarkable degree of fine-tuning that appears to be necessary both for the Universe we know to exist and also to support life. Such anthropic views do not prove the presence of a creator; as has often been pointed out, the data that appear to support this comforting premise may be refuted in the future. Still, from the present perspective they are at least consistent with a theistic view, albeit one far removed from the astounding claims of Christianity.

Apart from the quirkiness of the way carbon forms in the interiors of stars, as well as the peculiar properties of water (see above), the role of biology in this discussion, on the other hand, has been muted. But if the species that builds churches, temples, and mosques is genuinely an accidental sprig from the bush of ape evolution – one that in turn arose from a shrew-like animal that long, long ago clambered onto land, and so on – then the sensible thing might be to leave any theistic argument to the astronomers and cosmologists. Give us the Universe, let evolution run, and let's not worry too much about how we got here.

Yet it is we who are conscious, and we who believe, even if it is in nothing. What, then, might ring true for a biological creation? Let us not quibble that the biosphere is here through anything but through evolutionary processes: as

Ruse (2001) and many others have pointed out, that is the only way the world can work. The basis of life is simple, yet the building blocks of life not only combine in remarkable ways, but also do so through pathways that are highly specific. *A posteriori* we admire them, but *a priori* we blunder repeatedly in our attempts to simulate even the least complex: think of the innumerable origin-of-life experiments whose success is not only limited, but depends on a laboratory chemistry far removed from any credible prebiotic world (Shapiro 1987; Horgan 1991). Life is also pervaded by inherencies, by which I mean that much of the template of complex life is assembled long before the structures themselves evolve; as explained below, the eye is a good example. Following on, we can also see that, far from being a contingent muddle, life is pervaded with directionality; by no means everything is possible, but what is possible will evolve repeatedly. Finally, life constantly surprises us with its elegance and economy of construction, yet all of it emerges from a common and seemingly unremarkable substrate. None of this can be used to prove the activity of the Creator, but it seems consistent.

What is inevitable?

Let me now posit that the evolution of humans – by which I mean a sentient species with a religious instinct – is an inevitability. This seems scarcely compatible with our being an evolutionary accident, as strange in our way as an orchid or penguin (not to mention a bombardier beetle or a naked mole rat). But being human means much more than having descended from the jungle trees of Africa. Humans are a species with free will (see O'Hear 1997). What matters, therefore, is the emergence via evolution of certain biological properties that seem to be integral to humanity. If *Homo sapiens* had not evolved then, I would argue, something similar, sooner or later, would have emerged. The fact that we have a unique history does not rule out the emergence of common end-points.

How might we test this supposition? Broadly, there are two questions we might ask. First, can we conceive of an organism that "ought" in principle to exist but has failed to evolve on Earth? Second, are the examples of evolutionary convergence, such as the similarity between our eyes and those of the octopus, simply interesting coincidences, or are they indicative of something more significant?

Of imaginary animals the medieval bestiaries will provide little helpful guidance, being mostly chimeras or fantasy. They are important for our imagination, but not for biology. Of seemingly biologically credible animals, various people have wondered why the natural analogue of an airship, a floating dirigible scooping up food as it drifts across the skies, has never evolved (see Conway Morris 1999). In terms of construction, such an organism seems to be unproblematic. Gas-filled bladders are already present; think of the fish swim-bladder. This ingenious device controls leakage of gas both by rendering the bladder wall less permeable (by the deposition of thin and flexible crystals of guanine) and by a counter-current system whereby the gases that dissolve into the blood vessels as they leave the bladder are transferred back to the blood supply that is going to feed the bladder. Fish, of course, use their swim-bladders to achieve neutral buoyancy in the water, not to float clear of predators and fishermen. But, if needed, the lighter-than-air gas hydrogen should be readily available by the simple expedient of splitting the water molecule. In fact, a number of single-celled organisms contain specialized structures, the hydrogenosomes, which, as their name indicates, produce hydrogen (e.g. Embley *et al.* 1995). To be sure, they only operate in the exclusion of oxygen, but a gas-secreting organ with an effective method of protecting the hydrogen-generating area from the excess oxygen seems possible. So why aren't our skies full not only of birds, bats, insects, and now airplanes, but also of floating bladders? There seem to be two problems. First, the air is far too thin to support the equivalent of aerial plankton. Unlike viscous water where suspension feeders (e.g. serpulid worms, bivalve mollusks) are common, the nearest equivalent on land are spider webs. Scooping up airborne plankton is a recipe for starvation. Second, as Steve Vogel (1998) points out, in the evolutionary history of this hypothetical bladder, the bladder would have to start out at a very small size. Yet in such a case the area-to-volume ratio is decidedly unfavorable, with only a small contained volume and a relatively large surface area. Moreover, as the bladder wall would not be completely impermeable, especially given the small size of the hydrogen molecule, any ability to float would be compromised.

It seems, therefore, that the inhabitants of another world are no more likely to see a fleet of floating bladders in the sky than we are. And what about these sentient extraterrestrials? Will they be vaguely humanoid or perhaps repulsively reptilian? Might they be a colony of telepathic ant-like creatures? Not surprisingly, what we see here on Earth is likely to color our expectations, but in any event the humanoid option is regarded with wide suspicion. In an influential paper entitled "The non-prevalence of humanoids" the evolutionary biologist and paleontologist George Gaylord Simpson (1964) argued forcibly against the idea of any extraterrestrial looking remotely human. Simpson thought inhabited planets would be rare enough, but he argued that as and when life got a foothold, the twists and turns of history could never be replicated in the way that happened on Earth. Sentience may arise, but even an intelligence comparable to ours is regarded by some as a quirky byproduct of the evolutionary process (e.g. Calvin 1978; Diamond 1995). And humanoids? Absolutely not. Although this remains the prevailing opinion, others have looked to the prevalence of evolutionary convergence to suggest that if the biology of form is constrained, then, no matter how far away the worlds might be, there too will be humanoids. Thus, writing in the same year as Simpson, Bieri remarked, "If we ever succeed in communicating with conceptualizing beings in outer space, they won't be spheres, pyramids, cubes or pancakes. In all probability they will look an awful lot like us" (1964: 457).

I think Bieri is correct. Even so, I am surprised that as I have trawled the literature for examples of convergence the biologists describing a particular instance can seldom resist mentioning that it is "remarkable," "striking," or even "uncanny." But why should they be so surprised? By no means everything is possible through evolution. Examples of convergence appear in nearly all textbooks of evolution, but they are usually treated in a piecemeal fashion. So far as convergence has a wider significance it is usually because it represents a profound irritant to those engaged in reconstructing phylogenies. This is especially true of those wedded to cladistic methodologies, where similarity of form arising without common descent is referred to as homoplasy. To the cladist, homoplasy is equivalent to pouring sand into the machine, leading to horrible noises and overheating, and sometimes bringing their contrivance to a shuddering and smoking halt. Diligently the errant homoplasies are picked out and placed in neat heaps, where they cause no further mischief. In other words, they are neglected. Yet homoplasy is rife, and as such represents a profoundly interesting problem in its own right.

Focusing on convergence

At present, however, the many examples of convergence are seldom dealt with systematically, but are more often treated as anecdotes. And they certainly keep the audience interested, be it the example of convergence between the raptorial forelimbs of the praying mantis and the neuropteran *Mantispa* (Ulrich 1965), the huge stabbing canines of the saber-tooth cats and the marsupial thylacosmilids (Churcher 1985), the profound and extensive convergence between the moles and the many other burrowing animals (Nevo 1999), the close similarity in the societal structure of sperm whales and elephants (Weilgart *et al.* 1996), and even the independent evolution of a protein that prevents ice-crystal formation in the tissues of unrelated Arctic and Antarctic fish (Chen *et al.* 1997). All very interesting, but if convergence is going to provide a guiding principle, then it is now necessary to look at the evolution of complex systems, ideally those on the route to sentience.

Let us consider the eye. So often has this organ been used in evolutionary discussions that one might wonder what is worth repeating. The emphasis, however, has tended to focus more on the apparent perfection of design and, related to this, the incremental steps by which a dioptric eye might evolve (Nilsson and Pelger 1994). These are important points, especially as simulations suggest that with respect to geological time a complex eye could evolve rapidly, that is in less than a million years. What is not quite as widely appreciated is that the principal building blocks of the eye, notably the lens proteins (known as crystallins) and rhodopsin (which has a key role in the sensory retina), evolved long before the first eye and have been co-opted from previous functions. The example of crystallins is particularly informative because they are recruited from quite a wide variety of proteins, but typically ones involved in stress tolerance, such as the heat-shock proteins (e.g. Wistow 1993). The apparently fortuitous fact is that not only are such proteins stable against environmental insult, but, by being small and capable of close packing in a watery matrix, such proteins are preadapted for lens construction. Nor is a nervous system an essential prerequisite for the eye, because various single-celled organisms possess eye-spots. In the case of one group, the warnowiacean dinoflagellates, the eye is remarkably complex and consists of a bulbous lens, capable of focusing, above a light-sensitive cup (Greuet 1978). Quite how this tiny cell, swimming through the water, interprets the visual signal without a brain is an interesting question, but these tiny dinoflagellates are hunters. Some Italian investigators (Piccinni and Omodeo 1975) uneasily rejected the suggestion that these organisms "knew" what they were doing in pursuit of their prey.

Eyes, therefore, seem to be very much on the cards in any history of life. But, of course, there are several types of eye. As often as not, the alien intelligence is envisaged as remorseless, scanning the terrified humans with its empathy-free compound eyes, while its robotic limbs assemble the laser cannon. On Earth the bug-eyed monsters are the arthropods, but their type of eye has evolved independently several times. Our eyes, in contrast, show the so-called camera-design, with a large lens separating two chambers, the rear of which is lined with the retina. As is well known, the vertebrate eye is strongly convergent on that of the squid and octopus, advanced representatives of the cephalopod mollusks. This is a textbook example of evolutionary convergence, and although there are differences the overall similarity is impressive. It is much less often remarked that the camera-eye has evolved independently several other times, notably in the alciopid polychaetes (Hermans and Eakin 1974), as well as in at least three groups of snails: heteropods, littoriniids (Hamilton *et al.* 1983), and strombids (Gillary and Gillary 1979). For the most part these animals are agile, fast-moving, and often predatory. Good vision is essential for such activities. Even the mere presence of vision in the slow-moving littoriniids and strombids is less odd than first appears, because the former critically depend on recognizing landscapes on the tidal flats that provide local refuges during high water (Hamilton and Winter 1982), while when attacked the herbivorous strombids are exceptionally agile, exhibiting a remarkable jumping motion (Berg 1974). But is it a toss-up between the compound and camera-eye? After all, the former is used by the immensely successful insects, including such star turns as the navigating bees and migrating butterflies. For such small animals the compound eye suffices, but in comparison with the camera-eye it suffers serious drawbacks. Kuno

Kirschfeld has demonstrated that if we humans were to rely on a compound eye that provided the same degree of acuity as the eyes we possess, we would require a gigantic structure, at least a meter across (Kirschfeld 1976). Perhaps the Universe is full of eyes, but any sentient species gazing at the stars is much more likely to be using a camera-eye.

Converging brains?

Eyes in animals presuppose a brain, and to the first approximation brain size is proportional to that of the animal: big brains usually means big animals. However, a number of animals have brains that are disproportionately large, and none more spectacularly so than ours. Indeed, it is sometimes claimed that the dramatic increase in brain size in *Homo* is without parallel elsewhere, with the adjunct possibility that all the fruits of mentality, such as music and painting, are just accidental spin-offs of this cerebral mass. Yet so far as increase in brain size is concerned, there are striking parallels with the dolphins (Marino 2002). Here too the trend towards bigger brains was apparently rapid, and interestingly may have been spurred by oceanographic changes, notably a cooling in the southern ocean (Fordyce 1980). But it occurred about 20 million years ago, and in terms of brain size it was only about 2 million years ago, with *Homo erectus*, that the hominins overtook the dolphins (Marino 1996). Here too the major increase in hominin brain size was possibly driven by environmental factors, this time increasing aridity in Africa.

Big brains are often regarded as an evolutionary extravagance, a rococo embellishment, ruinously expensive metabolically, and, in our case, a cause for a post-Edenic pessimism, fuelled by existentialist doubts, which concludes that big brains have led, not to a surfeit of happiness, but to more ingenuity in the pursuit of malice. The solution to that problem lies elsewhere, specifically with a Man tracing his finger in the dust of Palestine, but so far as the dolphins are concerned there must be a substantial advantage in maintaining the swollen cerebral organ, given the millions of years that have elapsed since the size increase began.

Dolphin brains show some important differences from ours, including their cellular structure and a unique paralimbic lobe (Marino 1996, 2002). This is hardly surprising, seeing as these aquatic animals occupy a rather different environment from ours, a habitat where such features as echolocation and maneuverability are particularly important. In certain respects the dolphin brain still shows its primitive antecedents, but even so this group of toothed whales is remarkably versatile. Of particular interest are their communicative abilities (Janik 2000), a memory system convergent on ours (Thompson and Herman 1977), brain laterality (Goley 1999), at least hints at a capacity for abstraction (to judge from various experiments, including the comprehension of artificial languages – sign or auditory – whereby different word orders are intelligibly distinguished and executed) (Herman *et al.* 1993), and clear evidence for self-recognition (Reiss and Marino 2001). Dolphin social structure is also interesting, not only because of its long-appreciated similarity to chimp societies (the so-called fission–fusion arrangement), but also because recent work suggests the network of recognition by a dolphin extends to at least one hundred other individuals (Mann *et al.* 2000). Dolphins are sophisticated and intelligent creatures, but their aquatic habitat and flipper-like limbs suggest it would be difficult for them to take the step towards establishing complex material cultures, even though they display limited tool use (by perching conical sponges on their rostra to rootle in the seabed). Their importance, though, lies in the evolution of a large brain – if twice, why not many times?

Many animals use tools, such as the New Caledonian crows (Hunt 2000), or build structures – sometimes relatively simple ones like the tents some bats make by folding together large leaves (Kunz and McCracken 1995), sometimes remarkably elaborate ones such as those constructed by the bower birds (Madden 2001). Advanced technologies, however, are only seen in the hominin group, although even here a general survey would probably emphasize more the cultural conservatism rather than the endless rounds of innovation. Living as we do in a world of relentless and sometimes destructive technological innovation we find it difficult to understand those societies from Bronze-Age times onwards that failed – or did they decline? – to pursue technological leads, nor indeed why in any case the process took so long given there has been little change in brain size for more than 100,000 years. Nevertheless, the discovery of sophisticated javelins from 400,000-year-old peat in Germany (Thieme 1997) warns against excessive reliance on the evidence from lithic technologies, while still-controversial evidence from Africa (McBrearty and Brooks 2000) suggests that the development of more advanced Paleolithic stone cultures may have been earlier (circa 90,000 years ago) and more protracted than the European record of artifacts would suggest. Even so, the appearance of sophisticated technologies that presuppose not only an aesthetic dimension in tool production but also associated artifacts with unequivocal evidence of symbolic representation only approaches its full expression from about 50,000 years ago. And this event, the Paleolithic Revolution, has been interpreted as a unique development, restricted to our species.

The co-occurring Neandertals, which are generally but not universally identified as a separate species, are widely regarded as the runners-up. So close, so very close, with evidence for care of the disabled, use of fire, and burial of the dead, but lacking that special spark of creativity. This seems especially true in the last stages of Neandertal history, before their extinction about 28,000 years ago, with the development of the so-called Châtelperronian cultures. Here a step-change in sophistication occurs, not only in tool development but also with the manufacture of "useless" items such as necklaces. This culture is widely regarded as imitative, achieved perhaps by scavenging our rubbish sites, by trading, or by simply observing the clever chaps ("Very good, Arthur; not bad at all. Now if we can just hold the flint a bit higher...well, never mind, I expect we can use it for something; now if you would like to bring a flint from that pile over there, we'll continue the lesson..."). A reappraisal of the stratigraphy and dating of the often complex infill of cave sites has led to an overthrow of this view, although it needs to be pointed out that by no means everyone accepts the new view (Zilhao and d'Errico 1999), which is that, far from being imitative, the Châtelperronian culture was an independent development, convergent on our cultural explosion. The message throughout seems to be plain: once you get to a certain stage of evolution, further developments become very likely, if not inevitable.

To be sure, mass extinctions may postpone or divert the path, but their overall influence has been greatly exaggerated. Short of utter devastation, such as might be inflicted by a supernova exploding nearby, the emergence of various biological properties during the course of evolution is virtually guaranteed, at least to judge from the ubiquity of convergence. Above, I have given just a few examples, emphasizing convergence in the camera-eye, large brains, and culture. Many other cases, such as the convergent evolution of warm-bloodedness, "mammalness," and agriculture, are discussed elsewhere (Conway Morris, 2003). All you need is a habitable planet in a reasonably safe part of some galaxy. But that turns out to be much more difficult than might be imagined.

Rare Earth?

Confidence that there is a plurality of worlds has oscillated, although optimists have usually outnumbered skeptics in positing a Galaxy full of solar systems, and many with inhabited planets. In recent years a series of ingenious and very sensitive measurements has confirmed the reality of the so-called extra-solar planets (Lunine 1999). Whether life is also present, let alone prevalent, on these distant planets cannot be ascertained with the available methods, although plans are afoot to build deep-space telescopes that could resolve planets and analyze their atmospheres to see whether they are out of equilibrium in a manner comparable to Earth, where biological processes impose a strong impact, notably in the production of oxygen and methane. Even so, much has already been learned. More than ninety extra-solar systems have been detected, but with surprising results. The planets detected are enormous, often several times more massive than Jupiter. In itself this is not surprising, because only large planets exert sufficient gravitational pull on their stars for the effects (on the stellar spectra) to be detectable by our telescopes, and even so the perturbations are minuscule. Smaller planets no doubt exist, and as the range of detection expands we will get some sense of whether most solar systems do in fact consist of one or two gigantic planets. To date those detected are close to their suns, sometimes so near that they complete their orbits every few days. This too is an inevitable consequence of the methods used by astronomers, because planets further away from their suns take proportionally longer to orbit. It might take years of observation to confirm their presence. Jupiter, for example, takes almost twelve years to orbit the Sun. Accordingly, the detection methods are biased towards finding huge planets whirling close to their suns.

A potential implication, however, is that an arrangement like the Solar System may be much more unusual than was once thought. In itself, that might not matter too much, so long as the planet has a moderate mass – that is, a mass such that the gravitational field is not cripplingly high – and occupies the so-called "habitable zone" where liquid water can persist for thousands of millions of years. There may, however, be some problems. First, it seems rather likely that at least some of the giant extra-solar planets did not form in their present positions. Rather, they accreted much further out and were then moved towards their stars. As this happened, so their immensely strong gravitational fields would have drastically perturbed the orbits of any other planets, probably slingshotting them out into the wilderness of interstellar space (e.g. Rasio and Ford 1996; Weidenschilling and Marzori 1996). Paradoxically, so far as the Earth is concerned, the huge Jupiter is very much the "good neighbor." This is because its gravitational field helps to deflect the comets that might otherwise plow on towards the inner Solar System. Jupiter is not, of course, a perfect "goalkeeper" and some get through, potentially with catastrophic results. Yet such events are very rare, and although the post-disaster ecosystems take several million years fully to recover (Kirchner and Weil 2000), this is a substantially shorter period of time than is expected to elapse between giant collision. In the absence of Jupiter, however, it has been calculated that the Earth would experience a major impact approximately every 100,000 years (Wetherill 1994). The repeatedly traumatized biosphere would, no doubt, survive, but it is difficult to imagine species with cultures benefiting from these immense disruptions.

The possibility of the Earth forming at a substantially greater distance from the Sun and only subsequently moving to its present position is also intriguing, for the following reasons. As the Solar System forms, the bulk of the material falls into the newly forming star. This is a rapid process, and when the star is sufficiently massive it begins to shine as the thermonuclear reactions begin to operate. One result is that the lighter elements, known as the volatiles, still remaining in the swirling accretion disc are driven outwards by the Sun's radiation until they condense at the point known as the "snow line," which is approximately at the present position of Jupiter. If the Earth formed in much the same position as it occupies today, it would have lain well within the snow line. The outward displacement of the volatiles would, therefore, have left our planet with neither an ocean nor atmosphere, and hence no possibility of life. If, however, the Earth accreted beyond the snow line, the problem is potentially solved, although it would have to be followed by some perturbation that drove it closer to the Sun and into an orbit conveniently parked within the habitable zone. The alternative, and currently more popular, hypothesis is that the Earth accreted in its present position, but was resupplied with the volatiles necessary for the atmosphere and ocean by a massive bombardment of volatile-rich comets and water-rich asteroids (Morbidelli *et al.* 2000) early on in the history of the Solar System. This is appealing for two reasons. First, not withstanding the presence of Jupiter and its goalkeeping role as a "comet sink," it is sometimes objected that it was the formation of Jupiter that precluded the accretion of a planet between Jupiter and Mars, and thus led to the fragments remaining as the asteroid belt, members of which sometimes collide with the Earth. However, if the main water supply for the early Earth came from this asteroid belt, then the formation of a planet between Mars and Jupiter in fact would have reduced the influx of water to our planet. Second, if, as seems likely, the comets were an important source of primitive organic material, then they too may have been essential to kick-start the chemistry leading to the origin of life. The Solar System is richly endowed with these volatile-rich comets, but there are reasons to believe that equivalents to the so-called Oort cloud, the main repository for the billions of comets beyond the farthest planets, may be much less common in other solar systems (Sekanina 1976). If there were no comets, maybe there would be oceans, but there would be an insufficient supply of organics on planets in the habitable zone, and definitely no life.

So producing an Earth-like planet may be far more difficult than might be supposed (see also Ward and Brownlee 2000). And even the apparently trivial, such as the size of the planet or the presence of a moon, may be important factors. On a larger Earth, gravity would be much stronger; recall that a planet only 20 percent wider than ours would be five times more massive. In such a world topographies would be subdued, and much, if not all, of the surface would be covered by an ocean (Lissauer 1999). Such conditions are fine for the dolphins and maybe the rise of intelligence, although we must remember that back in the Eocene the ancestors of the dolphins and other whales were prowling along the seashore. Without extensive land surfaces and access to fire, however, it is difficult to imagine technology arising. If there were any land animals, they would be massively constructed, probably adapted to ground-hugging activities. What if the Earth were smaller? Mountains would be precipitous, but more importantly the weaker gravitational field would lead to a substantially thinner atmosphere and probably no large oceans. Land animals would be much more lightly built, but flight might be problematic, given the thinness of the air. Intelligence could arise, but it would be a very different world, and to our eyes much harsher.

Why might a moon-like satellite be important? Principally because of the lock between our daughter satellite and Earth that stabilizes the system and prevents the chaotic shifts in inclination that evidently characterize the other planets. In addition, the transfer of angular momentum to the Moon gradually slows the speed of the Earth's rotation, whereas without the Moon the rate of spinning would lead to a very different distribution of the Sun's heat and a surface characterized by hurricane-force winds (Comins 1993) – not a pleasant prospect.

The world without the Moon would, therefore, be a very different place, and arguably one in which advanced cultures might have had difficulty emerging. Yet its formation was apparently fortuitous. It originated early in the history of the Solar System, evidently the result of a gigantic collision between the proto-Earth and another rocky body, probably about the size of Mars. This catastrophic event led to a cloud of debris being ejected from the colliding bodies and eventually these coalesced into the Moon. To guarantee the Moon, however, requires precise conditions of impact, and in most reasonable instances the material either falls back onto the Earth or aggregates into a series of small moonlets. As with so much of the Solar System, it seems that very special circumstances gave us our Moon (Cameron and Benz 1991).

Inevitably lonely?

We are faced with a paradox. For all its richness, life is strongly constrained. Convergence is ubiquitous, and many of the apparent differences across the diversity of life are only skin-deep. If we ever visit an extraterrestrial biota, at first sight it will look pretty alien, but I am sure a closer look will reveal startling similarities. Convergence does not mean it will be identical – why should it? – but we can be confident that there are underlying biological properties that will emerge repeatedly. But will we ever find an alien biosphere? On this issue there are broadly three lines of thought. Two are positive, but with one of them supposing that while primitive life will be very widespread, very special conditions are necessary for the emergence of intelligence (Ward and Brownlee 2000), and the other regarding advanced technologies as inevitable. This runs into the so-called Fermi paradox – if extraterrestrials are so common, then where are they? – to which there are a variety of responses, of which the most likely is that our attempts at detection are still in their infancy. Even so, the absence of acceptable scientific evidence for visitations is puzzling. The third line of thought is seemingly much more pessimistic. This posits that the origin of life is actually highly fortuitous, and to compound the problem a very special type of solar system is needed for intelligence to evolve.

From this point of view, the evolution of humans is inevitable, but we live in a very lonely universe. Presently we have no reason to think otherwise, so let us suppose that this is true. It has some interesting theological ramifications. If we think of ourselves as castaways, then this has an echo of the Fall. One response is proud self-sufficiency, but history indicates clearly enough that this is the royal road to disaster. Another response is to take seriously both the promise made to Abraham and the startling and seemingly absurd truth of the Incarnation. Science has for too long been regarded as the enemy of religion, but strangely astronomy and evolutionary biology now suggest that we are something very special – so long as we remember who we are and how we came to be here.[1]

Notes

1 My thanks to Neil A. Manson for inviting me to join this enterprise. Also special thanks to Sandra Last for typing [Cambridge Earth Sciences Publication 6534].

References

Berg, C.J. (1974) "A comparative ethological study of strombid gastropods," *Behaviour* 51: 274–322.

Bieri, R. (1964) "Huminoids on other planets?" *American Scientist* 52: 452–8.

Calvin, W.H. (1978) "Fast tracks to intelligence (considerations from neurobiology and evolutionary biology)," in G. Marx (ed.) *Bioastronomy: The Next Steps*, Dordrecht: Kluwer, pp. 237–45.

Cameron, A.G.W. and Benz, W. (1991) "The origin of the moon and the single impact hypothesis IV," *Icarus* 92: 204–16.

Chen, L., DeVries, A.L. and Cheng, C.-H.C. (1997) "Convergent evolution of antifreeze glycoproteins in Antarctic notothenioid fish and Arctic cod," *Proceedings of the National Academy of Sciences, USA* 94: 3,817–22.

Chesterton, G.K. (1999) "The shadow of God," *The Chesterton Review* 25: 265–7.

Churcher, C.S. (1985) "Dental functional morphology in the marsupial sabretooth *Thylacosmilus atrox* (Thylacosmilidae) compared to that of felid sabretooths," *Australian Mammalogy* 8: 201–20.

Comins, N.F. (1993) *What if the Moon Didn't Exist? Voyages to Earth that Might Have Been*, New York: HarperCollins.

Conway Morris, S. (1999) "Palaeodiversifications: Mass extinctions, 'clocks', and other worlds," *GeoBios* 32: 165–74.

Conway Morris, S. (2003, in press) *Life's solution: Inevitable humans in a lonely universe*, Cambridge: Cambridge University Press.

Denton, M.J. (1998) *Nature's Destiny: How the Laws of Biology Reveal Purpose in the Universe*, New York: The Free Press.

Diamond, J. (1995) "Alone in a crowded universe," in B. Zuckerman and M.H. Hart (eds) *Extraterrestrials: Where Are They?*, Cambridge: Cambridge University Press, pp. 157–64.

Embley, T.M., Finley, B.J., Dyal, P.L., Hirt, R.P., Wilkinson, M., and Williams, A.G. (1995) "Multiple origins of anaerobic ciliates with hydrogenosomes within the radiation of aerobic ciliates," *Proceedings of the Royal Society of London B* 262: 87–93.

Eschenmoser, A. (1999) "Chemical etiology of nucleic acid structure," *Science* 284: 2,118–24.

Fordyce, R.E. (1980) "Whale evolution and Oligocene southern ocean environments," *Palaeogeography, Palaeoclimatology, Palaeoecology* 31: 319–36.

Freeland, S.J. and Hurst, L.D. (1998) "The genetic code is one in a million," *Journal of Molecular Evolution* 47: 238–48.

Freeland, S.J., Knight, R.D., and Landweber, L.F. (2000) "Measuring adaptation within the genetic code," *Trends in Biochemical Sciences* 25: 44–5.

Gillary, H.L. and Gillary, E.W. (1979) "Ultrastructural features of the retina and optic nerve of *Strombus luhuanus*, a marine gastropod," *Journal of Morphology* 159: 89–116.

Goley, P.D. (1999) "Behavioral aspects of sleep in Pacific white-sided dolphins (*Lagenorhynchus obliquidens*, Gill 1865)," *Marine Mammal Science* 15: 1,054–64.

Greuet, C. (1978) "Organisation ultrastructurale de l'ocelloide de *Nematodinium*. Aspect phylogénétique de l'evolution du photorécepteur des Péridiniens Warnowiidae Lindemann," *Cytobiologie* 17: 114–36.

Hamilton, P.V. and Winter, M.A. (1982) "Behavioural response to visual stimuli by the snail *Littorina irrorata*," *Animal Behaviour* 30: 757–60.

Hamilton, P.V., Ardizzoni, S.C., and Penn, J.S. (1983) "Eye structure and optics in the intertidal snail, *Littorina irrorata*," *Journal of Comparative Physiology A* 152: 435–45.

Henderson, L.J. (1913) *The Fitness of the Environment: An Enquiry into the Biological Significance of the Properties of Matter*, New York: Macmillan.

Herman, L.M., Kuczaj, S.A., and Holder, M.D. (1993) "Responses to anomalous gestural sequences by a language-trained dolphin: Evidence for processing of semantic relations and syntactic information," *Journal of Experimental Psychology: General* 122: 184–94.

Hermans, C.O. and Eakin, R.M. (1974) "Fine structure of the eyes of an alciopid polychaete, *Vanadis tagensis* (Annelida)," *Zeitschrift für Morphologie der Tiere* 79: 245–67.

Horgan, J. (1991) "In the beginning...," *Scientific American* 264(2): 100–9.

Hunt, G.R. (2000) "Tool use by the New Caledonian crow *Corvus moneduloides* to obtain Cerambycidae from dead wood," *Emu* 100: 109–14.

Janik, V.M. (2000) "Whistle matching in wild bottlenose dolphins (*Tursiops truncatus*)," *Science* 289: 1,355–7.

Kirchner, J.W. and Weil, A. (2000) "Delayed biological recovery from extinctions throughout the fossil record," *Nature* 404: 177–80.

Kirschfeld, K. (1976) "The resolution of lens and compound eyes," in F. Zettler and R. Weiler (eds) *Neural Principles in Vision*, Berlin: Springer, pp. 354–70.

Kunz, T.H. and McCracken, G.F. (1995) "Tents and harems: Apparent defence of foliage roosts by tent-making bats," *Journal of Tropical Ecology* 12: 121–37.

Lissauer, J.J. (1999) "How common are habitable planets?," *Nature* 402 (supplement): C11–14.

Lunine, J.I. (1999) "In search of planets and life around other stars," *Proceedings of the National Academy of Sciences, USA* 96: 5,353–5.

McBrearty, S. and Brooks, A.S. (2000) "The revolution that wasn't: A new interpretation of the origin of modern human behavior," *Journal of Human Evolution* 39: 453–63.

Madden, J. (2001) "Sex, bowers and brains," *Proceedings of the Royal Society of London B* 268: 833–8.
Mann, J., Connor, R.C., Tyack, P.L., and Whitehead, H. (eds) (2000) *Cetacean Societies: Field Studies of Dolphins and Whales*, Chicago: University of Chicago Press.
Marino, L. (2002) "Convergence of complex cognitive abilities in cetaceans and primates," *Brain, Behavior and Evolution* 5: 21–32.
—— (1996) "What can dolphins tell us about primate evolution?" *Evolutionary Anthropology* 5: 81–5.
Morbidelli, A., Chambers, J., Lunine, J.I., Petit, J.M., Robert, F., Valsecchi, G.B., and Cyr, K.E. (2000) "Source regions and timescales for the delivery of water to the earth," *Meteoritics and Planetary Science* 35: 1,309–20.
Nevo, E. (1999) *Mosaic Evolution of Subterranean Mammals: Regression, Progression and Global Convergence*, Oxford: Oxford University Press.
Nilsson, D.-E. and Pelger, S. (1994) "A pessimistic estimate of the time required for an eye to evolve," *Proceedings of the Royal Society of London B* 256: 53–8.
O'Hear, A. (1997) *Beyond Evolution: Human Nature and the Limits of Evolutionary Explanation*, Oxford: Oxford University Press.
Pace, N.R. (2001) "The universal nature of biochemistry," *Proceedings of the National Academy of Sciences, USA* 98: 805–8.
Piccinni, E. and Omodeo, P. (1975) "Photoreceptors and phototactic programs in Protista," *Bollettino di Zoologia* 42: 57–79.
Rasio, F.A. and Ford, E.B. (1996) "Dynamical instabilities and the formation of extrasolar planetary systems," *Science* 274: 954–6.
Reiss, D. and Marino, L. (2001) "Mirror self-recognition in the bottlenose dolphin: A case of cognitive convergence," *Proceedings of the National Academy of Sciences, USA* 98: 5,937–42.
Ruse, M. (2001) *Can a Darwinian be a Christian? The Relationship between Science and Religion*, Cambridge: Cambridge University Press.
Sekanina, Z. (1976) "A probability of encounter with interstellar comets and the likelihood of their existence," *Icarus* 27: 123–33.
Shapiro, R. (1987) *Origins: A Skeptic's Guide to the Creation of Life on Earth*, New York: Bantam.
Shapiro, R. and Feinberg, G. (1995) "Possible forms of life in environments very different from the earth," in B. Zuckerman and M.H. Hart (eds) *Extraterrestrials: Where Are They?*, Cambridge: Cambridge University Press, pp. 113–21.
Simpson, G.G. (1964) "The non-prevalence of humanoids," *Science* 143: 769–75.
Thieme, H. (1997) "Lower Palaeolithic hunting spears from Germany," *Nature* 385: 808–810.
Thompson, R.K. and Herman, L.M. (1977) "Memory for lists of sounds by the bottle-nosed dolphin: Convergence of memory processes with humans," *Science* 195: 501–3.
Ulrich, H. (1965) "Der Fang- und Greifapparat von *Mantispa* ein Vergleich mit *Mantis*," *Natur und Museum* 95: 499–508.
Vogel, S. (1998) *Cat's Paws and Catapults: Mechanical Worlds of Nature and People*, New York: Norton.
Wald, G. (1974) "Fitness in the Universe: Choices and necessities," *Origins of Life and Evolution of the Biosphere* 5: 7–27.

—— (1973) in R. Berendzen (ed.) *Life beyond Earth and the Mind of Man*, SP 328, Washington, DC: NASA Scientific and Technical Information Office.

Ward, P.D. and Brownlee, D. (2000) *Rare Earth: Why Complex Life is Uncommon in the Universe*, New York: Copernicus.

Weidenschilling, S.J. and Marzori, F. (1996) "Gravitational scattering as a possible origin for giant planets at small distances," *Nature* 384: 619–21.

Weilgart, L., Whitehead, H., and Payne, K. (1996) "A colossal convergence," *American Scientist* 84: 278–87.

Wetherill, G.W. (1994) "Possible consequences of absence of 'Jupiters' in planetary systems," *Astrophysics and Space Sciences* 212: 23–32.

Wistow, G. (1993) "Lens crystallins: Gene recruitment and evolutionary dynamism," *Trends in Biochemical Sciences* 18: 301–6.

Zilhao, J. and d'Errico, F. (1999) "The chronology and taphonomy of the earliest Aurignacian and its implications for the understanding of Neandertal extinction," *Journal of World Prehistory* 13: 1–68.

19

THE COMPATIBILITY OF DARWINISM AND DESIGN

Peter van Inwagen

It is often said, both by Darwinians and anti-Darwinians, that Darwin's account of evolution is incompatible with the thesis that living organisms (or any of their features or any aspect whatsoever of the biological world) are products of intelligent design.[1]

This thesis must be carefully distinguished from the following thesis: Darwin's account of evolution refutes the argument from design. I reject the former thesis and will argue against it in this essay. I have a great deal of sympathy with the latter, however, although my sympathy is tempered by two reservations. First, I do not think it is altogether clear what it means to "refute" an argument. Secondly, there is more than one design argument, and Darwin's account of evolution is more damaging to some of them than to others. I will concede, however, that the very existence of Darwin's account – whether or not it is true – renders all versions of the design argument considerably less cogent than they would have been if no one had ever thought of it.[2] Despite my concession, I do think that there are some versions of the design argument that are not too bad as philosophical arguments – it is, of course, a philosophical argument – go. But that, frankly, is not a very rigorous standard. (What would be an example of a really *good* philosophical argument that could set the standard against which the design argument could be measured?)

The first thesis, the topic of this essay, is an entirely different sort of thesis from the second. The difference between them is all the difference between the thesis that a particular argument for a certain conclusion lacks force and the thesis that that conclusion is false – all the difference between "We should not believe that Richard murdered the princes in the Tower simply because Shakespeare wrote a play in which he did" and "Richard did not murder the princes in the Tower."

For my part, I do not think that any of my beliefs about God, even my belief that living things are products of intelligent design, are based on the design argument – any more than I think that any of my beliefs about my wife are based on the analogical argument for the existence of other minds (which, in my view, is not too bad an argument – as philosophical arguments go). As Newman says somewhere (I paraphrase from uncertain memory), "I do not believe in God because I look at nature and see design; rather, I look at nature and see design because I believe in God." For me, the design argument is an object of philosophical, not religious, interest. A definitive refutation of the argument would trouble me as a lover and child of God no more than a definitive refutation of the analogical argument for the existence of other minds would trouble me as a husband. If, however, the first thesis were established, I should have to admit that Darwin's account of evolution was fundamentally incompatible with beliefs that are a part of the fabric of my life. As a Christian, I should have to look at Darwin's account of evolution in more or less the way that a committed feminist would look at an allegedly scientific account of sexual dimorphism that entailed that women were, of biological necessity, intellectually inferior to men. This is perhaps obvious enough. As a Christian, I am committed to the thesis that the biological world is a product of intelligent design. I am not committed to any very specific thesis about this design, to any thesis about the exact nature of the connection between the mind of God and the structures of organisms. I am not committed to the thesis that God molded human beings out of the dust of the Earth. I am not committed to the thesis that organisms are designed by God in a way that is at all like the way in which machines are designed by a human engineer ("this will have to fit into that space, so the bit on the left will have to be just a little smaller"). Darwinism is no doubt incompatible with some of these very specific theses, theses about the exact nature of the connection between the mind of God and the structures of organisms. What I am committed to is the thesis that the terrestrial biosphere exists because it is God's will that it exist, and to the thesis that it has various of its large-scale features because it is God's will that there be a biosphere having those features. It is in this sense that I should be understood when I proceed to argue for the thesis that Darwin's account of evolution is compatible with the thesis that organisms – the components of the biosphere – are the product of intelligent design. I will call this thesis "compatibilism."[3] Unlike many religious believers who have argued for compatibilism, I can claim that my arguments are disinterested, for I do not think that the Darwinian account of evolution is *true*. (I do not exactly think it is false, but I do find it highly implausible.) And most theists who have argued for compatibilism have been motivated to do so by a belief that the Darwinian account is true, or at least very likely to be true. But, from my point of view, not a lot hangs on the question whether compatibilism is right or wrong: after all, any theist must admit, as a simple matter of logic, that there are lots of false propositions that are incompatible with theism. One more would be no great thing.

I am not going to devote any space in this chapter to explaining why I believe, or am strongly inclined to believe, that Darwin's account of evolution is false.[4] For the curious, however, I will present a quotation from a recent book by the biologist Brian Goodwin that sums up my own views nicely:

> [D]espite the power of molecular genetics to reveal the hereditary essences of organisms, the large-scale aspects of evolution remain unexplained, including the origin of species. There is "no clear evidence…for the gradual emergence of any evolutionary novelty," says Ernst Mayr, one of the most eminent of contemporary evolutionary biologists. New types of organisms simply appear on the evolutionary scene, persist for various periods of time, and then become extinct. So Darwin's assumption that the tree of life is a consequence of the gradual accumulation of small hereditary differences seems to be without significant support. Some other process is responsible for the emergent properties of life, those distinctive features that separate one group of organisms from another – fishes and amphibians, worms and insects, horsetails and grasses. Clearly something is missing from biology. It appears that Darwin's theory works for the small-scale aspects of evolution: it can explain the variations and the adaptations within species that produce fine-tuning of varieties to different habitats. The large-scale differences of form between types of organism that are the foundation of biological classification systems seem to require another principle than natural selection operating on small variations, some process that gives rise to distinctly different forms of organism. This is the problem of emergent order in evolution, the origins of novel structures in organisms, which has always been one of the primary foci of attention in biology.
>
> (Goodwin 1994: viii–ix)

Let us now turn to the question that is our primary concern. Why is it held that the Darwinian account of evolution is incompatible with intelligent design?

I will first make what seems to me to be an important logical point (logical as opposed to scientific or metaphysical or epistemological – not as opposed to illogical). The word "incompatibility," in its central, logical sense, names a relation that holds between *propositions* (or theses or statements or assertions or beliefs). If, therefore, Darwin's account of evolution is incompatible with design, this must mean that some *proposition* (or some set of propositions) is incompatible with *the proposition that* the biological world is a product of design. But what proposition, or set of propositions, would that be? What proposition or set of propositions are phrases like "The Darwinian account of evolution" or "Darwin's theory of evolution" or "Darwinism" names for? It is remarkably hard to find an explicit answer to this question in the literature on evolution, or at least in the minuscule segment of it that I have read. I will make a proposal. If anyone thinks that my proposal misrepresents the propositional content of Darwinism, I would at least like to see some other equally specific suggestion. Unless some reasonably specific proposal is on the table, no investigation of what Darwin's account of evolution is or is not compatible with can be usefully undertaken.

I begin with the word "evolution," which I will understand as a name for a thesis rather than a phenomenon – the thesis that people commit themselves to when they say things like "I believe in evolution." I will set out a series of propositions – four in all – that I mean to express the content of the proposition that evolution is real or that evolution has actually occurred. (Since I confine the scope of my remarks to our planet, some may prefer to call the thesis I shall lay out "the thesis of *terrestrial* evolution.") When I have set out the four propositions that constitute this thesis, I will add one proposition to the list and I will claim that the resulting set of propositions comprises "The Darwinian account of evolution." (Later, however, I will argue that this definition needs to be revised.)

Here are the first two of my propositions:

(1) Any two living organisms, past or present, have a common ancestor.
(2) There have been living organisms for a very long time, not just for a few thousand years but for *millions* of thousands of years (perhaps since a few hundreds of millions of years after the Earth's surface was cool enough to support life).

These two propositions taken together make up a rather weak thesis. For one thing, it is weak because it says nothing about biological diversity. It could be true if the only organisms there had ever been were a particular sort of bacterium that had persisted unchanged for billions of years. It is also weak because it says almost nothing about causation – although "ancestor" is a causal concept. It is compatible, for example, with the statement that God has been responsible for a vast array of miraculous innovations in the history of life, and it is compatible with the statement that intelligent extraterrestrials have been dropping in on the Earth every 10 million years or so to perform prodigies of genetic engineering in aid of some mysterious agenda involving terrestrial life. To get a more interesting thesis to associate with the word "evolution," let us add some propositions about diversity and causation:

(3) Life exhibits (and has exhibited for a very long time now) enormous taxonomic diversity.
(4) Only natural causes have been at work in the production of all this diversity.

And what does "natural" mean? Well, the word can be opposed both to "miraculous" or "supernatural" on the one hand, and to "artificial" on the other. Let us understand "natural" in this context as carrying both implications. The thesis of evolution implies that only the laws of physics (operating of course under an enormously complex set of boundary conditions) have been at work in the terrestrial biosphere during the course of the diversification of life. It also implies that the only extraterrestrial influences on terrestrial life have been things that are in no way the instruments of intelligence or purpose: light from the Sun, cosmic rays, falls of meteor dust, asteroid strikes, and the like.

I think it is useful to regard these four propositions as together constituting the thesis of evolution. (Should there be something here suggestive of the notion of "progress," or, at any rate, of increasing complexity? Anyone who thinks so may add a clause to the effect that, in the very long run, the complexity of both the biosphere and of the most complex organisms in the biosphere tends to increase. I should not object to the addition. This seems to be a part of what a lot of people mean by "evolution," and it seems to be true.)

Now I will turn to "Darwin's account of evolution" or "Darwin's theory of evolution" or "Darwinism." I take Darwinism to be an identification of the "natural causes" referred to in the last of the four propositions. I take Darwinism to be a specification of a mechanism, a single mechanism, which explains taxonomic diversification. This mechanism is the operation of natural selection on random small hereditable variations that come about in the course of reproduction. Although I shall later have something to say about the word "random," I am not going to try to give an exposition of what lies behind the slogan "the operation of natural selection on random small hereditable variations." I know that there is considerable diversity of opinion among those who describe themselves as Darwinians as to how the reality behind the slogan should be spelled out in detail, but I do not think that these disagreements have much to do with what I want to say. At any rate, I take it that we all have some idea of what these words mean. Even the slogan is too cumbersome for frequent repetition, so I'll call the mechanism simply "natural selection." Darwinism, then, is the thesis of evolution plus the further thesis that the sole mechanism behind the enormous taxonomic diversity displayed by terrestrial life – behind the existence of all of those vastly different phyla and orders and classes – is natural selection. (I am aware that Darwin was probably not a Darwinian in this sense – at least not always and consistently – and I am aware that he sometimes opposed natural selection to sexual selection. As to the former point, I am trying to capture at least something close to the most usual sense the word "Darwinism" has in current debates. As to the latter point, unless I am mistaken, most people today use the term "natural selection" in such a way that what Darwin called sexual selection is a special case of natural selection.)

Now, why is this proposition – "Darwinism" or "Darwin's account of evolution" – supposed to be inconsistent with the proposition that the biological world is a product of design? One might well ask why, if unaided natural selection really is capable of producing the ordered diversity we see in the terrestrial biosphere today, a God – or any intelligent being capable of working on such a scale – who wanted such ordered diversity should not have used this very elegant mechanism. If I myself doubt whether God did use this mechanism, it is only because I doubt whether unaided natural selection could do the job. I think that other mechanisms would be required and that He therefore cannot have used natural selection alone. But if unaided natural selection would work – well, why shouldn't God use something that would work? And if God (or any intelligent being) did establish an environment in which this mechanism could operate, and if its operations in due course produced a biosphere having certain features, and if God foresaw and intended the existence of a biosphere having these features – then why would it not be correct to say that these features were products of intelligent design?

I think that these are excellent questions, questions that have never been properly answered – that have never in fact been properly addressed. It seems to be a widespread opinion that there is something about natural selection that unfits it for use as a divine instrument. I have never been able to see this. When I was an agnostic, I was a Darwinian. When I became a Christian – a very old-fashioned, orthodox one – I was a Darwinian still. And although I have experienced many intellectual difficulties with my faith, my belief in Darwinism never caused me the least intellectual discomfort. (My doubts about Darwinism began only when I discovered that the "smoothness" of the fossil record that I had always believed in was not there.) I should add, in this connection, that I do not regard the difficulties that I believe Darwinism faces – the difficulties summarized in my quotation from Brian Goodwin – as constituting any sort of evidence for theism. I think that the truth or falsity of Darwinism has no more to do with the truth or falsity of theism than does, say, the hypothesis of continental drift. But many people do not see things this way. I could quote both Darwinians and anti-Darwinians, both atheists and theists, to this effect. Here is a famous quotation from Monod that will do as well as any. Speaking of the events that have been identified as the sources of mutations, he says:

> We call these events accidental; we say that they are random occurrences. And since they constitute the *only* possible source of modifications in the genetic text, itself the *sole* repository of the organism's hereditary structure, it necessarily follows that chance alone is at the source of every innovation, of all creation in the biosphere. Pure chance, absolutely free but blind, at the very root of

> the stupendous edifice of evolution: this central concept of modern biology…is today the *sole* conceivable hypothesis, the only one that squares with observed and tested fact.
>
> (Monod 1971: 112–13)

He goes on to make it clear that he understands chance in Aristotle's sense, as arising from the coincidence of independent lines of causation. (Thus, it is due to chance that Shakespeare and Cervantes died on the same day, as it would not be if they had killed each other in a duel. In this sense, chance can exist even in a fully deterministic world.) He identifies the source of this chance with imperfections in the fundamental mechanisms of molecular invariance in living organisms. He mentions only the causes of mutations, but he might have mentioned other sorts of events that are of evolutionary significance and can with equal plausibility be ascribed to chance: the flood that happened to destroy a certain herd of ruminants, the raising by geological forces of a land bridge that enabled representatives of certain species to move into a new environment, the intersection of the trajectories of the Earth and a certain comet, and so on.

I do not quite see how it is that the hypothesis that all such events are due to chance is the only conceivable hypothesis. (Is the hypothesis that the motions of the air and water molecules in the sky over Dunkirk in late May and early June 1940 were due entirely to chance the only *conceivable* hypothesis?) But let us suppose that this hypothesis is at any rate *true*. Does it follow that the general features of the biosphere are products of chance? It does not. To suppose that it did would be to commit what logicians call the fallacy of composition. It would be as if one reasoned that because a cow is entirely composed of quarks and electrons, and quarks and electrons are non-living and invisible, a cow must therefore be non-living and invisible.

There is a marvelous device for calculating the areas surrounded by irregular closed curves. It is an electronic realization of what is sometimes called the dartboard technique. To simplify somewhat: you draw the curve on a screen; then the device selects points on the screen at random, and looks at each point to see whether it falls inside or outside the curve; as the number of points chosen increases, the ratio of the chosen points that fall inside the curve to the total number of points chosen tends to the ratio of the area enclosed by the curve to the area of the screen. For a large class of curves, including all that you could draw by hand, and probably all that would be of practical interest to scientists or engineers, the convergence of ratios is quite rapid. Because of this, such devices are useful and have been built.

Now the properties of each point that is chosen – its co-ordinates – are products of chance in just Monod's sense. But the whole assemblage of points chosen in the course of solving a given area problem has an important property that is not due to chance: its capacity to represent the area of a curve that had been drawn before any of the points were chosen. Indeed, since the device was built by purposive beings, there can be no objection to saying that the whole assemblage of points has the *purpose* of representing the area of that curve – despite the fact that the co-ordinates of each individual point have no purpose whatsoever. It is also true that the fact that each point has co-ordinates that are due to chance is not due to chance and has a purpose: its purpose is the elimination of bias, to insure that the probability of a given point's falling inside the curve depends on the proportion of the screen enclosed by the curve and on nothing else.

Suppose that every mutation that has ever occurred is, as Monod says, due to chance. Suppose, in fact, that every individual event of any kind that is a part of the causal history of the biosphere is due to chance. It does not follow that every aspect of the biosphere is due to chance. And if none of these individual events has a purpose, it does not follow that the biosphere has no purpose. To make either inference is to commit the fallacy of composition.

Now this reasoning shows at most that the thesis that some features of the biosphere are not due to chance (and likewise the stronger thesis that they have a purpose) is logically consistent with Darwinism. It could still be that the conditional probability of the thesis that there are features of the biosphere that are not due to chance is very low, even negligible, on the hypothesis of Darwinism. But the reasoning does show that if someone wants to construct an *argument* for the conclusion that Darwinism is in any sense incompatible with the thesis that some features of the biosphere are not products of chance, he will have to employ some premise in addition to "Darwinism implies that all events of evolutionary significance are due to chance." (And, as I have implied, I do not find that premise itself indisputable.)

How might an advocate of the thesis that Darwinism is incompatible with design respond to these points? One way might be to argue that the features of the biosphere are in a very important respect unlike the features of an assemblage of points produced by our area-measuring device. Each time we draw a curve on the screen of the area-measurer and turn the thing on, it is for all practical purposes determined, foreordained, that the assemblage of points it produces will have the property of representing the area enclosed by the curve. But, it might be argued, the properties of the biosphere are not like that. There used to be a popular thesis called Biochemical Predestination, according to which they *were* like that. According to Biochemical Predestination, you just take a lifeless planet that satisfies certain conditions (conditions the Earth satisfied before there was any life on it, and which are undemanding enough that it would be reasonable to suppose that a pretty fair number of planets in a given galaxy satisfied them) and in due course you will "automatically" have life, eukaryotic life, multicellular life, sexually dimorphic life, highly differentiated life, and, finally, intelligent life – the whole *Star Trek* scenario. Biochemical Predestination does not seem to be very popular among the practitioners of the life sciences these days, although belief in it seems to be common among physicists and astronomers, and nearly universal among university undergraduates, who believe that Vulcans and Klingons await us among the stars with the same unreflective assurance that attended the belief of their twenty-times-great grandparents that elves and trolls awaited them in the woods. But if Biochemical Predestination is not true, if the main features of the biosphere did not fall into place automatically, but are rather due to remote chances that just happened to come off, then how can it be that these features are due to the purposes of a divine being – or any intelligent being? In short, the failure of Biochemical Predestination shows that, since the evolutionary process has no determinate "output," it is not the kind of thing that could be anyone's instrument.[5] It can no more be used for that purpose than a flamingo can be used as a croquet mallet.

This is an interesting and important argument. It deserves a more careful formulation. I offer the following.

It seems plausible to suppose that if any features of the biosphere are products of intelligent design, then some very *particular* features of the biosphere are products of intelligent design: this one if no other: the existence of rational beings like ourselves, creatures made "in God's image and in His likeness." If natural selection cannot be used (even by an omnipotent and omniscient being) as an instrument to produce living things with "special" characteristics like rationality (or binocular vision or opposable thumbs or pentadactyl limbs; but let us use rationality as our primary example), then it is unreasonable to suppose that any intelligence has been using it as its (sole) instrument in imparting features to the biosphere.

Advocates of the argument we are considering hold that natural selection is indeed unusable for this purpose, owing to the radical contingency of its output. The concept of radical contingency may be explained as follows. Consider the Earth as it was at some very early stage after the emergence of life – when, say, there was only a single type of organism, some bacterium-like prokaryote. Let us say that we are considering the Earth as it was at a time called "t_0." Consider all the physically possible sets of subsequent trajectories of the particles whose precise arrangement at t_0 constituted this "initial state." (We suppose a given set of diachronic boundary conditions, that is, a given, predetermined "schedule" of extraterrestrial "inputs" into terrestrial conditions: sunlight, meteors, and so on.) A complete set of these particle trajectories may be called a history. Consider a space each of whose points is a history. Postulate a numerical measure, a measure of proportion, defined on this space. The idea behind this measure is that it should allow a sufficiently knowledgeable being – a being of the epistemological order of Laplace's Intelligence – to make judgments like this: in 70 percent of the space of histories, the Earth has feature F at t_0 + 1 billion years. If we suppose that each history is exactly as probable as any other, and if the space of histories satisfies a few unexciting formal conditions, our measure is a probability measure, and the above judgment may be read as, "Given the way things were at t_0, the probability that the Earth would have feature F at $t_0 + 1$ billion years was 70 percent." [6]

The thesis that rationality is radically contingent is this: the set of histories that contain rational beings comprises only a small proportion of the total space; that is, the probability of rational beings was small, given the way the Earth was at t_0. The thesis that opposable thumbs or pentadactyl limbs are radically contingent is, of course, to be explained in the same way. The rather more vague general thesis of "radical contingency *simpliciter*" is that the existence of all, or at least most, of the specific features of living organisms are radically contingent. ("Gouldian contingency" may be defined as the thesis that the existence of every phylum that exists today is radically contingent.)

Now a moment's thought will show that there is an annoying technical difficulty that must be faced by anyone who thinks that the existence of rationality, or anything else, is radically contingent. If the physical world is strictly deterministic, there is only one history, and, therefore, in a strictly deterministic world, nothing is radically contingent. (If the world is strictly deterministic, God, or the Laplacian Intelligence, could have produced *every* feature of the present biosphere simply by seeing to it that the world was as it in fact was at any time in the past – after all, the world of the remote past did in fact manage to "turn into" the present world and, if strict determinism is true, it could not have turned into a world having any features but those of the present world.) There are various ways this technical problem might be solved. To discuss them, however, would take us away from our discussion of radical contingency and Darwinism. Let us, therefore, simply assume that there is enough indeterminacy in the world (rooted in quantum indeterminacy, perhaps) that the proponent of the radical contingency of the special characteristics of the biological world need not attend to this problem.

Let the argument continue. If rationality is radically contingent, then the processes of the natural world cannot be anyone's instrument for producing rationality. Of course, this does not show that *natural selection* could not be anyone's instrument for producing rationality – not unless the thesis that rationality is a product of natural selection entails or somehow requires that the genesis of rationality be a matter of radical contingency. I am not sure how one would argue for this conclusion. We have seen that it does not follow logically from the premise that all the individual events that collectively compose the course of evolution are due to chance. It may be, however, that it does follow from this thesis in conjunction with some set of true statements about the conditions under which natural selection has actually operated. If a set of statements having this feature could actually be produced, and if they were known to be true, it would be pedantic to insist that it had not been demonstrated that Darwin's account of evolution *per se* was incompatible with design, but only the conjunction of Darwin's account of evolution with certain other statements – statements that were known to be true. Let us, in order to give the proponents of the incompatibility of Darwinism and design as strong a case as possible, assume that Darwinism commits its adherents to the thesis that certain features of the world, features that it is reasonable to suppose have been conferred on the world by God if *any* features have been conferred on the world by God (the existence of rational beings, for example), are radically contingent.

If we do suppose this, some of us may find the world a bit suspicious. If the existence of rational beings is of a very low order of probability, given that all the features of the biosphere are due to natural selection, and there in fact are rational beings, doesn't that provide some reason to doubt whether all the features of the biosphere are due to natural selection? "Of course not. Given the general thesis of radical contingency, whatever reasonably specific features chance happens to endow the biosphere with will be radically contingent. That the world of living things exhibits many features that are radically contingent is therefore not itself a matter of radical contingency. There is no more reason for you to be astonished by the existence of rational beings than there is for you to be astonished by your own existence – which is, in almost anyone's view, radically contingent." Such exchanges as this are very tricky. Those who think that the existence of rational beings is evidence for the falsity of natural selection will reply by arguing that the existence of rational living organisms (unlike the existence of any particular rational living organism, such as you or me) is highly probable on the hypothesis that the world has been created by God, and, therefore, that the fact that there are such beings favors this hypothesis over any hypothesis on which their probability is low. There are, of course, ways of replying to this reply, and there are ways of replying to the replies to the reply – and so on, for all practical purposes, *ad infinitum*. I do not propose to enter into the ins and outs of a debate on this topic (it would be similar to debates about whether the "fine-tuning for life" that the cosmos apparently exhibits requires an explanation). I will only observe that the contention that the existence of rational beings counts against any theory according to which their existence is extremely improbable has sufficient plausibility that it deserves to be discussed seriously and at length.

As to whether or not this is correct, however, haven't I conceded that Darwinism is incompatible with design if Darwinism commits its adherents to the thesis that certain features of the world that a designer would want are radically contingent? And doesn't Darwinism carry this commitment – if not evidently, then at least for all anyone knows? The point is well taken. If you define Darwinism as I have and if you assume that Darwinism, so defined, entails the radical contingency of some features of the world such that God (or any designer) would create a world only if He, or it, could ensure that it had those features...*then* you have got a thesis that is incompatible with design. (And I will concede that "radical contingency" either is a consequence of Darwinism or could for all anyone knows be a consequence of Darwinism.)

There is, however, more to be said. It is time to re-examine our statement of Darwinism. Most compatibilists, or so I would judge, think that the biosphere has assumed its present form owing to God's guidance of a (generally speaking) Darwinian world. This thesis is not compatible with our statement of Darwinism, owing to the fourth clause in our statement of the thesis of evolution:

(4) Only natural causes have been at work in the production of all this diversity.

If you think about it, however, this would seem to be a metaphysical thesis. And, one may well ask, what business has a scientific theory making pronouncements on metaphysical matters? Let us grant that a theory that postulates supernatural causes is *ipso facto* not a scientific theory. Let us grant that it is an essential part of the methodology of natural science always to search for purely natural causes – and always to assume that our failure so far to find an explanation in terms of natural causes of any event reflects only the limits of our present theoretical knowledge and experimental technique. Some would dispute these assumptions, but let us grant them for the sake of the argument; let us grant them to see what follows (or, more importantly, does not follow) from them. What does not follow is that it is proper for a scientific theory to include, to have as a part of its propositional content, the thesis that the phenomena of which it treats never have supernatural causes. That may be true, but if it is true, establishing it would require some sort of argument. I do not know how the argument would go. Newton's laws of motion and his law of universal gravitation tell us (at least to a good approximation in many circumstances) how massive bodies move when the only forces that are acting on them are gravitational. But they no more contain within themselves the statement "Supernatural agencies never affect the motions of massive bodies" than they contain within themselves the statement "Electromagnetic forces never affect the motions of massive bodies." The obvious position to take on this question, it seems to me, is that the laws of nature have no more to say about the operation of supernatural agencies in the physical world than the laws of gravitational mechanics have to say about the operations of electromagnetic forces. This obvious thesis could be wrong, but I will accept it till someone shows me why I should not.

Do the best meteorological theories (those that are embodied in computer programs for predicting the weather) have as a part of their content that no supernatural agency ever affects the weather? Is someone who believes that God had a special hand in the way the weather was at Dunkirk in the position of rejecting the best meteorological theories? I do not see why I should think so. And I do not see why anyone who thinks that God had a hand in the way evolution went can properly be said – just in virtue of having that very general belief, and not some much more specific belief (as it may be: a belief that each species is a special creation) – to reject any theory of evolution that could properly be called scientific.

Still, it might be argued that the Darwinian account of evolution is a special case. It says that every event of evolutionary significance is due to chance; and if an event is chosen, if it is deliberately brought about by a rational agent in order to serve that agent's ends, then that event is *not* due to chance.

There is certainly a sense of "chance" in which this is true. But the word is a tricky one with many senses. Consider the closely related word "random" (I in fact used "random" and not "chance" in my statement of Darwinism). In one sense of the word, a sequence of things – numbers, say – each of which is individually and deliberately chosen by a rational being, is not "random." Nevertheless, if the members of some odd sect claimed to have in their possession a book of mystically significant numbers, numbers chosen by God, you could not refute their belief by applying statistical tests to show that the book (despite its fancy calligraphy and illuminated capitals) was in fact a table of random numbers, for there is no inconsistency in saying both that a sequence of numbers satisfies all the statistical tests necessary for it to be pronounced "random" and that it was chosen by God for some purpose. Like its near relation "random," the word "chance" has more than one sense, and some of its senses are compatible with "deliberately chosen." If Darwinism is to be a scientific theory, a theory that treats only of the natural world, and if it is to incorporate the concept of chance, that concept must be understood in a way that can be spelled out entirely by reference to the natural world.

Is there such a sense? Of course there is. It can be found in any textbook discussion of Darwinian theory. Its statement is a commonplace of Darwinism. Let us consider mutations, the most important class of events to which Darwinists apply the word "chance." It is of the essence of Darwinism to insist that mutations do not occur in response to changes in the environmental perils or opportunities that confront individuals or species. There is – Darwinians insist – simply no correlation whatsoever between the "usefulness" to a particular species of a possible mutation and the likelihood that it will occur. Suppose, for example, that a certain species of toad is slowly dying out, owing to some gradual environmental change. Suppose that three possible mutations in the genome of that species are equally likely from the point of view of molecular biology. Suppose that one of the mutations, if it became established, would enable the species to cope with its changing environment, one would have no significant effect at all, and one would be lethal. If Darwinism is correct, then these facts about the "usefulness" of the three mutations have no effect whatsoever on the probability of any of the three mutations turning up in some toad of the coming generation. The probability of each is a matter of biochemistry and, apart from the fact that radiation or chemical mutagens in the environment can cause mutations, is independent of the toads' environment – is independent of the species' needs with respect to coping with or exploiting the features of its environment. This thesis entails that, in a perfectly good sense of the word, mutations are due to chance: the Aristotelian sense that I mentioned earlier in connection with Monod. That is to say: Changes in an organism's DNA (as opposed to the transmission of such changes to the organism's descendants) and the features of the organism's environment that are relevant to its success in having descendants are causally independent of each other.

It is, however, consistent with the thesis that all mutations (and, more generally, all events of evolutionary significance) are due to chance in this Aristotelian sense that God has been guiding evolution – by deliberately causing certain mutations (and other events of evolutionary significance). If God has been doing this, it does not follow that the history of terrestrial life would reveal anything inconsistent with the Darwinian thesis that all mutations are due to chance. Suppose that God has in fact been guiding evolution in this way. Suppose also that there is a record of all the uncounted billions of mutations that have ever occurred. Is there any reason to suppose that a statistical analysis of all these mutations and the circumstances under which they occurred (perhaps Laplace's Intelligence could be pressed into service to perform the analysis) would have to uncover some significant correlation between the potential usefulness to species of various mutations and the likelihood of their occurrence?

If there is such a reason, I don't see it. If the course of Darwinian evolution would indeed have to be radically contingent, then a theist who accepts Darwinism (and who accepts the thesis that radical contingency is a consequence of Darwinism) might speculate that God has directed it down the path it has in fact taken by a judicious choice of mutations (and of climatic changes and of events of many other types). And the atheistic Darwinian will have to admit that nothing in the history of life, no possible paleontological discovery, could be inconsistent with, or even cast doubt on, this thesis. After all, the atheistic Darwinian thinks that the actual course of evolutionary history *was* produced by a sequence of events that was due to chance in the Aristotelian sense. Therefore, he must admit that if God chose the actual course of evolutionary history, God chose – produced, created – a course of events that was due to chance in the Aristotelian sense. And this is something that an omnipotent and omniscient being would find no more difficulty in doing than He would in creating a table of random numbers.

It does seem as if there are a lot of people who, even if they are willing to admit that God *could* have done this, think it's at least very unlikely that God – if He existed – *would* have done anything of the sort. Presumably they think that if the biological world were the creation of an infinite being, a being whose power and knowledge were absolutely without limit, the biological world would look very different. (I'm not talking about the existence of suffering, which is an entirely different problem, and quite unrelated to Darwinism.) But how, then, would it look? When I actually talk to people who think this and ask them this question, I do not generally get answers – or I get ones that I (frankly) regard as simple-minded. I think an answer is in order, and one that is the product of a little thought and at least some familiarity with theology. Anyone who thinks that the history of terrestrial life is inconsistent with its being the vehicle by which God's purposes have unfolded in time really should have something to say about how the history of life *would* look if it were the vehicle of God's unfolding purposes.

Notes

1 That is, of *non-human* intelligent design. Obviously, a few of the features of a few species are – very recent – products of human intelligent design.

2 It is arguable that Aquinas's Fifth Way depends on the premise that if an object always or for the most part acts in such a way as to produce some "useful" end, then intelligence *must* have been in some way among the causes of that object's action. I think it should be evident that the very existence of Darwin's account of evolution shows that this premise is false. (But do I not, as a theist, believe that it is a necessary truth that intelligence is in some way among the causes of *everything*? Yes. A more careful statement of my thesis would be this: Darwin has shown that it is false that Aquinas's premise is a *conceptual* truth, a proposition that can be seen to be true by anyone who understands the concepts it involves. If I am right in thinking that the Fifth Way depends on this premise, Darwin may be said to have "refuted" the Fifth Way in a very straightforward sense of the word. But not all versions of the design argument depend on this premise.)

3 This is Phillip Johnson's use of the term. I do not know whether he originated it. It should perhaps be noted that the term "compatibilism" has a standard and quite unrelated meaning in philosophy: it is used as a name for the thesis that free will is compatible with determinism.

4 My reasons are presented in the following two essays: "Genesis and evolution" (Stump 1993: 93–127, reprinted in van Inwagen 1995); and "Doubts about Darwinism" (Buell and Hearn 1994: 177–91).

5 Curiously enough, Biochemical Predestination was said by those who believed in it to show that the evolutionary process was not anyone's instrument.

6 If the number of histories is finite, each should have probability 1/N, where N is the number of histories; if the number of histories is infinite, each should have probability 0 (or, if infinitesimal probabilities are allowed, each should have the same infinitesimal probability).

References

Buell, J. and Hearn, V. (eds) (1994) *Darwinism: Science or Philosophy?*, Richardson, Texas: Foundation for Thought and Ethics.

Goodwin, B. (1994) *How the Leopard Changed its Spots: The Evolution of Complexity*, New York: Charles Scribner's Sons.

Monod, J. (1971) *Chance and Necessity: An Essay on the Natural Philosophy of Modern Biology*, trans. Austryn Wainhouse, New York: Vintage Books.
Stump, E. (ed) (1993) *Reasoned Faith*, Ithaca, New York: Cornell University Press.
van Inwagen, P. (1995) *God, Knowledge, and Mystery: Essays in Philosophical Theology*, Ithaca: Cornell University Press.

INDEX

Contact Us!
YourVoiceMinistry.com